Lecture Notes in Computational Science and Engineering

87

Editors:

For further volumes:
http://www.springer.com/series/3527

Shaun Forth • Paul Hovland • Eric Phipps
Jean Utke • Andrea Walther
Editors

Recent Advances in Algorithmic Differentiation

Springer

Editors
Shaun Forth
Applied Mathematics
and Scientific Computing
Cranfield University
Shrivenham
Swindon
United Kingdom

Paul Hovland
Mathematics and Computer Science
Division
Argonne National Laboratory
Argonne
Illinois
USA

Eric Phipps
Sandia National Laboratory
Albuquerque
New Mexico
USA

Jean Utke
Mathematics and Computer Science
Division
Argonne National Laboratory
Argonne
Illinois
USA

Andrea Walther
Department of Mathematics
University of Paderborn
Paderborn
Germany

ISSN 1439-7358
ISBN 978-3-642-30022-6 ISBN 978-3-642-30023-3 (eBook)
DOI 10.1007/978-3-642-30023-3
Springer Heidelberg New York Dordrecht London

Library of Congress Control Number: 2012942187

Mathematics Subject Classification (2010): 65D25, 90C30, 90C31, 90C56, 65F50, 68N20, 41A58, 65Y20

Printed on acid-free paper

Springer is part of Springer Science+Business Media (www.springer.com)

Preface

The Sixth International Conference on Automatic Differentiation (AD2012) held July 23–27, 2012, in Fort Collins, Colorado (USA), continued this quadrennial conference series. While the fundamental idea of differentiating numerical programs is easy to explain, the practical implementation of this idea for many nontrivial numerical computations is not. Our community has long been aware of the discrepancy between the aspiration of an *automatic* process suggested by the name automatic differentiation and the reality of its practical use, which often requires substantial effort from the user. New algorithms and methods implemented in differentiation tools improve their usability and reduce the need for user intervention. On the other hand, the demands to compute derivatives for numerical models on parallel hardware, using a wide variety of libraries and having components implemented in different programming languages, pose new challenges, particularly for the efficiency of the derivative computation. These challenges, as well as new applications, have been driving research for the past four years and will continue to do so. Despite retaining automatic differentiation in the conference name, the editors purposely switched to *algorithmic differentiation* (AD) in the proceedings title. Thus, the conference proceedings follow somewhat belatedly the more appropriate naming chosen by Andreas Griewank for the first edition of his seminal monograph covering our subject area. This name better reflects the reality of AD usage and the research results presented in the papers collected here.

The 31 contributed papers cover the application of AD to many areas of science and engineering as well as aspects of AD theory and its implementation in tools. For all papers the referees, selected from the program committee and the wider AD community, as well as the editors have emphasized accessibility of the presented ideas also to non-AD experts.

In the AD tools arena new implementations are introduced covering, for example, Java and graphical modeling environments, or join the set of existing tools for Fortran. New developments in AD algorithms target: efficient derivatives for matrix-operation, detection and exploitation of sparsity, partial separability, the treatment of nonsmooth functions, and other high-level mathematical aspects of the numerical computations to be differentiated.

Applications stem from the Earth sciences, nuclear engineering, fluid dynamics, and chemistry, to name just a few. In many cases the applications in a given area of science or engineering share characteristics that require specific approaches to enable AD capabilities or provide an opportunity for efficiency gains in the derivative computation. The description of these characteristics and of the techniques for successfully using AD should make the proceedings a valuable source of information for users of AD tools.

The image on the book cover shows the high-harmonic emission spectrum of a semiconductor quantum dot for different excitation conditions. To favor specific frequencies one has to find an appropriate input pulse within a large parameter space. This was accomplished by combining a gradient-based optimization algorithm with AD. The data plots were provided by Matthias Reichelt.

Algorithmic differentiation draws on many aspects of applied mathematics and computer science and ultimately is useful only when users in the science and engineering communities become aware of its capabilities. Furthering collaborations outside the core AD community, the AD2012 program committee invited leading experts from diverse disciplines as keynote speakers. We are grateful to Lorenz Biegler (Carnegie Mellon University, USA), Luca Capriotti (Credit Suisse, USA), Don Estep (Colorado State University, USA), Andreas Griewank (Humboldt University, Germany), Mary Hall (University of Utah, USA), Barbara Kaltenbacher (University of Klagenfurt, Austria), Markus Püschel (ETH Zurich, Switzerland), and Bert Speelpenning (MathPartners, USA) for accepting the invitations.

We want to thank SIAM and the NNSA and ASCR programs of the US Department of Energy for their financial support of AD2012.

Albuquerque, Chicago
Paderborn, Shrivenham
April 2012

Shaun Forth
Paul Hovland
Eric Phipps
Jean Utke
Andrea Walther

Program Committee AD2012

Brad Bell, University of Washington (USA)
Martin Berz, Michigan State University (USA)
Christian Bischof, TU Darmstadt (Germany)
Martin Bücker, RWTH Aachen (Germany)
Bruce Christianson, University of Hertfordshire (UK)
David Gay, AMPL Optimization Inc. (USA)
Andreas Griewank, Humboldt University Berlin (Germany)
Laurent Hascoët, INRIA (France)
Patrick Heimbach, Massachusetts Institute of Technology (USA)
Koichi Kubota, Chuo University (Japan)
Kyoko Makino, Michigan State University (USA)
Jens-Dominik Müller, Queen Mary University of London (UK)
Uwe Naumann, RWTH Aachen (Germany)
Boyana Norris, Argonne National Laboratory (USA)
Trond Steihaug, University of Bergen (Norway)

Contents

Contributors

Hany S. Abdel-Khalik Department of Nuclear Engineering, North Carolina State University, Raleigh, NC, USA, abdelkhalik@ncsu.edu

Johan Åkesson Department of Automatic Control, Faculty of Engineering, Lund University, Lund, Sweden, johan.akesson@control.lth.se

Juan J. Alonso Department of Aeronautics and Astronautics, Stanford University, Stanford, CA, USA, jjalonso@stanford.edu

Joel Andersson Electrical Engineering Department (ESAT) and Optimization in Engineering Center (OPTEC), K.U. Leuven, Heverlee, Belgium, joel.andersson@esat.kuleuven.be

Mihai Anitescu Mathematics and Computer Science, Argonne National Laboratory, Argonne, IL, USA, anitescu@mcs.anl.gov

Frédéric Baret INRA, Avignon, France, baret@avignon.inra.fr

Paul I. Barton Process Systems Engineering Laboratory, Department of Chemical Engineering, Massachusetts Institute of Technology, Cambridge, MA, USA, pib@mit.edu

Markus Beckers STCE, RWTH Aachen University, Aachen, Germany, beckers@stce.rwth-aachen.de

Christian H. Bischof Scientific Computing Group, TU Darmstadt, Darmstadt, Germany, christian.bischof@sc.tu-darmstadt.de

Torsten Bosse Humboldt-Universität zu Berlin, Berlin, Germany, bosse@math.hu-berlin.de

H. Martin Bücker Institute for Scientific Computing, RWTH Aachen University, Aachen, Germany, buecker@sc.rwth-aachen.de

David C. Carothers James Madison University, Harrisonburg, USA, carothdc@jmu.edu

Jieqiu Chen Mathematics and Computer Science, Argonne National Laboratory, Argonne, IL, USA, jieqchen@mcs.anl.gov

Bruce Christianson School of Computer Science, University of Hertfordshire, Hatfield, UK, b.christianson@herts.ac.uk

Thomas F. Coleman Department of Combinatorics and Optimization, University of Waterloo, Ontario, Canada, tfcoleman@uwaterloo.ca

Heather Cole-Mullen Argonne National Laboratory, The University of Chicago, Chicago, IL, USA, hlcm@mcs.anl.gov

Benoit Delinchant Grenoble Electrical Engineering Laboratory, Saint-Martin d'Hères, France, benoit.delinchant@grenoble-inp.fr

Moritz Diehl Electrical Engineering Department (ESAT) and Optimization in Engineering Center (OPTEC), K.U. Leuven, Heverlee, Belgium, moritz.diehl@esat.kuleuven.be

Beth Drewniak Environmental Science, Argonne National Laboratory, Argonne, IL, USA, bbye@anl.gov

Jeffrey A. Fike Department of Aeronautics and Astronautics, Stanford University, Stanford, CA, USA, jfike@stanford.edu

Shaun A. Forth Applied Mathematics and Scientific Computing, Cranfield University, Swindon, UK, S.A.Forth@cranfield.ac.uk

Nicolas R. Gauger Computational Mathematics Group, CCES, RWTH Aachen University, Aachen, Germany, gauger@mathcces.rwth-aachen.de

Assefaw Gebremedhin Department of Computer Science, Purdue University, West Lafayette, IN, USA, agebreme@purdue.edu

Ralf Giering FastOpt GmbH, Lerchenstrasse 28a, 22767 Hamburg, Germany, Ralf.Giering@FastOpt.com

Vaibhava Goel IBM, TJ Watson Research Center, Yorktown Heights, NY, USA, vgoel@us.ibm.com

Andreas Griewank Humboldt-Universität zu Berlin, Berlin, Germany, griewank@math.hu-berlin.de

Abhishek Kr. Gupta Department of Electrical Engineering, IIT Kanpur, Kanpur, India, g.kr.abhishek@gmail.com

Laurent Hascoët INRIA, Sophia-Antipolis, France, Laurent.Hascoet@inria.fr

Paul Hovland Mathematics and Computer Science, Argonne National Laboratory, Argonne, IL, USA, hovland@mcs.anl.gov

John Junkins Aerospace Engineering, Texas A&M University, College Station, TX, USA, junkins@tamu.edu

Kamil A. Khan Process Systems Engineering Laboratory, Department of Chemical Engineering, Massachusetts Institute of Technology, Cambridge, MA, USA, kamil@mit.edu

Kshitij Kulshreshtha Institut für Mathematik, Universität Paderborn, Paderborn, Germany, kshitij@math.upb.de

Claire Lauvernet Irstea, Lyon, France, claire.lauvernet@irstea.fr

François-Xavier Le Dimet Université de Grenoble, Grenoble, France, Francois-Xavier.Le-Dimet@imag.fr

Benjamin Letschert Universität Paderborn, Institut für Mathematik, Paderborn, Germany, Benny.Letschert@web.de

Johannes Lotz STCE, RWTH Aachen University, Aachen, Germany, lotz@stce.rwth-aachen.de

Stephen K. Lucas James Madison University, Harrisonburg, VA, USA, lucassk@jmu.edu

Andrew Lyons Dartmouth College, Hanover, NH, USA, lyonsam@gmail.com

Manoranjan Majji Mechanical and Aerospace Engineering, University at Buffalo, Buffalo, NY, USA, majjineo@gmail.com

Azamat Mametjanov Mathematics and Computer Science, Argonne National Laboratory, Argonne, IL, USA, azamat.mametjanov@gmail.com

Jan Marburger Fraunhofer-Institut für Techno- und Wirtschaftsmathematik, Kaiserslautern, Germany, jan.marburger@itwm.fhg.de

Viktor Mosenkis STCE, RWTH Aachen University, Aachen, Germany, mosenkis@stce.rwth-aachen.de

Todd Munson Mathematics and Computer Science, Argonne National Laboratory, Argonne, IL, USA, tmunson@mcs.anl.gov

Sri Hari Krishna Narayanan Mathematics and Computer Science, Argonne National Laboratory, Argonne, IL, USA, snarayan@mcs.anl.gov

Uwe Naumann STCE, RWTH Aachen University, Aachen, Germany, naumann@stce.rwth-aachen.de

Marco Nehmeier Institute of Computer Science, University of Würzburg, Würzburg, Germany, nehmeier@informatik.uni-wuerzburg.de

Anil Nemili Computational Mathematics Group, CCES, RWTH Aachen University, Aachen, Germany, nemili@mathcces.rwth-aachen.de

Duc Nguyen Department of Computer Science, Purdue University, West Lafayette, IN, USA, nguyend@purdue.edu

Boyana Norris Mathematics and Computer Science, Argonne National Laboratory, Argonne, IL, USA, norris@mcs.anl.gov

Peder A. Olsen IBM, TJ Watson Research Center, Yorktown Heights, NY, USA, pederao@us.ibm.com

Emre Özkaya Computational Mathematics Group, CCES, RWTH Aachen University, Aachen, Germany, ozkaya@mathcces.rwth-aachen.de

G. Edgar Parker James Madison University, Harrisonburg, VA, USA, parkerge@jmu.edu

Valérie Pascual INRIA, Sophia-Antipolis, Sophia-Antipolis, France, Valerie.Pascual@inria.fr

Roger Pawlowski Sandia National Laboratories, Multiphysics Simulation Technologies Department, Albuquerque, NM, USA, rppawlo@sandia.gov

Barak A. Pearlmutter Department of Computer Science and Hamilton Institute, National University of Ireland, Maynooth, Ireland, barak@cs.nuim.ie

Phuong Pham-Quang CEDRAT S.A., Meylan Cedex, France, phuong.phamquang@cedrat.com

Eric Phipps Sandia National Laboratories, Optimization and Uncertainty Quantification Department, Albuquerque, NM, USA, etphipp@sandia.gov

Alex Pothen Department of Computer Science, Purdue University, West Lafayette, IN, USA, apothen@purdue.edu

Alexey Radul Hamilton Institute, National University of Ireland, Maynooth, Ireland, alexey.radul@nuim.ie

James A. Reed Department of Nuclear Engineering, North Carolina State University, Raleigh, NC, USA, jareed3@ncsu.edu

Steven J. Rennie IBM, TJ Watson Research Center, Yorktown Heights, NY, USA, sjrennie@us.ibm.com

Daniel R. Reynolds Mathematics, Southern Methodist University, Dallas, TX, USA, reynolds@smu.edu

Joseph D. Rudmin James Madison University, Harrisonburg, VA, USA, rudminjd@jmu.edu

Ravi Samtaney Mechanical Engineering, Division of Physical Scienceand Engineering, King Abdullah University of Science and Technology, Thuwal, Saudi Arabia, Ravi.Samtaney@kaust.edu.sa

Jeffrey Mark Siskind Electrical and Computer Engineering, Purdue University, West Lafayette, IN, USA, qobi@purdue.edu

James S. Sochacki James Madison University, Harrisonburg, VA, USA, sochacjs@jmu.edu

Hamdi A. Tchelepi Department of Energy Resources Engineering, Stanford University, Stanford, CA, USA, tchelepi@stanford.edu

Roger J. Thelwell James Madison University, Harrisonburg, VA, USA, thelwerj@jmu.edu

Anthony Tongen James Madison University, Harrisonburg, VA, USA, tongenal@jmu.edu

James Turner Aerospace Engineering, Texas A&M University, College Station, TX, USA, turner@aero.tamu.edu

Jörn Ungermann Institute of Energy and Climate Research – Stratosphere (IEK-7), Research Center Jülich GmbH, Jülich, Germany, j.ungermann@fz-juelich.de

Jean Utke Argonne National Laboratory, The University of Chicago, Chicago, IL, USA, utke@mcs.anl.gov

Michael Voßbeck FastOpt GmbH, Lerchenstrasse 28a, 22767 Hamburg, Germany, Michael.Vossbeck@FastOpt.com

Andrea Walther Institut für Mathematik, Universität Paderborn, Paderborn, Germany, andrea.walther@uni-paderborn.de

Paul G. Warne James Madison University, Harrisonburg, VA, USA, warnepg@jmu.edu

Johannes Willkomm Scientific Computing Group, TU Darmstadt, Darmstadt, Germany, johannes.willkomm@sc.tu-darmstadt.de

Xin Xiong Department of Combinatorics and Optimization, University of Waterloo, Ontario, Canada, x6xiong@uwaterloo.ca

Wei Xu Department of Mathematics, Tongji University, Shanghai, China, wdxu@tongji.edu.cn

Ahmad Bani Younes Aerospace Engineering, Texas A&M University, College Station, TX, USA, olalahmad@gmail.com

Rami M. Younis Department of Energy Resources Engineering, Stanford University, Stanford, CA, USA, ryounis@stanford.edu

Xiaoyan Zeng Mathematics and Computer Science, Argonne National Laboratory, Argonne, IL, USA, zeng@mcs.anl.gov

A Leibniz Notation for Automatic Differentiation

Bruce Christianson

Abstract Notwithstanding the superiority of the Leibniz notation for differential calculus, the dot-and-bar notation predominantly used by the Automatic Differentiation community is resolutely Newtonian. In this paper we extend the Leibniz notation to include the reverse (or adjoint) mode of Automatic Differentiation, and use it to demonstrate the stepwise numerical equivalence of the three approaches using the reverse mode to obtain second order derivatives, namely forward-over-reverse, reverse-over-forward, and reverse-over-reverse.

Keywords Leibniz • Newton • Notation • Differentials • Second-order • Reverse mode

1 Historical Background

Who first discovered differentiation?[1] Popular European[2] contenders include Isaac Barrow, the first Lucasian Professor of Mathematics at Cambridge [5]; Isaac Newton, his immediate successor in that chair [21]; and Godfrey Leibniz, a librarian employed by the Duke of Brunswick [19]. The matter of priority was settled in Newton's favour by a commission appointed by the Royal Society. Since the report

[1] Archimedes' construction for the volume of a sphere probably entitles him to be considered the first to discover integral calculus.

[2] Sharaf al-Din al-Tusi already knew the derivative of a cubic in 1209 [1], but did not extend this result to more general functions.

B. Christianson (✉)
School of Computer Science, University of Hertfordshire, College Lane, Hatfield, England Europe
e-mail: b.christianson@herts.ac.uk

S. Forth et al. (eds.), *Recent Advances in Algorithmic Differentiation*, Lecture Notes in Computational Science and Engineering 87, DOI 10.1007/978-3-642-30023-3_1,

of the commission [2] was written by none other than Isaac Newton himself[3] we may be assured of its competence as well as its impartiality. Cambridge University thenceforth used Newton's notation exclusively, in order to make clear where its loyalties lay.

However, if instead we ask, who first discovered *automatic* differentiation, then Leibniz has the best claim. In contrast with Newton's geometric and dynamical interpretation, Leibniz clearly envisaged applying the rules of differentiation to the numerical values which the coefficients represented, ideally by a mechanical means, as the following excerpts [18, 19] respectively show:

> Knowing thus the *Algorithm* (as I may say) of this calculus, which I call *differential* calculus, all other differential equations can be solved by a common method. ... For any other quantity (not itself a term, but contributing to the formation of the term) we use its differential quantity to form the differential quantity of the term itself, not by simple substitution, but according to the prescribed Algorithm. The methods published before have no such transition.[4]

> When, several years ago, I saw for the first time an instrument which, when carried, automatically records the number of steps taken by a pedestrian, it occurred to me at once that the entire arithmetic could be subjected to a similar kind of machinery ...

Although Leibniz did devise and build a prototype for a machine to perform some of the calculations involved in automatic differentiation [18], the dream of a mechanical device of sufficient complexity to perform the entire sequence automatically had to wait until 1837, when Charles Babbage completed the design of his programmable analytical engine [20]. Babbage, who was eventually to succeed to Newton's chair, had while still an undergraduate been a moving force behind the group of young turks[5] who forced the University of Cambridge to change from the Newton to the Leibniz notation for differentiation. Babbage described this as rescuing the University from its dot-age [3].

There is no doubt that by the time of Babbage the use of Newton's notation was very badly hindering the advance of British analysis,[6] so it is ironic to reflect that we in the automatic differentiation community continue to use the Newton notation almost exclusively, for example by using a dot to denote the second field of an active variable.

---

[3]Although this fact did not become public knowledge until 1761, nearly 50 years later.

[4]The word Algorithm derives from the eponymous eighth century mathematician Al-Khwarizmi, known in Latin as Algoritmi. Prior to Leibniz, the term referred exclusively to mechanical arithmetical procedures, such as the process for extraction of square roots, applied (by a human) to numerical values rather than symbolic expressions. The italics are in the Latin original: "Ex cognito hoc velut *Algorithmo*, ut ita dicam, calculi hujus, quem voco *differentialem*."

[5]The Analytical Society was founded by Babbage and some of his friends in 1812. So successful was their program of reform that 11 of thc 16 original members subsequently became professors at Cambridge.

[6]Rouse Ball writes [4] "It would seem that the chief obstacle to the adoption of analytical methods and the notation of the differential calculus arose from the professorial body and the senior members of the senate, who regarded any attempt at innovation as a sin against the memory of Newton."

2 The Leibniz Notation

Suppose that we have independent variables w, x and dependent variables y, z given by the system

$$y = f(w, x) \qquad z = g(w, x)$$

2.1 The Forward Mode

In Newton notation we would write the forward derivatives as

$$\dot{y} = f'_w \dot{w} + f'_x \dot{x} \qquad \dot{z} = g'_w \dot{w} + g'_x \dot{x}$$

It is quite straightforward to turn this into a Leibniz notation by regarding the second field of an active variable as a differential, and writing dx, dy etc in place of $\dot{x}, \dot{y}$, etc.

In Leibniz notation the forward derivatives become[7]

$$dy = \frac{\partial f}{\partial w} dw + \frac{\partial f}{\partial x} dx \qquad dz = \frac{\partial g}{\partial w} dw + \frac{\partial g}{\partial x} dx$$

where dw, dx are independent and dy, dz are dependent differential variables.[8]

2.2 The Reverse Mode

For the reverse mode of automatic differentiation, the backward derivatives are written in a Newton style notation as

$$\bar{w} = \bar{y} f'_w + \bar{z} g'_w \qquad \bar{x} = \bar{y} f'_x + \bar{z} g'_x$$

This can be turned into a Leibniz form in a similar way to the forward case. We introduce a new notation, writing by, bz in place of the independent barred variables $\bar{y}, \bar{z}$, and bw, bx in place of the dependent barred variables $\bar{w}, \bar{x}$.

---

[7] Since $y \equiv f(w, x)$ we allow ourselves to write $\frac{\partial f}{\partial x}$ interchangeably with $\frac{\partial y}{\partial x}$.

[8] Actually the tradition of treating differentials as independent variables in their own right was begun by d'Alembert as a response to Berkeley's criticisms of the infinitesimal approach [6], but significantly he made no changes to Leibniz's original notation for them. Leibniz's formulation allows for the possibility of non-negligible differential values, referring [19] to "the fact, until now not sufficiently explored, that dx, dy, dv, dw, dz can be taken *proportional* [my italics] to the momentary differences, that is, increments or decrements, of the corresponding x, y, v, w, z", and Leibniz is careful to write $d(xv) = xdv + vdx$, without the term $dxdv$.

$$bw = by\frac{\partial f}{\partial w} + bz\frac{\partial g}{\partial x} \qquad bx = by\frac{\partial f}{\partial w} + bz\frac{\partial g}{\partial x}$$

We refer to quantities such as bx as *barientials*. Note that the bariential of a dependent variable is independent, and vice versa. Differentials and barientials will collectively be referred to as *varientials*.

The barientials depend on all the dependent underlying variables so, as always with the reverse mode, the full set of equations must be explicitly given before the barientials can be calculated.

2.3 Forward over Forward

Repeated differentiation in the forward mode (the so-called forward-over-forward approach) produces the Newton equation

$$\ddot{y} = f''_{ww}\dot{w}\dot{w} + 2f''_{wx}\dot{w}\dot{x} + f''_{xx}\dot{x}\dot{x} + f'_w\ddot{w} + f'_x\ddot{x}$$

and similarly for $\ddot{z}$. This has the familiar[9] Leibniz equivalent

$$d^2y = \frac{\partial^2 f}{\partial w^2}dw^2 + 2\frac{\partial^2 f}{\partial w \partial x}dwdx + \frac{\partial^2 f}{\partial x^2}dx^2 + \frac{\partial f}{\partial w}d^2w + \frac{\partial f}{\partial x}d^2x$$

and similarly for d^2z.

2.4 Forward over Reverse

Now consider what happens when we apply forward mode differentiation to the backward derivative equations (the so-called forward-over-reverse approach). Here are the results in Newton notation

$$\dot{\bar{w}} = \dot{\bar{y}}f'_w + \bar{y}f''_{ww}\dot{w} + \bar{y}f''_{wx}\dot{x} + \dot{\bar{z}}g'_w + \bar{z}g''_{ww}\dot{w} + \bar{z}g''_{wx}\dot{x}$$

and here is the Leibniz equivalent

[9] The familiarity comes in part from the fact that this is the very equation of which Hademard said [15] "que signifie ou que représente l'égalité? A mon avis, rien du tout." ["What is meant, or represented, by this equality? In my opinion, nothing at all."] It is good that the automatic differentiation community is now in a position to give Hadamard a clear answer: (y, dy, d^2y) is the content of an active variable.

$$dbw = dby\frac{\partial f}{\partial w} + by\frac{\partial^2 f}{\partial w^2}dw + by\frac{\partial^2 f}{\partial w \partial x}dx + dbz\frac{\partial g}{\partial w} + bz\frac{\partial^2 g}{\partial w^2}dw + bz\frac{\partial^2 f}{\partial w \partial x}dx$$

with similar equations for $\dot{\bar{x}}$ and dbx respectively.

What happens when we repeatedly apply automatic differentiation in other combinations?

3 Second Order Approaches Involving Reverse Mode

For simplicity, in this section we shall consider the case[10] of a single independent variable x and a single dependent variable $y = f(x)$.

3.1 Forward over Reverse

Here are the results in Newton notation for forward-over-reverse in the single variable case. The reverse pass gives

$$y = f(x) \qquad \bar{x} = \bar{y} f'$$

and then the forward pass, with independent variables x and $\bar{y}$, gives

$$\dot{y} = f'\dot{x} \qquad \dot{\bar{x}} = \dot{\bar{y}} f' + \bar{y} f'' \dot{x}$$

The Leibniz equivalents are

$$y = f(x) \qquad bx = by\frac{\partial f}{\partial x}$$

and

$$dy = \frac{\partial f}{\partial x}dx \qquad dbx = dby\frac{\partial f}{\partial x} + by\frac{\partial^2 f}{\partial x^2}dx$$

[10] The variables x and y may be vectors: in this case the corresponding differential dx and bariential by are respectively a column vector with components dx^j and a row vector with components by_i; f' is the matrix $J^i_j = \partial_j f^i = \partial f^i / \partial x^j$, and f'' is the mixed third order tensor $K^i_{jk} = \partial^2_{jk} f^i = \partial^2 f^i / \partial x^j \partial x^k$.

3.2 Reverse over Forward

Next, the corresponding results for reverse-over-forward. First the forward pass in Newton notation

$$y = f(x) \qquad \dot{y} = f'\dot{x}$$

then the reverse pass, applying the rules already given, and treating both y and $\dot{y}$ as dependent variables. We use a long bar to denote ADOL-C style reverse mode differentiation [13], starting from $\dot{y}$ and y

$$\overline{x} = \overline{y}\, f' + \overline{\dot{y}}\, f''\dot{x} \qquad \overline{\dot{x}} = \overline{\dot{y}}\, f'$$

In Leibniz notation the forward pass gives

$$y = f(x) \qquad dy = \frac{\partial f}{\partial x} dx$$

and for the reverse pass we treat y and dy as the dependent variables. We denote the bariential equivalent of the long bar by the letter p for the moment, although we shall soon see that this notation can be simplified. This gives

$$px = py\frac{\partial f}{\partial x} + pdy\frac{\partial^2 f}{\partial x^2} dx \qquad pdx = pdy\frac{\partial f}{\partial x}$$

3.3 Reverse over Reverse

Finally we consider reverse over reverse. The first reverse pass gives

$$y = f(x) \qquad \bar{x} = \bar{y}\, f'$$

the dependent variables are y and $\bar{x}$. We denote the adjoint variables on the second reverse pass by a long bar

$$\overline{x} = \overline{y}\, f' + \bar{y}\, f''\overline{\bar{x}} \qquad \overline{\bar{y}} = f'\,\overline{\bar{x}}$$

and we shall see shortly that the use made here of the long bar is consistent with that of the previous subsection. In Leibniz notation, the first reverse pass corresponds to

$$y = f(x) \qquad bx = by\frac{\partial f}{\partial x}$$

with the dependent variables being y and bx. Denoting the barientials for the second reverse pass by the prefix p, we have

$$px = py\frac{\partial f}{\partial x} + by\frac{\partial^2 f}{\partial x^2}pbx \qquad\qquad pby = \frac{\partial f}{\partial x}pbx$$

In general we write differentials on the right and barientials on the left, but pbx is a bariential of a bariential, and so appears on the right.[11]

4 The Equivalence Theorem

By collating the equations from the three previous subsections, we can immediately see that all three of the second-order approaches involving reverse differentiation produce structurally equivalent sets of equations, in which certain pairs of quantities correspond. In particular, where v is any dependent or independent variable,

$$\overline{\;v\;} = \dot{\bar{v}} \qquad\qquad \overline{\dot{v}} = \bar{v} \qquad\qquad \overline{\bar{v}} = \dot{v}$$

or, in Leibniz notation

$$pv = dbv \qquad\qquad pdv = bv \qquad\qquad pbv = dv$$

allowing the use of p-barientials to be eliminated.

However, we can say more than this. Not only are the identities given above true for dependent and independent varientials,[12] the correspondences also hold for the varientials corresponding to all the intermediate variables in the underlying computation. Indeed, the three second-order derivative computations themselves are structurally identical.

This can be seen by defining the intermediate variables v_i in the usual way [14] by the set of equations

$$v_i = \phi_i(v_{j:j \prec i})$$

and then simulating the action of the automatic differentiation algorithm, by using the rules in the preceding subsections to successively eliminate the varientials corresponding to the intermediate variables, in the order appropriate to the algorithm being used.

In all three cases, we end up computing the varientials of each intermediate variable with exactly the same arithmetical steps

$$pbv_i = dv_i = \sum_{j:j \prec i} \frac{\partial \phi_i}{\partial v_j} dv_j \qquad\qquad pdv_i = bv_i = \sum_{k:i \prec k} bv_k \frac{\partial \phi_k}{\partial v_i}$$

[11] If x is a vector then pbx is a column vector.

[12] Recall that this term includes all combinations of differentials and barientials.

and

$$pv_i = dbv_i = \sum_{k:i \prec k} \left\{ dbv_k \frac{\partial \phi_k}{\partial v_i} + bv_k \sum_{j:j \prec k} \frac{\partial^2 \phi_k}{\partial v_i \partial v_j} dv_j \right\}$$

We therefore have established the following

Theorem 1. *The three algorithms forward-over reverse, reverse-over-forward, and reverse-over-reverse are all numerically stepwise identical, in the sense that they not only produce the same numerical output values, but at every intermediate stage perform exactly the same floating point calculations on the same intermediate variable values.*

Although the precise order in which these calculations are performed may depend on which of the three approaches is chosen, each of the three algorithms performs exactly the same floating point arithmetic. Strictly speaking, this statement assumes that an accurate inner product is available as an elemental operation to perform accumulations, such as those given above for dv_i, bv_i, dbv_i, in an order-independent way.

A final caveat is that the statement of equivalence applies only to the floating point operations themselves, and not to the load and store operations which surround them, since a re-ordering of the arithmetic operations may change the contents of the register set and cache.

Historically, all three of the second-order methods exploiting reverse were implemented at around the same time in 1989 [11]: reverse-over-reverse in PADRE2 by Iri and Kubota [16, 17]; reverse-over-forward in ADOL-C by Griewank and his collaborators [12, 13]; and forward-over-reverse by Dixon and Christianson in an Ada package [7, 10]. The stepwise equivalence of forward-over-reverse with reverse-over-reverse was noted in [9] and that of forward-over-reverse with reverse-over-forward in [8].

The stepwise equivalence of the three second order approaches involving the reverse mode nicely illustrates the new Leibniz notation advanced in this paper, but also deserves to be more widely known than is currently the case.

References

1. Al-Tusi, S.A.D.: Treatise on Equations. Manuscript, Baghdad (1209)
2. Anonymous: An account of the book entitled commercium epistolicum collinii et aliorum, de analysi promota; published by order of the Royal Society, in relation to the dispute between Mr. Leibnitz [sic] and Dr. Keill, about the right of invention of the method of fluxions, by some called the differential method. Philosophical Transaction of the Royal Society of London **342**, 173–224 (January and February 1714/5)
3. Babbage, C.: Passages from the Life of a Philosopher. London (1864)
4. Ball, W.W.R.: A History of the Study of Mathematics at Cambridge. Cambridge (1889)

5. Barrow, I.: Lectiones Opticae et Geometricae. London (1669)
6. Berkeley, G.: The Analyst; or, A Discourse Addressed to an Infidel Mathematician, Wherein it is examined whether the Object, Principles, and Inferences of the modern Analysis are more distinctly conceived, or more evidently deduced, than Religious Mysteries and Points of Faith. London (1734)
7. Christianson, B.: Automatic Hessians by reverse accumulation. Technical Report NOC TR228, Numerical Optimisation Centre, Hatfield Polytechnic, Hatfield, United Kingdon (1990)
8. Christianson, B.: Reverse accumulation and accurate rounding error estimates for Taylor series coefficients. Optimization Methods and Software **1**(1), 81–94 (1991). Also appeared as Tech. Report No. NOC TR239, The Numerical Optimisation Centre, University of Hertfordshire, U.K., July 1991
9. Christianson, B.: Automatic Hessians by reverse accumulation. IMA Journal of Numerical Analysis **12**(2), 135–150 (1992)
10. Dixon, L.C.W.: Use of automatic differentiation for calculating Hessians and Newton steps. In: Griewank and Corliss [11], pp. 114–125
11. Griewank, A., Corliss, G.F. (eds.): Automatic Differentiation of Algorithms: Theory, Implementation, and Application. SIAM, Philadelphia, PA (1991)
12. Griewank, A., Juedes, D., Srinivasan, J., Tyner, C.: ADOL-C: A package for the automatic differentiation of algorithms written in C/C++. Preprint MCS-P180-1190, Mathematics and Computer Science Division, Argonne National Laboratory, Argonne, Illinois (1990)
13. Griewank, A., Juedes, D., Utke, J.: Algorithm 755: ADOL-C: A package for the automatic differentiation of algorithms written in C/C++. ACM Transactions on Mathematical Software **22**(2), 131–167 (1996). URL http://doi.acm.org/10.1145/229473.229474
14. Griewank, A., Walther, A.: Evaluating Derivatives: Principles and Techniques of Algorithmic Differentiation, 2nd edn. No. 105 in Other Titles in Applied Mathematics. SIAM, Philadelphia, PA (2008). URL http://www.ec-securehost.com/SIAM/OT105.html
15. Hadamard, J.: La notion de différentiel dans l'enseignement. Mathematical Gazette **XIX**(236), 341–342 (1935)
16. Kubota, K.: PADRE2 Version 1 Users Manual. Research Memorandum RMI 90-01, Department of Mathematical Engineering and Information Physics, Faculty of Engineering, University of Tokyo, Tokyo, Japan (1990)
17. Kubota, K.: PADRE2, a Fortran precompiler yielding error estimates and second derivatives. In: Griewank and Corliss [11], pp. 251–262
18. Leibniz, G.W.: Machina arithmetica in qua non additio tantum et subtractio sedet multiplicatio nullo, division veropaene nullo animi labore peragantur. [An arithmetic machine which can be used to carry out not only addition and subtraction but also multiplication with no, and division with really almost no, intellectual exertion.]. Manuscript, Hannover (1685). A translation by Mark Kormes appears in 'A Source Book in Mathematics' by David Eugene Smith, Dover (1959)
19. Leibniz, G.W.: Nova methodvs pro maximis et minimis, itemque tangentibus, quae nec fractas, nec irrationales quantitates moratur, et singulare pro illis calculi genus. [A new method for maxima and minima as well as tangents, which is impeded neither by fractional nor irrational quantities, and a remarkable type of calculus for them.]. Acta Erutitorium (October 1684)
20. Menabrea, L.F.: Sketch of the analytical engine invented by Charles Babbage, with notes by the translator Augusta Ada King, Countess of Lovelace. Taylor's Scientific Memoirs **3**, 666–731 (1842)
21. Newton, I.: Philosophiae Naturalis Principia Mathematica. London (1687)

Sparse Jacobian Construction for Mapped Grid Visco-Resistive Magnetohydrodynamics

Daniel R. Reynolds and Ravi Samtaney

Abstract We apply the automatic differentiation tool OpenAD toward constructing a preconditioner for fully implicit simulations of mapped grid visco-resistive magnetohydrodynamics (MHD), used in modeling tokamak fusion devices. Our simulation framework employs a fully implicit formulation in time, and a mapped finite volume spatial discretization. We solve this model using inexact Newton-Krylov methods. Of critical importance in these iterative solvers is the development of an effective preconditioner, which typically requires knowledge of the Jacobian of the nonlinear residual function. However, due to significant nonlinearity within our PDE system, our mapped spatial discretization, and stencil adaptivity at physical boundaries, analytical derivation of these Jacobian entries is highly nontrivial. This paper therefore focuses on Jacobian construction using automatic differentiation. In particular, we discuss applying OpenAD to the case of a spatially-adaptive stencil patch that automatically handles differences between the domain interior and boundary, and configuring AD for reduced stencil approximations to the Jacobian. We investigate both scalar and vector tangent mode differentiation, along with simple finite difference approaches, to compare the resulting accuracy and efficiency of Jacobian construction in this application.

Keywords Forward mode • Iterative methods • Sparse Jacobian construction

---

D.R. Reynolds (✉)
Department of Mathematics, Southern Methodist University, Dallas, TX 75275, USA
e-mail: reynolds@smu.edu

R. Samtaney
Department of Mechanical Engineering, Division of Physical Science and Engineering, King Abdullah University of Science and Technology, Thuwal, Saudi Arabia
e-mail: Ravi.Samtaney@kaust.edu.sa

S. Forth et al. (eds.), *Recent Advances in Algorithmic Differentiation*, Lecture Notes in Computational Science and Engineering 87, DOI 10.1007/978-3-642-30023-3_2,

1 Introduction

In this paper, we examine application of the Automatic Differentiation (AD) tool OpenAD [12–14] toward fully implicit simulations of mapped grid visco-resistive magnetohydrodynamics (MHD). These simulations are used to study tokamak devices for magnetically-confined fusion plasmas. However, such problems are indicative of a much more expansive class of large-scale simulations involving multi-physics systems of partial differential equations (PDEs), and most of the work described herein will apply in that larger context. We note that similar efforts have been made in Jacobian construction within the context of compressible fluid dynamics [6, 11], and the current study complements that work through our investigation of an increasingly complex PDE system with more significant nonzero Jacobian structure.

This paper addresses using OpenAD to generate Jacobian components required within iterative solvers for nonlinear implicit equations arising from our PDE model. We begin by describing the model (Sect. 1.1), and our discretization and solver framework (Sect. 1.2). We then describe three competing approaches for Jacobian construction (Sect. 2): scalar mode AD, vector mode AD, and simple finite difference approximation, as well as the variety of code modifications that were required to enable these techniques. We then describe our experimental tests on these approaches and the ensuing results (Sect. 3), and conclude with some proposed optimizations for Jacobian construction in similar applications.

1.1 Model

We study visco-resistive MHD in cylindrical, (r, φ, z), coordinates [15],

$$\partial_t \mathbf{U} + \frac{1}{r}\partial_r(r\mathbf{F}(\mathbf{U})) + \partial_z \mathbf{H}(\mathbf{U}) + \frac{1}{r}\partial_\varphi \mathbf{G}(\mathbf{U}) = \mathbf{S}(\mathbf{U}) + \nabla \cdot \mathbf{F}_d(\mathbf{U}), \tag{1}$$

where $\mathbf{U} = \left(\rho, \rho u_r, \rho u_\varphi, \rho u_z, B_r, B_\varphi, B_z, e\right)$, with plasma density ρ, velocity $\mathbf{u} = \left(u_r, u_\varphi, u_z\right)$, magnetic induction $\mathbf{B} = \left(B_r, B_\varphi, B_z\right)$, total energy e, and radial location r. Here, the hyperbolic fluxes are given by

$$\mathbf{F} = \begin{aligned}&\Big(\rho u_r,\ \rho u_r^2 + \tilde{p} - B_r^2,\ \rho u_r u_\varphi - B_r B_\varphi,\ \rho u_r u_z - B_r B_z,\ 0,\\ &\quad u_r B_\varphi - u_\varphi B_r,\ u_r B_z - u_z B_r,\ (e + \tilde{p})u_r - (\mathbf{B}\cdot\mathbf{u})B_r\Big),\end{aligned} \tag{2}$$

$$\mathbf{G} = \begin{aligned}&\Big(\rho u_\varphi,\ \rho u_r u_\varphi - B_r B_\varphi,\ \rho u_\varphi^2 + \tilde{p} - B_\varphi^2,\ \rho u_z u_\varphi - B_z B_\varphi,\\ &\quad u_\varphi B_r - u_r B_\varphi,\ 0,\ u_\varphi B_z - u_z B_\varphi,\ (e + \tilde{p})u_\varphi - (\mathbf{B}\cdot\mathbf{u})B_\varphi\Big),\end{aligned} \tag{3}$$

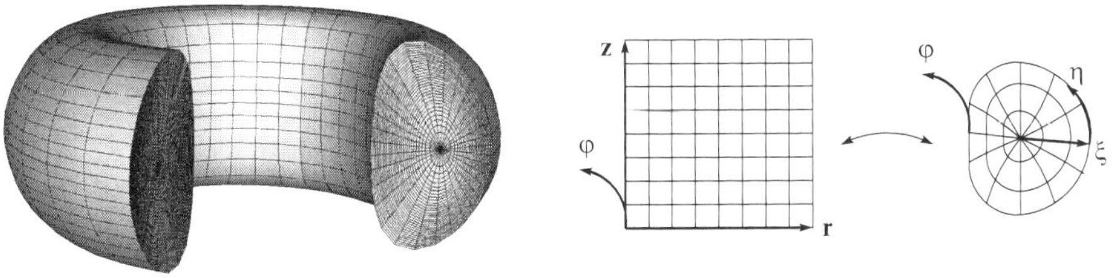

Fig. 1 *Left*: tokamak domain (a slice has been removed to show the poloidal cross-section). Note the coordinate singularity at the torus core. Cells near the core exhibit a loss of floating-point accuracy in evaluation of $\mathscr{J}$ in (6). *Right*: mapping between cylindrical and shaped domains

$$\mathbf{H} = \Big(\rho u_z,\ \rho u_r u_z - B_r B_z,\ \rho u_z u_\varphi - B_z B_\varphi,\ \rho u_z^2 + \tilde{p} - B_z^2,\ u_z B_r - u_r B_z,\ u_z B_\varphi - u_\varphi B_z,\ 0,\ (e + \tilde{p})u_z - (\mathbf{B}\cdot\mathbf{u})B_z\Big), \tag{4}$$

where $\tilde{p} = p + \frac{\mathbf{B}\cdot\mathbf{B}}{2}$ and pressure $p = \frac{2e}{3} - \frac{\rho\mathbf{u}\cdot\mathbf{u}}{3} - \frac{\mathbf{B}\cdot\mathbf{B}}{3}$. In this model, $\mathbf{S}(\mathbf{U})$ is a local source term resulting from the cylindrical coordinate system,

$$\mathbf{S} = \Big(0,\ B_z^2 - \rho u_z^2 - \tilde{p},\ 0,\ \rho u_r u_z - B_r B_z,\ 0,\ 0,\ u_z B_r - u_r B_z,\ 0\Big)/r. \tag{5}$$

A similar cylindrical divergence is applied to the diffusive terms $\nabla\cdot\mathbf{F}_d(\mathbf{U})$,

$$\nabla\cdot\mathbf{F}_d(\mathbf{U}) = \Big(0,\ \nabla\cdot\overline{\tau},\ \nabla\cdot(\overline{\tau}\mathbf{u} + \kappa\nabla T + \mathbf{B}\times(\eta(\nabla\times\mathbf{B}))),\ -\nabla\times(\eta(\nabla\times\mathbf{B})),\ 0\Big),$$

where stress tensor $\overline{\tau} = \mu(\nabla\mathbf{u} + (\nabla\mathbf{u})^T) - \frac{2}{3}\mu(\nabla\cdot\mathbf{u})\overline{I}$, temperature $T = \frac{2p}{\rho}$, μ, η and κ are input parameters for the plasma viscosity, resistivity and heat conductivity.

We map (1) to a shaped grid corresponding to the toroidal tokamak geometry (see Fig. 1). These mappings are encoded in the functions

$$\begin{aligned} &\xi = \xi(r,z), \quad \eta = \eta(r,z), \quad \varphi = \varphi \qquad \text{(cylindrical} \rightarrow \text{mapped)},\\ &r = r(\xi,\eta), \quad z = z(\xi,\eta), \quad \varphi = \varphi \qquad \text{(mapped} \rightarrow \text{cylindrical)},\\ &\mathscr{J} = (\partial_\xi r)(\partial_\eta z) - (\partial_\eta r)(\partial_\xi z), \qquad \mathscr{J}^{-1} = (\partial_r \xi)(\partial_z \eta) - (\partial_r \eta)(\partial_z \xi). \end{aligned} \tag{6}$$

Under this mapping, we rewrite the visco-resistive MHD system as

$$\partial_t \mathbf{U} + \frac{1}{\mathscr{J} r}\Big[\partial_\xi(r\tilde{\mathbf{F}}(\mathbf{U})) + \partial_\eta(r\tilde{\mathbf{H}}(\mathbf{U})) + \partial_\varphi(\tilde{\mathbf{G}}(\mathbf{U}))\Big] = \mathbf{S}(\mathbf{U}) + \nabla\cdot\tilde{\mathbf{F}}_d(\mathbf{U}), \tag{7}$$

$$\text{where} \quad \begin{aligned} \tilde{\mathbf{F}} &= \mathscr{J}(\partial_r\xi\,\mathbf{F} + \partial_z\xi\,\mathbf{H}) = \partial_\eta z\,\mathbf{F} - \partial_\eta r\,\mathbf{H}, \qquad \tilde{\mathbf{G}} = \mathscr{J}\mathbf{G},\\ \tilde{\mathbf{H}} &= \mathscr{J}(\partial_r\eta\,\mathbf{F} + \partial_z\eta\,\mathbf{H}) = -\partial_\xi z\,\mathbf{F} + \partial_\xi r\,\mathbf{H}. \end{aligned}$$

Similar transformations are performed on the diffusive fluxes $\nabla\cdot\tilde{\mathbf{F}}_d(\mathbf{U})$. We also employ a 2D version of this model for simulations within the poloidal, (ξ,η), plane.

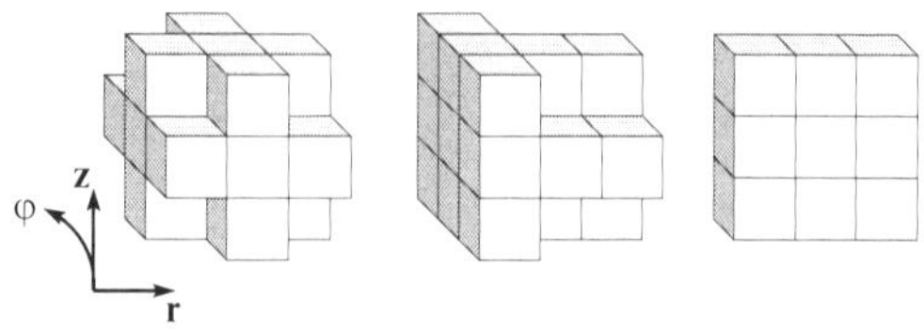

Fig. 2 Stencils used in difference calculations: centered 19-point stencil for 3D domain interior (*left*), one-sided 19-point stencil for r-left 3D domain boundary (*center*), 9-point 2D stencil (*right*)

To approximate solutions to (7), we follow a method-of-lines approach to split the time and space dimensions. In space, we use a second-order accurate finite volume discretization. Due to the coordinate mappings (6), this discretization results in a 19-point stencil. Additionally, at the boundaries $\xi = 0$ and $\xi = \xi_{max}$, the centered stencil must be modified to a one-sided 19-point approximation. For the two-dimensional approximation, similar differencing requires a 9 point stencil. Schematics of these stencils are shown in Fig. 2.

1.2 Implicit Solver Framework

For time discretization, we condense notation to write the spatially semi-discretized equation (7) as $\partial_t \mathbf{U} = \mathscr{R}(\mathbf{U})$. Due to fast waves (corresponding to fast magnetosonic and Alfvén modes) in MHD, we employ a fully implicit discretization to ensure numerical stability when evolving at time steps of interest. To this end, we update the solution from time step t^n to t^{n+1}, with $\Delta t^n = t^{n+1} - t^n$, using a θ-method,

$$\mathbf{U}^{n+1} - \mathbf{U}^n - \Delta t^n \left[\theta\, \mathscr{R}(\mathbf{U^{n+1}}) + (\mathbf{1} - \theta)\, \mathscr{R}(\mathbf{U^n})\right] = 0, \qquad 0.5 \leq \theta \leq 1. \quad (8)$$

We compute the time-evolved $\mathbf{U}^{n+1}$ as the solution to the nonlinear algebraic system

$$\mathbf{f(U)} \;\equiv\; \mathbf{U}^{n+1} - \theta \Delta t^n \mathscr{R}(\mathbf{U}^{n+1}) - \mathbf{g} = 0, \qquad \mathbf{g} \equiv \mathbf{U}^n + \Delta t^n\,(1-\theta)\,\mathscr{R}(\mathbf{U^n}). \quad (9)$$

We solve (9) using inexact Newton-Krylov methods from the SUNDIALS library [5,9,10]. At each step of these methods we solve a linear system, $J(\mathbf{U})\mathbf{V} = -\mathbf{f}(\mathbf{U})$, with Jacobian matrix $J(\mathbf{U}) = \frac{\partial \mathbf{f}}{\partial \mathbf{U}}$. For parallel efficiency we solve these linear systems using iterative Krylov methods, a critical component of which is the *preconditioner*, $P \approx J^{-1}$, that accelerates and robustifies the Krylov iteration [8]. A focus of our research program is the derivation of scalable and efficient preconditioners P. In this vein, so-called "physics-based" preconditioners often approximate J so that P^{-1} has desirable nonzero structure, e.g. many fusion codes decouple the poloidal (ξ, η)-plane from the toroidal direction. Hence, the focus of this paper is the construction of a flexible approach for computing Jacobian approximations, allowing us to use either the full stencils from Fig. 2, or to approximate J based on reduced stencils: 11 or 7 point in 3D, and 5 point in 2D, as shown in Fig. 3.

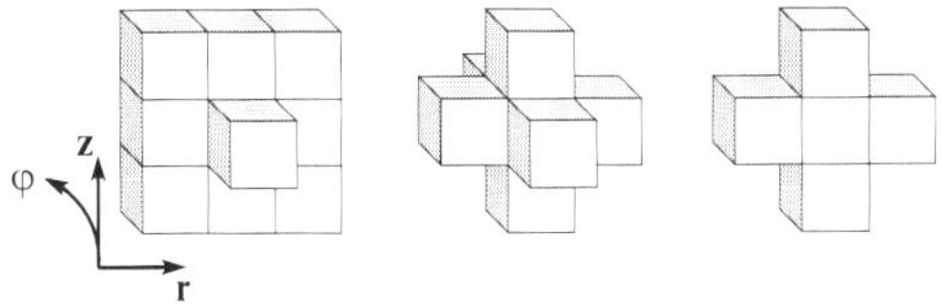

Fig. 3 Modified stencils used in approximate preconditioners: 11-point 3D stencil (*left*), 7-point 3D stencil (*middle*), 5-point 2D stencil (*right*)

We note that SUNDIALS solvers approximate the directional derivatives in the Krylov method using one-sided differences,

$$J(\mathbf{U})\mathbf{V} = [\mathbf{f}(\mathbf{U} + \sigma\mathbf{V}) - \mathbf{f}(\mathbf{U})]/\sigma + \mathcal{O}(\sigma),$$

but preconditioners typically require direct access to the preconditioner entries. It is in the construction of these entries of P^{-1} that we employ AD.

2 Preconditioner Construction

Due to the complexity of (7), the changing spatial stencil between domain interior and boundary, and our desire to explore different reduced stencil approximations within P^{-1}, automated approaches for preconditioner construction were desired.

To this end, a variety of automated approaches based on graph coloring were considered. Such approaches perform dependency analysis on the entire code, to allow for automatic calculation of the sparsity patterns in $J(\mathbf{U})$ and storage of the resulting nonzero entries [2–4]. However, for our application we are not concerned with the matrix J, instead preferring approximations to J that result from reduced stencils (see Fig. 3). To our knowledge, none of the standard graph coloring approaches allow for customization of these dependency graphs, and so we chose to follow a more manual approach to AD.

We therefore focus on three competing approaches which first compute the local Jacobian of the residual associated with a single finite volume cell with respect to the elements of $\mathbf{U}$ in its stencil – in Sect. 2.1 we term this the stencil patch Jacobian. As we further detail in Sect. 2.1 these patch Jacobians are then assembled to form the sparse preconditioner P^{-1}. Two of our approaches use the AD tool OpenAD [13, 14], for forward (tangent) differentiation in both scalar (one column of the patch Jacobian) and vector (all columns of the patch Jacobian) mode – in scalar mode we combine the columns of the patch Jacobian in a strip-mining approach [1]. Our third approach employs a simple finite difference approximation to these derivatives. Moreover, due to our decision to manually control the nonzero structure when approximating $P^{-1} \approx J$, these approaches required a multi-stage process: reconfiguration of our current code, OpenAD usage for the scalar and vector mode cases, and integration of the resulting routines into our code base.

2.1 Code Reconfiguration

Due to our compact stencils and eight variables in each cell, the function $\mathbf{f}$ at any given spatial location depends on a relatively small number of unknowns: a maximum of 152 for the full 3D stencil, or as little as 40 for the 5-point 2D stencil. However, since all of our unknowns $\mathbf{U}$ on a processor are stored in a single array (typically 8×64^3 entries), naïve vector mode AD of $\mathbf{f}(\mathbf{U})$ would compute dependencies on the entire processor-local mesh, resulting in a dense matrix of approximate size 2 million $\times$ 2 million, even though over 99% of these entries will be zero. Therefore, our first step in preconditioner generation consisted of separating out the nonlinear residual calculation at each spatial location $\mathbf{x}_i$, creating a routine of the form $\hat{\mathbf{f}}_i(\hat{\mathbf{U}}_i)$, where the input "patch" $\hat{\mathbf{U}}_i$ contains only the spatially-neighboring values of $\mathbf{U}$ that contribute to the nonlinear residual function at $\mathbf{x}_i$. Moreover, this "patch" of unknowns adapts based on whether $\mathbf{x}_i$ is in the domain interior or boundary (i.e. requires a centered vs. one-sided stencil). Additionally, through the use of preprocessor directives we also control whether this patch is 3D or 2D, and whether the desired Jacobian approximation uses a full or reduced stencil approximation. The result was a new routine in which all inputs $\hat{\mathbf{U}}_i$ contribute to the output $\hat{\mathbf{f}}_i(\hat{\mathbf{U}}_i)$, eliminating the possiblility of computing unnecessary derivatives.

We note that this new patch-based residual routine was also critical for developing an efficient finite-difference approximation routine for the Jacobian entries, since simple FD strip mining approaches would also compute predominantly zero-valued entries. In that routine, we employed the one-sided difference formula

$$[J(\mathbf{U})]_{\mathrm{i,j}} \approx \left[\hat{\mathbf{f}}_{\mathrm{i}}(\hat{\mathbf{U}}_{\mathrm{i}} + \sigma \mathbf{e}_{\mathrm{j}}) - \hat{\mathbf{f}}_{\mathrm{i}}(\hat{\mathbf{U}}_{\mathrm{i}})\right] / \sigma, \tag{10}$$

with a fixed parameter $\sigma = 10^{-8}$, since solution values are unit normalized [7].

A further code modification involved our use of Fortran 90 modules to store grid and boundary condition information. While these modules do not perform calculations, they hold a multitude of parameters that help define the simulation. Fortunately, OpenAD supports F90 modules, but it required copies of these modules to be included in the source code files prior to processing. To this end, we included duplicate declarations for all included modules, surrounded by preprocessor directives that ignore the additional code when used within our main simulation framework.

2.2 OpenAD Usage

In constructing our patch-based residual routine $\hat{\mathbf{f}}_i(\hat{\mathbf{U}}_i)$, we made significant use of preprocessor directives to handle problem dimensionality, different physical modules, and even choose between spatial discretizations at compile time.

Unfortunately, this multitude of options within a single source file is not amenable to AD, since every combination would need to be differentiated separately. Hence preprocessor directives are not retained by OpenAD. Therefore, prior to processing with OpenAD we performed a Makefile-based pre-processing stage to generate separate source code routines corresponding to our typical usage scenarios.

We then used OpenAD on these pre-processed files to generate source code for calculating entries of P^{-1}. As we wanted to determine the most efficient approach possible, we generated versions for both scalar and vector mode AD to compare their efficiency. Scalar mode forward differentiation was straightforward, requiring the simple "`openad -c -m f <file>`" call. Vector mode forward differentiation was almost as easy, however it required modification of the internal OpenAD parameter `max_deriv_vec_len`, in the file

```
$OPENADROOT/runTimeSupport/vector/OAD_active.f90
```

from the default of 100 up to 152, corresponding to our maximum domain of dependence. Once this change was made, it again required the simple command "`openad -c -m fv <file>`", to produce the desired forward mode routine.

2.3 Code Integration

The resulting routines required only minor modification before being reintroduced to our main code base. First, since OpenAD retains the original subroutine name on differentiated routines, and each of our preprocessed routines had the same subroutine function names (but in separate files), we modified these names to be more indicative of their actions, including specification of dimensionality and approximate stencil structure. Second, since the OpenAD-generated code copied all module declarations into the produced code, but F90 compilers typically do not tolerate multiple module declarations, we manually removed the extraneous declarations from the generated files. Lastly, since we applied different types of AD to the same routines, we wrote simple code wrappers to differentiate between the scalar and vector mode differentiated code.

Finally, our preconditioner construction routine merely declares the input and output arguments to the OpenAD-generated files using the supplied Fortran90 "active" type. Depending on the spatial location within the domain, and the choice of reduced spatial approximation, this routine filled in the input patch $\tilde{\mathbf{U}}_i$ with the relevant differentials set to 1. After calling the dimension/approximation dependent differentiation function described above, the routine then sifts through the resulting Jacobian entries to copy them back into a block-structured Jacobian matrix data structure, used within our preconditioning routines. We do not use the resulting function values $\hat{\mathbf{f}}_i(\hat{\mathbf{U}}_i)$, only their differentials.

3 Results

Given these three competing approaches for constructing our preconditioning matrix, we investigated each based on computational efficiency and numerical accuracy. We therefore set up a suite of tests at a variety of mesh sizes to estimate the average wall-clock time required for Jacobian construction per spatial cell:

- 2D discretizations: 64^2, 128^2 and 256^2 grids,
- 3D discretizations: 32^3, 64^3 and 128^3 grids.

For each of these meshes, we constructed the sparse preconditioner matrix based on full and reduced stencil approximations:

- 2D stencils: 9 and 5 point,
- 3D stencils: 19, 11 and 7 point.

Finally, for each of these discretization/stencil combinations, we used the three previously-described construction methods: vector mode OpenAD, scalar mode OpenAD, and finite difference approximation. All computations were performed on a linux workstation with two quad-core 64-bit Intel Xeon 3.00 GHz processors and 48 GB of RAM. No attempt was made at multi-threading the differentiated code produced by OpenAD, though all code is inherently amenable to distributed-memory parallelism, since the patches $\tilde{\mathbf{U}}_i$ are formed using an underlying data structure that contains ghost zone information from neighboring processes.

Results from these tests are provided in Table 1. Tests with the same stencil at different mesh sizes are combined to generate averaged statistics. All times represent average wall-clock time in seconds required on a per-mesh-cell basis, which remained relatively uniform over the different mesh sizes. The reported accuracy is given in the maximum norm for the relative error, computed as

$$\left(\max_{i,j} |J_{i,j}^{AD,vector} - J_{i,j}^{FD}|\right) \Big/ \left(\max_{i,j} |J_{i,j}^{AD,vector}|\right).$$

From these results, we see that the fastest approach used the simple finite difference approximation (10). Of the OpenAD-produced code, the vector mode routine outperformed the scalar mode routine, as is expected within strip mining approaches [14]. While these efficiency differences were statistically significant, they were not dramatic. For the full 19 and 9 point stencils in 3D and 2D, the vector mode OpenAD code was less than a factor of 2 slower than the finite difference routine, and even for the reduced 7 point stencil it only took a factor of 5 more of wall clock time. This slowdown of the vector vs. finite difference routines as the stencil approximation shrinks can be understood through our use of full 152-entry vectors within the OpenAD calculations. While finite difference calculations only compute with the relevant nonzero entries in the stencil patch, the vector mode OpenAD code performs array operations on all possible dependencies in the `max_deriv_vec_len` length array. Therefore, due to our current desire for

Table 1 Average wall-clock times and numerical accuracy for Jacobian construction approaches (*S* scalar, *V* vector, *FD* finite difference). All times are reported in seconds, and correspond to the average wall clock time required per spatial cell over the various grid sizes. Finite difference accuracy values are averaged over the test grids for each stencil

Dimension	Stencil (pt)	S time	V time	FD time	FD accuracy
3	19	3.484e−3	4.728e−4	2.868e−4	9.996e−5
3	11	1.515e−3	4.201e−4	1.452e−4	1.579e−4
3	7	8.946e−4	3.947e−4	8.085e−5	1.259e−4
2	9	1.586e−3	2.476e−4	1.528e−4	5.015e−6
2	5	5.127e−4	2.165e−4	4.887e−5	1.652e−5

flexibility in preconditioner construction, based on either full or reduced stencil Jacobian approximations, all codes set this globally-defined OpenAD parameter to 152. Hence the vector mode tests on reduced stencils end up wasting cycles performing unnecessary calculations.

Additionally, $\hat{\mathbf{f}}_i(\hat{\mathbf{U}}_i)$ computes a number of intermediate quantities that are reused between different output values in the same spatial cell. As a result, the scalar mode AD routines must recompute these values at each call, whereas the vector mode and FD routines naturally reuse this information. Resultingly, the scalar approach was the slowest of the three methods.

However, in typical nonlinear solvers the overall computational speed is determined more by the number of iterations required to converge than by the preconditioner construction time. It is in this metric that the accuracy of the Jacobian entries becomes relevant, since inaccuracies in P^{-1} can slow convergence. Moreover, it is here that the OpenAD-generated routines easily surpass the finite difference approximation. The accuracy of a one-sided finite difference calculation (10) is $O(\sigma)$, which in double precision and for ideally normalized units is at best the chosen value of 10^{-8} [7]. Furthermore, floating-point inaccuracies in the evaluation of $\hat{\mathbf{f}}_\mathbf{i}(\hat{\mathbf{U}}_\mathbf{i})$ can further deteriorate the approximation accuracy. As noted in Fig. 1, our mapping $(r, z) \rightarrow (\xi, \eta)$ results in an increase in floating-point evaluation error near the plasma core. As a result, the finite difference accuracy reported in Table 1 shows that these approximations only retain from 3 to 5 significant digits. Meanwhile, since the OpenAD-generated routines analytically differentiate $\hat{\mathbf{f}}(\tilde{\mathbf{U}}_i)$, the resulting Jacobian error is orders of magnitude more accurate.

3.1 Conclusions and Future Work

In this paper, we have explored a somewhat straightforward use of AD, but in a rather complex application resulting from a system of mapped-grid nonlinear PDEs for visco-resistive magnetohydrodynamics. Through development of a highly flexible patch-based approach to modification of our code, we were able to apply the OpenAD tool to generate relatively efficient source code that allows exploration of a wide variety of preconditioning options in our application.

We are currently exploring optimal preconditioners for our problem, which will help determine which reduced stencil approximation, if any, we wish to use in a production code. Once this determination has been made, further optimizations in our use of OpenAD are possible. Specifically, our choice of stencil will determine the optimum value of the `max_deriv_vec_len` parameter within OpenAD's vector mode, which we can use to eliminate unnecessary derivative calculations. Additionally, we may apply techniques similar to [11] to compute Jacobian blocks only once per finite-volume face, however unlike in that work our use of cylindrical coordinates and spatial mapping require additional manipulations of these blocks before adding their contributions to the overall Jacobian matrix.

Acknowledgements The work of D. Reynolds was supported by the U.S. Department of Energy, in grants DOE-ER25785 and LBL-6925354. R. Samtaney acknowledges funding from KAUST for the portion of this work performed at KAUST.

We would also like to thank Steve Jardin for insightful discussions on preconditioning techniques for toroidal plasmas, and Mike Fagan for help in building and using OpenAD for this work.

References

1. Bischof, C.H., Green, L., Haigler, K., Knauff, T.: Calculation of sensitivity derivatives for aircraft design using automatic differentiation. In: Proceedings of the 5th AIAA/-NASA/USAF/ISSMO Symposium on Multidisciplinary Analysis and Optimization, AIAA 94-4261, pp. 73–84. American Institute of Aeronautics and Astronautics (1994). Also appeared as Argonne National Laboratory, Mathematics and Computer Science Division, Preprint MCS-P419-0294.
2. Coleman, T.F., Garbow, B.S., Moré, J.J.: Software for estimating sparse Jacobian matrices. ACM Trans. Math. Software **10**(3), 329–345 (1984)
3. Gebremedhin, A.H., Manne, F., Pothen, A.: What color is your Jacobian? graph coloring for computing derivatives. SIAM Review **47**(4), 629–705 (2005). DOI 10.1137/S0036144504444711. URL http://link.aip.org/link/?SIR/47/629/1
4. Griewank, A., Walther, A.: Evaluating Derivatives: Principles and Techniques of Algorithmic Differentiation, 2nd edn. No. 105 in Other Titles in Applied Mathematics. SIAM, Philadelphia, PA (2008). URL http://www.ec-securehost.com/SIAM/OT105.html
5. Hindmarsh, A.C. et al.: SUNDIALS, suite of nonlinear and differential/algebraic equation solvers. ACM Trans. Math. Softw. **31**(3), 363–396 (2005)
6. Hovland, P.D., McInnes, L.C.: Parallel simulation of compressible flow using automatic differentiation and PETSc. Tech. Rep. ANL/MCS-P796-0200, Mathematics and Computer Science Division, Argonne National Laboratory (2000). To appear in a special issue of *Parallel Computing* on "Parallel Computing in Aerospace"
7. Kelley, C.T.: Iterative Methods for Linear and Nonlinear Equations. SIAM, Philadelphia (1995)
8. Knoll, D.A., Keyes, D.E.: Jacobian-free Newton-Krylov methods: a survey of approaches and applications. J. Comput. Phys. **193**, 357–397 (2004)
9. Reynolds, D., Samtaney, R., Woodward, C.: A fully implicit numerical method for single-fluid resistive magnetohydrodynamics. J. Comput. Phys. **219**, 144–162 (2006)
10. Reynolds, D., Samtaney, R., Woodward, C.: Operator-based preconditioning of stiff hyperbolic systems. SIAM J. Sci. Comput. **32**, 150–170 (2010)

11. Tadjouddine, M., Forth, S., Qin, N.: Elimination AD applied to Jacobian assembly for an implicit compressible CFD solver. Int. J. Numer. Meth. Fluids **47**, 1315–1321 (2005)
12. Utke, J.: OpenAD. http://www.mcs.anl.gov/OpenAD
13. Utke, J., Naumann, U., Fagan, M., Tallent, N., Strout, M., Heimbach, P., Hill, C., Wunsch, C.: OpenAD/F: A modular, open-source tool for automatic differentiation of Fortran codes. ACM Transactions on Mathematical Software **34**(4), 18:1–18:36 (2008). DOI 10.1145/1377596.1377598
14. Utke, J., Naumann, U., Lyons, A.: OpenAD/F: User Manual. Tech. rep., Argonne National Laboratory. Latest version available online at http://www.mcs.anl.gov/OpenAD/openad.pdf
15. Woods, L.C.: Principles of Magnetoplasma Dynamics. Clarendon Press, Oxford (1987)

Combining Automatic Differentiation Methods for High-Dimensional Nonlinear Models

James A. Reed, Jean Utke, and Hany S. Abdel-Khalik

Abstract Earlier work has shown that the efficient subspace method can be employed to reduce the effective size of the input data stream for high-dimensional models when the effective rank of the first-order sensitivity matrix is orders of magnitude smaller than the size of the input data. Here, the method is extended to handle nonlinear models, where the evaluation of higher-order derivatives is important but also challenging because the number of derivatives increases exponentially with the size of the input data streams. A recently developed hybrid approach is employed to combine reverse-mode automatic differentiation to calculate first-order derivatives and perform the required reduction in the input data stream, followed by forward-mode automatic differentiation to calculate higher-order derivatives with respect only to the reduced input variables. Three test cases illustrate the viability of the approach.

Keywords Reverse mode • Higher-order derivatives • Low-rank approximation

1 Introduction

As is the case in many numerical simulations in science and engineering, one can use derivative information to gain insight into the model behavior. Automatic differentiation (AD) [7] provides a means to efficiently and accurately compute such derivatives to be used, for example, in sensitivity analysis, uncertainty propagation,

J.A. Reed (✉) · H.S. Abdel-Khalik
Department of Nuclear Engineering, North Carolina State University, Raleigh, NC, USA
e-mail: jareed3@ncsu.edu; abdelkhalik@ncsu.edu

J. Utke
Argonne National Laboratory, The University of Chicago, Chicago, IL, USA
e-mail: utke@mcs.anl.gov

S. Forth et al. (eds.), *Recent Advances in Algorithmic Differentiation*, Lecture Notes in Computational Science and Engineering 87, DOI 10.1007/978-3-642-30023-3_3,

and design optimization. The basis for AD is the availability of a program that implements the model as source code. Transforming or reinterpreting the source code enables the derivative computation. Given the complexity of the numerical simulations the derivative computation, however, can remain quite costly, despite the efficiency gains made possible by AD techniques.

Exploiting model properties that are known at a higher mathematical level but are not easily recognizable at the source code level in an automatic fashion is a major factor for improving the efficiency of derivative based methods. Problems in nuclear engineering provide a good example for such higher-level properties. Detailed nuclear reactor simulations involve high-dimensional input and output streams. The effective numerical rank r of such models, however, is known to be typically much lower than the size of the input and output streams naively suggests. By reducing the higher-order approximation of the model to r (pseudo) variables, one can significantly reduce the approximation cost while maintaining reasonable approximation errors. This approach, the efficient subspace method (ESM), is discussed in Sect. 2. The implementation with AD tools is described in Sect. 3, and three test cases are presented in Sect. 4.

2 Methodology

For simplicity we begin with constructing a low-rank approximation to a matrix operator. Let $\boldsymbol{A} \in \mathbb{R}^{m \times n}$ be the unknown matrix, and let the operator provide matrix vector products with $\boldsymbol{A}$ and $\boldsymbol{A}^T$. The following steps yield a low-rank approximation of $\boldsymbol{A}$:

1. Form k matrix vector products $\boldsymbol{y}^{(i)} = \boldsymbol{A}\boldsymbol{x}^{(i)}, i = 1, \ldots, k$ for randomly chosen Gaussian vectors $\boldsymbol{x}^{(i)}$ (assume stochastic independence for all random vectors).
2. QR factorize the matrix of responses: $\left[\boldsymbol{y}^{(1)} \ldots \boldsymbol{y}^{(k)}\right] = \boldsymbol{QR} = \left[\boldsymbol{q}^{(1)} \ldots \boldsymbol{q}^{(k)}\right]\boldsymbol{R}$.
3. Determine the effective rank r by using the rank finding algorithm (RFA):
   (a) Choose a sequence of k random Gaussian vectors $\boldsymbol{w}^{(i)}$.
   (b) Compute $z^{(i)} = (\boldsymbol{I} - \boldsymbol{QQ}^T)\boldsymbol{A}\boldsymbol{w}^{(i)}$.
   (c) Test for any i if $||z^{(i)}|| > \epsilon$, if true, increment k and go back to step 1 else set $r = k$ and continue.
4. Calculate $\boldsymbol{p}^{(i)} = \boldsymbol{A}^T\boldsymbol{q}^{(i)}$ for all $i = 1, \ldots, k$.
5. Using the $\boldsymbol{p}^{(i)}$ and $\boldsymbol{q}^{(i)}$ vectors, calculate a low-rank approximation of the form $\boldsymbol{A} = \boldsymbol{USV}^T$, as shown in the appendix of [1].

Halko et al. showed in [8] that with at least $1–10^{-k}$ probability, one can determine a matrix $\boldsymbol{Q}$ of rank r such that the following error criterion is satisfied

$$||(\boldsymbol{I} - \boldsymbol{QQ}^T)\boldsymbol{A}|| \leq \epsilon/(10\sqrt{2/\pi}) ,$$

where ϵ is the user specified error allowance.

In real applications, these ideas can be applied by replacing the matrix operator with a computational model. Let the computational model of interest be described by a vector-valued function $\boldsymbol{y} = \Theta(\boldsymbol{x})$, where $\boldsymbol{y} \in \mathbb{R}^m$ and $\boldsymbol{x} \in \mathbb{R}^n$. The goal is to compute all derivatives of a given order by reducing the dimensions of the problem and thus reducing the computational and storage requirements. First we consider the case $m = 1$. A function $\Theta(\boldsymbol{x})$ can be expanded around a reference point $\boldsymbol{x}_0$. Bang et al. showed in [2] that an infinite Taylor-like series expansion may be written as follows (without loss of generality, assume $\boldsymbol{x}_0 = \boldsymbol{0}$ and $\Theta(\boldsymbol{x}_0) = 0$):

$$\Theta(\boldsymbol{x}) = \sum_{k=1}^{\infty} \sum_{j_1,\ldots,j_l,\ldots,j_k=1}^{n} \psi_1(\beta_{j_1}^{(k)T}\boldsymbol{x}) \ldots \psi_l(\beta_{j_l}^{(k)T}\boldsymbol{x}) \ldots \psi_k(\beta_{j_k}^{(k)T}\boldsymbol{x}), \tag{1}$$

where the $\{\psi_l\}_{l=1}^{\infty}$ can be any kind of scalar functions. The outer summation over the variable k goes from 1 to infinity. Each term represents one order of variation; $k = 1$ represents the first-order term and $k = 2$ the second-order terms. For the case of $\psi_l(\theta) = \theta$, the kth term reduces to the kth term in a multivariable Taylor series expansion. The inside summation for the kth term consists of k single-valued functions $\{\psi_l\}_{l=1}^{\infty}$ that are multiplying each other. The arguments for the $\{\psi_l\}_{l=1}^{\infty}$ functions are scalar quantities representing the inner products between the vector $\boldsymbol{x}$ and n vectors $\{\beta_{j_l}^{(k)}\}_{j_l=1}^{n}$ that span the parameter space. The superscript (k) implies that a different basis is used for each of the k terms, that is, one basis is used for the first-order term, another for the second-order term, and so on. Any input parameter variations that are orthogonal to the range formed by the collection of the vectors $\{\beta_{j_l}^{(k)}\}$ will not produce changes in the output response. If the $\{\beta_{j_l}^{(k)}\}$ vectors span a subspace of dimension r as opposed to n (i.e., $dim(span\{\beta_{j_l}^{(k)}\}) = r$), then the effective number of input parameters can be reduced from n to r. The mathematical range can be determined by using only first-order derivatives. Differentiating (1) with respect to $\boldsymbol{x}$ gives

$$\nabla\Theta(\boldsymbol{x}) = \sum_{k=1}^{\infty} \sum_{j_1,\ldots,j_l,\ldots,j_k=1}^{n} \left(\psi_l'(\beta_{j_l}^{(k)T}\boldsymbol{x})\beta_{j_l}^{(k)} \prod_{i=1,i\neq l}^{k} \psi_i(\beta_{j_i}^{(k)T}\boldsymbol{x}) \right), \tag{2}$$

where $\psi_l'(\beta_{jl}^{(k)T}\boldsymbol{x})\beta_{jl}^{(k)}$ is the derivative of the term $\psi_l(\beta_{jl}^{(k)T}\boldsymbol{x})$. We can reorder (2) to show that the gradient of the function is a linear combination of the $\{\beta_{j_l}^{(k)}\}$ vectors

$$\nabla\Theta(\boldsymbol{x}) = \sum_{k=1}^{\infty} \sum_{j_1,\ldots,j_l,\ldots,j_k=1}^{n} \chi_{j_l}^{(k)}\beta_{j_l}^{(k)} = \begin{bmatrix} \cdots & \beta_{j_l}^{(k)} & \cdots \end{bmatrix} \begin{bmatrix} \vdots \\ \chi_{j_l}^{(k)} \\ \vdots \end{bmatrix} = \boldsymbol{B}\chi,$$

where

$$\chi_{j_l}^{(k)} = \psi_l'(\beta_{j_l}^{(k)T}\boldsymbol{x}) \prod_{i=1,i\neq l}^{k} \psi_i(\beta_{j_i}^{(k)T}\boldsymbol{x}).$$

In a typical application, the $\boldsymbol{B}$ matrix will not be known beforehand. One need only to know the range of $\boldsymbol{B}$, which can be accomplished by using the rank finding algorithm; see above. After determining the effective rank, the function depends only on r effective dimensions and can be reduced to simplify the calculation. The reduced model requires only the use of the subspace that represents the range of $\boldsymbol{B}$, of which there are infinite possible bases.

This concept is now expanded to a vector-valued model. The qth response $\Theta_q(\boldsymbol{x})$ of the model and its derivative $\nabla\Theta_q(\boldsymbol{x})$ can be written like (1) and (2) with an additional index q in the vectors $\{\beta_{j_l,q}^{(k)}\}$. The active subspace of the overall model must contain the contributions of each individual response. The matrix $\boldsymbol{B}$ will contain the $\{\beta_{j_l,q}^{(k)}\}$ vectors for all orders and responses. To determine a low-rank approximation, a pseudo response Θ_{pseudo} will be defined as a linear combination of the m responses:

$$\Theta_{pseudo}(\boldsymbol{x}) = \sum_{q=1}^{m} \gamma_q \sum_{k=1}^{\infty} \sum_{j_1,\ldots,j_l,\ldots j_k=1}^{n} \psi_1(\beta_{j_1,q}^{(k)T}\boldsymbol{x}) \ldots \psi_l(\beta_{j_l,q}^{(k)T}\boldsymbol{x}) \ldots \psi_k(\beta_{j_k,q}^{(k)T}\boldsymbol{x}), \quad (3)$$

where γ_q are randomly selected scalar factors. The gradient of the pseudo response is

$$\nabla\Theta_{pseudo}(\boldsymbol{x}) = \sum_{q=1}^{m} \gamma_q \sum_{k=1}^{\infty} \sum_{j_1,\ldots,j_l,\ldots j_k=1}^{n} \left(\psi_l'(\beta_{j_l,q}^{(k)T}\boldsymbol{x})\beta_{j_l,q}^{(k)} \prod_{i=1,i\neq l}^{k} \psi_i(\beta_{j_i,q}^{(k)T}\boldsymbol{x}) \right).$$

Calculating derivatives of the pseudo response as opposed to each individual response provides the necessary derivative information while saving considerable computational time for large models with many inputs and outputs.

3 Implementation

In this section we discuss the rationale for the specific AD approach, tool independent concerns, and some aspects of applying the tools to the problem scenario.

3.1 Gradients with OpenAD

The numerical model Θ has the form $\boldsymbol{y} = \Theta(\boldsymbol{x})$, where $\boldsymbol{y} \in \mathbb{R}^m$ is the output and $\boldsymbol{x} \in \mathbb{R}^n$ the input vector. No additional information regarding Θ is required other than the program P implementing Θ. Following (3), we define the pseudo response $\tilde{y}$ as the weighted sum

$$\tilde{y} = \sum_{i=1}^{m} \gamma_i y_i . \tag{4}$$

This requires a change in P but is easily done in a suitable top-level routine. The source code (Fortran) for the modified program $\tilde{P}$ becomes the input to OpenAD [10], which facilitates the computation of the gradient $\nabla\tilde{y}$ using reverse-mode source transformation. The overall preparation of the model and the first driver was done following the steps outlined in [11].

The source code of MATWS (see Sect. 4) exhibited some of the programming constructs known to be obstacles for the application of source transformation AD. Among them is the use of `equivalence`, especially for the initialization of `common` blocks. The idiom there was to equivalence an array of length 1 with the first element in the common block. Then, the length-1 array was used to access the entire common block via subscript values greater than 1, which does not conform to the standard (though this can typically not be verified at compile time). Similar memory-aliasing patterns appear to be common in nuclear engineering models.

OpenAD uses *association by address* [5], that is an active type, as the means of augmenting the original program data to hold the derivative information. The usual activity analysis would ordinarily trigger the redeclaration of only a subset of common block variables. Because the access of the common block via the array enforces a uniform type for all common block variables to maintain proper alignment, all common block variables had to be activated. Furthermore, because the equivalence construct applied syntactically only to the first common block variable, the implicit equivalence of all other variables cannot be automatically deduced and required a change of the analysis logic for OpenAD to maintain alignment by conservatively overestimating the active variable set. The alternatively used *association by name* [5] would likely resort to the same alignment requirement.

Once the source transformation succeeds, suitable driver logic must be written to accommodate the steps needed for k evaluations of the gradient $\nabla\tilde{y}^{(j)}$ using random weights $\gamma_i^{(j)}$ and randomly set Gaussian inputs $x_i^{(j)}$. The k gradients form the columns of

$$\boldsymbol{G} = \left[\nabla\tilde{y}^{(1)}, \ldots, \nabla\tilde{y}^{(k)}\right]$$

$\boldsymbol{G}$ is QR factorized, $\boldsymbol{G} = \boldsymbol{QR} = [\boldsymbol{Q}_r\ \boldsymbol{Q}_2]\,\boldsymbol{R}$, where the submatrix $\boldsymbol{Q}_r \in \mathbb{R}^{n\times r}$ contains only the first r columns of $\boldsymbol{Q}$. The rank is selected to satisfy a user-defined error metric such that $||(\boldsymbol{I} - \boldsymbol{Q}_r\boldsymbol{Q}_r^T)\boldsymbol{G}|| < \epsilon$ The columns of $\boldsymbol{Q}_r$ are used to define the (reduced) pseudo inputs $\tilde{\boldsymbol{x}} = \boldsymbol{Q}_r^T\,\boldsymbol{x}$. Because of orthogonality we can simply prepend to the original program P the logic implementing $\boldsymbol{x} = \boldsymbol{Q}_r\,\tilde{\boldsymbol{x}}$ to have the $\tilde{\boldsymbol{x}}$ as our new reduced set of input variables for which derivatives will be computed. Similar to (4), this is easily done by adding code in a suitable top-level routine, yielding $\hat{P}(\tilde{\boldsymbol{x}}) = \boldsymbol{y}$, $\hat{P} : \mathbb{R}^r \mapsto \mathbb{R}^m$, which is the effective model differentiated by Rapsodia.

3.2 Higher-Order Derivatives with Rapsodia

Rapsodia [4] is used to compute all derivative tensor elements up to order o:

$$\left[\frac{\partial^{o} y_i}{\partial \tilde{x}_1^{o_1} \dots \partial \tilde{x}_r^{o_r}}\right], \quad \text{with multi-index } \boldsymbol{o}, \text{ where} \quad o = |\boldsymbol{o}| = \sum_{k=1}^{r} o_k, \tag{5}$$

for $\hat{P}$ following the interpolation approach in [6] supported by Rapsodia (see also [3]). Rapsodia is based on operator overloading for the forward propagation of univariate Taylor polynomials. Most other operator overloading-based AD tools have overloaded operators that are hand-coded, operate on Taylor coefficient arrays with variable length in loops with variable bounds to accommodate the derivative orders and numbers of directions needed by the application. In contrast, Rapsodia generates on demand a library of overloaded operators for a specific number of directions and a specific order. The generated code exhibits (partially) flat data structures, partially unrolled loops over the directions, and fully unrolled loops over the derivative order. This implies few array dereferences in the generated code, which in turn provides more freedom for compiler optimization yielding better performance than conventional overloaded operators even with fixed loop bounds. Because of the overall assumption that r, the reduced input dimension, is much smaller than m, the higher-order derivative computation in forward mode is feasible and appropriate.

Because overloaded operators are triggered by using a special type for which they are declared it now appears as a nice confluence of features that OpenAD for the gradient computation already does the data augmentation via association by address, i.e. via an active type. However, one cannot simply exchange the OpenAD and Rapsodia active types to use the operator overloading library. The following features of the OpenAD type change done for Sect. 3.1 can (partially) be reused.

Selective type change based on activity analysis: The main difference to Sect. 3.1 is the change of inputs from $\boldsymbol{x}$ to $\tilde{\boldsymbol{x}}$ and conversely $\tilde{\boldsymbol{y}}$ to $\boldsymbol{y}$. This requires merely changing the pragma declarations identifying the dependent and independent program variables in the top-level routine.

Type conversion for changing activity patterns in calls: The activity analysis intentionally does not yield matching activity signatures for all calling contexts of any given subroutine. Therefore, for a subroutine `foo(a,b,c)`, the formal parameters `a,c` may be determined as active while `b` remains passive. For a given calling context `call foo(d,e,f)` the type of the actual parameter `d` may be passive or `e` may be active, in which case pre- and postconversion calls to a type-matching temporary may have to be generated; see Fig. 1.

Default projections to the value component: The type change being applied to the program variables, arithmetic, and I/O statements referencing active variables is adapted to access the value component of the active type to replicate the original computation.

```
subroutine foo(a,b,c)     real :: d, t2; type(active):: e, f t1
  type(active)::a,c       !...
  real::b                 call cvrt_p2a(c,t1); call cvrt_a2p(d,t2)
  !....                   call foo(t1,t2,f)
end subroutine            call cvrt_a2p(t1,c); call cvrt_p2a(t2,d)
```

Fig. 1 Passive ↔ active type change conversions `cvrt_{p2a|a2p}` for a subroutine call `foo(d,e,f)` made by OpenAD for applying a Rapsodia-generated library (shortened names, active variables *underlined*)

These portions are implemented in the `TypeChange` algorithm stage in the OpenAD component `xaifBooster`. The last feature prevents triggering the overloaded operators, and the value component access needs to be dropped from the transformation. Following a common safety measure, there is no assignment operator or implicit conversion from active types to the passive floating-point types. Therefore, assignment statements to passive left-hand sides need to retain the value component access in the right-hand-side expressions. These specific modifications were implemented in OpenAD's postprocessor and are enabled by the `--overload` option. While manual type change was first attempted, it quickly proved a time-intensive task even on the moderately sized nuclear engineering source code, in particular because of the many I/O statements that would need adjustments and the fact that the Fortran source code given in fixed format made simple editor search and replaces harder. Therefore, the manual attempt was abandoned and this modification of the OpenAD source transformation capabilities proved useful.

Given the type change transformation, the tensors in (5) are computed with Rapsodia. The first-order derivatives in terms of $\boldsymbol{x}$ rather than $\tilde{\boldsymbol{x}}$ are recovered as follows.

$$\frac{\partial y_i}{\partial x_j} = \sum_{k=1}^{r} \underbrace{\frac{\partial y_i}{\partial \tilde{x}_k}}_{\in \widetilde{\boldsymbol{J}}} \frac{\partial \tilde{x}_k}{\partial x_j} = \sum_{k=1}^{r} \frac{\partial y_i}{\partial \tilde{x}_k} q_{jk}$$

In terms of the Jacobian this is $\boldsymbol{J} = \boldsymbol{Q}_r \widetilde{\boldsymbol{J}}$. Similarly for second order one has

$$\frac{\partial^2 y_i}{\partial x_j \partial x_g} = \sum_{k,l} \underbrace{\frac{\partial^2 y_i}{\partial \tilde{x}_k \partial \tilde{x}_l}}_{\in \widetilde{\boldsymbol{H}}_i} q_{jk} q_{gl},$$

which in terms of the Hessian $\boldsymbol{H}_i$ for ith output y_i is $\boldsymbol{H}_i = \boldsymbol{Q}_r \widetilde{\boldsymbol{H}}_i \boldsymbol{Q}_r^T$. The oth order derivatives are recovered by summing over the multi-indices $\boldsymbol{k}$.

$$\frac{\partial^o y_i}{\partial x_{j_1} \dots \partial x_{j_o}} = \sum_{|\boldsymbol{k}|=o} \frac{\partial^o y_i}{\partial \tilde{x}_{k_1} \dots \partial \tilde{x}_{k_o}} \prod_{l=1}^{o} q_{j_l k_l}$$

For all derivatives in increasing order, products of the q_{jk} can be incrementally computed.

4 Test Cases

Simple scalar-valued model. We consider an example model given as

$$y = \boldsymbol{a}^T\boldsymbol{x} + (\boldsymbol{b}^T\boldsymbol{x})^2 + \sin(\boldsymbol{c}^T\boldsymbol{x}) + \frac{1}{1 + e^{\boldsymbol{d}^T\boldsymbol{x}}},$$

where vectors $\boldsymbol{x}, \boldsymbol{a}, \boldsymbol{b}, \boldsymbol{c}$, and $\boldsymbol{d} \in \mathbb{R}^n$. The example model is implemented in a simple subroutine named `head` along with a `driver` main program that calls `head` and is used to extract the derivatives. Then `head` was transformed with OpenAD to compute the gradient of y with respect to the vector $\boldsymbol{x}$.

A Python script was written to execute the subspace identification algorithm with the compiled executable code. The script takes a guess k for the effective rank and runs the code for k random Gaussian input vectors $\boldsymbol{x}$. Within the Python script, the responses are collected into a matrix $\boldsymbol{G}$. Following the algorithm, a QR decomposition is then performed on $\boldsymbol{G}$, and the effective rank is found by using the RFA. The first r columns of $\boldsymbol{Q}$ are written to a file to be used as input to the Rapsodia code. With the model above with $n = 50$ and random input vectors with eight digits of precision for $\boldsymbol{a}, \boldsymbol{b}, \boldsymbol{c}$, and $\boldsymbol{d}$, with $\epsilon = 10^{-6}$, the effective rank was found to be $r = 3$.

The `driver` is then modified for use with Rapsodia, and the library is generated with the appropriate settings for the order and the number or directions. For first order the number of directions is simply the number of inputs. Once the derivatives $\left(\frac{dy}{d\tilde{x}}\right)$ are calculated , the full derivatives can be reconstructed by multiplying the Rapsodia results by the $\boldsymbol{Q}_r$ matrix used as input. With an effective rank of $r = 3$ and therefore a $\boldsymbol{Q}_r$ matrix of dimension 50×3, the reconstructed derivatives were found to have relative errors on the order of 10^{-13} compared with results obtained from an unreduced Rapsodia calculation.

Using Rapsodia to calculate second-order derivatives involves simply changing the derivative order (and the associated number of directions) to $o = 2$ and recompiling the code. The output can then be constructed into a matrix $\widetilde{\boldsymbol{H}}$ of size $r \times r$, and the full derivatives can be recovered by $\boldsymbol{Q}_r\tilde{\boldsymbol{H}}\boldsymbol{Q}_r^T$; the result is a $n \times n$ symmetric matrix. When the second-order derivatives are calculated for the example above, only six directions are required for an effective rank of 3, as opposed to 1,275 directions for the full problem. The relative errors of the reduced derivatives are on the order of 10^{-12}.

Third-order derivatives were also calculated using this example. The unreduced problem would require 22,100 directions, while the reduced problem requires only 10. Relative errors were much higher for this case but still at a reasonable order of 10^{-6}. The relative errors for each derivative order are summarized in Table 1.

We note here that in practice the derivatives are employed to construct a surrogate model that approximates the original function. Therefore, it is much more instructive to talk about the accuracy of the surrogate model employed rather than the accuracy of each derivative. This approach is illustrated in the third test case using an engineering model.

Table 1 Comparison of number of required directions for the unreduced and reduced model together with the relative error for simple vector test case

Derivative order	Unreduced directions	Reduced directions	Relative error
1	50	3	10^{-13}
2	1,275	6	10^{-12}
3	22,100	10	10^{-6}

Simple Vector-valued model. Problems with multiple outputs require a slightly different approach when determining the subspace. First we consider

$$y_1 = \boldsymbol{a}^T\boldsymbol{x} + (\boldsymbol{b}^T\boldsymbol{x})^2;\ y_2 = \sin(\boldsymbol{c}^T\boldsymbol{x}) + (1 + e^{-\boldsymbol{d}^T\boldsymbol{x}})^{-1};$$

$$y_3 = (\boldsymbol{a}^T\boldsymbol{x})(\boldsymbol{e}^T\boldsymbol{x});\ y_4 = 2^{\boldsymbol{e}^T\boldsymbol{x}};\ y_5 = (\boldsymbol{d}^T\boldsymbol{x})^{-3}.$$

Following (4), we compute the pseudo response $\tilde{y}$ in the modified version of the `head` routine, implementing the above example model with randomly generated factors γ_i that are unique for each execution of the code. The computed gradient is of $\tilde{y}$ with respect to $\boldsymbol{x}$. Then, following the same procedure as before, we ran the subspace identification script for $n = 50$ and $\epsilon = 10^{-6}$. The effective rank was found to be $r = 5$, and we found for similar accuracy the reduction of directions needed for the approximation from 250 to 5, 6,375 to 75, and 110,500 to 175 for first up to third order, respectively.

MATWS. A more realistic test problem was done with the MATWS (a subset of SAS4A [9]) Fortran code for nuclear reactor simulations. Single-channel calculations were performed by using as inputs the axial expansion coefficient, Doppler coefficient, moderator temperature coefficient, control rod driveline, and core radial expansion coefficient. The outputs of interest are temperatures within the channel in the coolant, the structure, the cladding, and the fuel. These give a 4×5 output for the first-order derivatives and 15 and 35 directions for second and third order, respectively. After applying the subspace identification algorithm, the effective rank was found to be r $=$ 3, giving 6 and 10 directions for second- and third-order derivative calculations. The results were evaluated by using the derivatives from reduced and unreduced cases to construct a surrogate model. This surrogate model was then used to approximate temperature values of the model with a 0.01% input perturbation. The surrogate model that calculates the temperature vector $\boldsymbol{t}$ from perturbations of the input coefficients $\Delta\alpha$ was constructed as follows:

$$\boldsymbol{t} = \boldsymbol{t}_0 + \boldsymbol{J}\Delta\alpha + \begin{bmatrix} \vdots \\ \Delta\alpha^T \boldsymbol{H}_i \Delta\alpha \\ \vdots \end{bmatrix} \quad \text{with rows} \quad i = 1, \ldots, 5,$$

Table 2 Comparison of number of required directions for unreduced and the reduced model together with the relative errors for MATWS test case with 0.01% input perturbations

Derivative order	Unreduced directions	Reduced directions	Relative error (AD)	Relative error (real)
1	5	3	9.216×10^{-5}	6.695×10^{-3}
2	15	6	1.252×10^{-4}	3.182×10^{-3}

where $\boldsymbol{J}$ is the 4×5 Jacobian and the $\boldsymbol{H}_i$ are the 5×5 Hessians that correspond to each output. The maximum relative errors between the approximate temperature values calculated with the unreduced and reduced surrogate models are given in the "relative error (AD)" column of Table 2. The "relative error (real)" column gives the maximum relative difference between the reduced surrogate temperature calculations and the real values that a normal MATWS run gives. The typical temperature values that MATWS gives as output are about 800°F, making the absolute differences on the order of single degrees. Future work will focus on implementing this method on larger test problems in order to more drastically illustrate the potential computational savings.

This manuscript has presented a new approach to increase the efficiency of automatic differentiation when applied to high-dimensional nonlinear models where high-order derivatives are required. The approach identifies few pseudo input parameters and output responses that can be related to the original parameters and responses via simple linear transformations. Then, AD is applied to the pseudo variables, resulting in significant computational savings.

Acknowledgements This work was supported by the U.S. Department of Energy, under contract DE-AC02-06CH11357.

References

1. Abdel-Khalik, H.: Adaptive core simulation. Ph.D. thesis (2004). URL http://books.google.com/books?id=5moolOgFZ84C
2. Bang, Y., Abdel-Khalik, H., Hite, J.M.: Hybrid reduced order modeling applied to nonlinear models. International Journal for Numerical Methods in Engineering (to appear)
3. Charpentier, I., Utke, J.: Rapsodia: User manual. Tech. rep., Argonne National Laboratory. Latest version available online at http://www.mcs.anl.gov/Rapsodia/userManual.pdf
4. Charpentier, I., Utke, J.: Fast higher-order derivative tensors with Rapsodia. Optimization Methods Software **24**(1), 1–14 (2009). DOI 10.1080/10556780802413769
5. Fagan, M., Hascoët, L., Utke, J.: Data representation alternatives in semantically augmented numerical models. In: Proceedings of the Sixth IEEE International Workshop on Source Code Analysis and Manipulation (SCAM 2006), pp. 85–94. IEEE Computer Society, Los Alamitos, CA, USA (2006). DOI 10.1109/SCAM.2006.11
6. Griewank, A., Utke, J., Walther, A.: Evaluating higher derivative tensors by forward propagation of univariate Taylor series. Mathematics of Computation **69**, 1117–1130 (2000)
7. Griewank, A., Walther, A.: Evaluating Derivatives: Principles and Techniques of Algorithmic Differentiation, 2nd edn. No. 105 in Other Titles in Applied Mathematics. SIAM, Philadelphia, PA (2008). URL http://www.ec-securehost.com/SIAM/OT105.html

8. Halko, N., Martinsson, P.G., Tropp, J.A.: Finding structure with randomness: Probabilistic algorithms for constructing approximate matrix decompositions. SIAM Review **53**(2), 217–288 (2011). DOI 10.1137/090771806. URL http://link.aip.org/link/?SIR/53/217/1
9. SAS4A: http://www.ne.anl.gov/codes/sas4a/
10. Utke, J., Naumann, U., Fagan, M., Tallent, N., Strout, M., Heimbach, P., Hill, C., Wunsch, C.: OpenAD/F: A modular, open-source tool for automatic differentiation of Fortran codes. ACM Transactions on Mathematical Software **34**(4), 18:1–18:36 (2008). DOI 10.1145/1377596.1377598
11. Utke, J., Naumann, U., Lyons, A.: OpenAD/F: User Manual. Tech. rep., Argonne National Laboratory. Latest version available online at http://www.mcs.anl.gov/OpenAD/openad.pdf

Application of Automatic Differentiation to an Incompressible URANS Solver

Emre Özkaya, Anil Nemili, and Nicolas R. Gauger

Abstract This paper deals with the task of generating a discrete adjoint solver from a given primal Unsteady Reynolds Averaged Navier-Stokes (URANS) solver for incompressible flows. This adjoint solver is to be employed in active flow control problems to enhance the performance of aerodynamic configurations. We discuss on how the development of such a code can be eased through the use of the reverse mode of Automatic/Algorithmic Differentiation (AD). If AD is applied in a black-box fashion then the resulting adjoint URANS solver will have prohibitively expensive memory requirements. We present several strategies to circumvent the excessive memory demands. We also address the parallelization of the adjoint code and the adjoint counterparts of the MPI directives that are used in the primal solver. The adjoint code is validated by applying it to the standard test case of a rotating cylinder by active flow control. The sensitivities based on the adjoint code are compared with the values obtained from finite differences and forward mode AD code.

Keywords Unsteady discrete adjoints • Optimal flow control • Reverse mode of AD • Checkpointing • Reverse accumulation

1 Introduction

In the past few decades, the usage of adjoint methods have gained popularity in the design optimization. After being introduced by Pironneau [18] in fluid mechanics, the adjoint methods received much attraction after they are successfully used by

E. Özkaya (✉) · A. Nemili · N.R. Gauger
Computational Mathematics Group, CCES, RWTH Aachen University, Schinkelstr. 2, 52062 Aachen, Germany
e-mail: ozkaya@mathcces.rwth-aachen.de; nemili@mathcces.rwth-aachen.de; gauger@mathcces.rwth-aachen.de

S. Forth et al. (eds.), *Recent Advances in Algorithmic Differentiation*, Lecture Notes in Computational Science and Engineering 87, DOI 10.1007/978-3-642-30023-3_4,

Jameson [15] for aerodynamic design optimization. Jameson had derived the continuous adjoint method for compressible Euler equations, which was later extended to the compressible Navier-Stokes by Jameson et al. [16]. In the continuous adjoint method, one first derives the optimality system from a given object function (e.g. drag coefficient) and the state partial differential equations (PDEs) that are to be satisfied (e.g. Euler equations or Navier-Stokes). From the optimality conditions, the resulting adjoint PDEs can be written. The adjoint PDEs are then discretized and solved with a numerical method. Although being computationally efficient both in terms of memory and run time, the continuous adjoint approach is known to suffer from consistency problems. For the derivation of the continuous adjoints, one assumes that the primal solution of the underlying state PDEs is exact. But in practice one uses an approximate numerical solution instead of the exact primal one, which might lead to an error in the adjoint system. However, by refining the grid the numerical solution can be theoretically converged towards its exact value. Further, one has to take care that the discretization of the adjoint equations is consistent with the primal ones. Otherwise, one ends up with a consistency error. Yet another source of inconsistency is due to the *constant eddy viscosity* or the so called *frozen turbulence* assumption. This inconsistency emanates from the fact that the non-differentiability of some turbulence models in the primal Reynolds Averaged Navier-Stokes (RANS) equations results in non-existence of the corresponding adjoint equations. Therefore, one often treats the eddy viscosity as constant in deriving and solving the continuous adjoint equations. Because of these inconsistencies, the continuous approach lacks robustness and accuracy in the computation of gradient vectors [2]. Although some effort has been done in the past in order to derive the continuous adjoint formulation of some turbulence models (e.g. [22]), still a correct treatment is missing for many turbulence models, which are used in design optimization.

In contrast to the continuous adjoint method, one can follow the discrete adjoint approach, in which the discretized state PDEs are used to derive the optimality conditions for a given object function [6, 17]. An advantage of this approach is that it guarantees the consistency between the primal and discrete adjoint solutions on any arbitrary grid. Also, it does not have any inconsistency due to the frozen turbulence assumption as the discrete realizations of the turbulence models are algorithmically differentiable. Further, automatic differentiation (AD) techniques can be used to ease the development of discrete adjoint codes [4, 9].

In this paper, we present the development of a discrete adjoint solver for incompressible URANS equations using AD techniques. In Sect. 2, we describe the governing equations and the basic structure of an incompressible URANS solver based on a pressure-velocity coupling algorithm. Section 3 presents the details of differentiating the primal flow solver using reverse mode of AD. Further, we discuss the strategies to adjoin the unsteady time iterations, the pressure-velocity coupling scheme in each time iteration and MPI function calls. Finally, in Sect. 4 numerical results are shown to validate the unsteady discrete adjoint solver.

2 Incompressible URANS Equations and Flow Solver

The incompressible unsteady RANS equations govern subsonic flows, in which compressibility effects can be neglected. For subsonic flows with a small Mach number ($Ma < 0.3$), the density ρ and the laminar dynamic viscosity μ can be assumed as constants. In tensor notation, the unsteady RANS equations can be written in the absence of body forces as follows:

$$\frac{\partial(\rho \overline{U}_i)}{\partial X_i} = 0 \tag{1}$$

$$\frac{\partial(\rho \overline{U}_i)}{\partial t} + \frac{\partial}{\partial X_j}(\rho \overline{U}_i \overline{U}_j + \rho \overline{U_i' U_j'}) = -\frac{\partial \overline{p}}{\partial X_i} + \frac{\partial \overline{\tau}_{ij}}{\partial X_j}, \tag{2}$$

where $\overline{U}_i$ and U_i' denote the mean and fluctuating velocity fields, $\overline{p}$ is the mean pressure and $\overline{\tau}_{ij}$ denotes the mean viscous stress tensor, given by

$$\overline{\tau}_{ij} = \mu \left(\frac{\partial \overline{U}_i}{\partial X_j} + \frac{\partial \overline{U}_j}{\partial X_i} \right) . \tag{3}$$

The unknown Reynolds stresses $\rho \overline{U_i' U_j'}$ are modeled by the eddy-viscosity model

$$-\rho \overline{U_i' U_j'} = \mu_t \left(\frac{\partial \overline{U}_i}{\partial X_j} + \frac{\partial \overline{U}_j}{\partial X_i} \right) . \tag{4}$$

The eddy viscosity μ_t can be modeled by various turbulence models. In the present work, numerical simulations are performed by using the pressure-based URANS solver ELAN [21]. The ELAN code, written in FORTRAN 77 has various state of the art features and is on par with other industry-standard CFD codes. For the incompressible system, momentum and energy equations are decoupled so that one need not necessarily solve the energy equation unless the temperature distribution is desired. However, one difficulty in solving the incompressible RANS equations is to do the coupling between the pressure and the velocity fields. Various pressure-velocity coupling schemes have been developed and the most frequently used is the SIMPLE algorithm [1, 7].

In order to understand the general structure of an incompressible solver based on the SIMPLE scheme, we present the following pseudo-code for solving $2D$ URANS equations with $k - \omega$ turbulence model [20]:

Initialize velocity and pressure fields
for $t = T_0, t \leq T_N$ **do** *//Time iterations*
    for $i = 0, i \leq imax$ **do** *//Outer iterations*
        for $j = 0, j \leq jmax$ **do** solve the x-momentum equation

for $j = 0, j \leq jmax$ **do** solve the y-momentum equation
Compute the uncorrected mass fluxes at faces
for $j = 0, j \leq jmax$ **do** solve the pressure correction equation
Correct pressure field, face mass fluxes and cell velocities
for $j = 0, j \leq jmax$ **do** solve scalar equation for k
for $j = 0, j \leq jmax$ **do** solve scalar equation for ω
Calculate eddy-viscosity μ_t
if $(\|U_i - U_{i-1}\| \leq \epsilon)$ **and** $(\|p_i - p_{i-1}\| \leq \epsilon)$) **break**
endfor
endfor

It is important to note that we have three main loops in this solution strategy. Each j loop corresponds to a system of linear equations, which is solved iteratively by SIP (Strongly Implicit Procedure) or Stone's method. Usually in practice, the linear system of equations for each state variable is not solved very accurately but only some moderate number of iterations are performed. These iterations are known as inner iterations in the CFD community. The outer i loop corresponds to iterations of the pressure-velocity coupling scheme, which are known as outer iterations. In the above algorithm, `imax` and `jmax` represent the maximum number of outer and inner iterations respectively. At each time t, the outer iterations are performed until convergence is achieved. On that level we have a fixed point solution of the state vector $y = (U_i, p, k, \omega)$, which we denote by $y_* = G(y_*)$. Here the fixed point iterator G includes all the steps in one outer iteration. For the time iterations, usually we do not have a fixed point solution, since the state vector might have an oscillatory behavior due to the unsteadiness of the fluid flow.

3 Generation of a Discrete Adjoint Solver

Consider the case where we want to compute the sensitivities of the average drag coefficient ($C_{d,ave}$) with respect to some control u over the time interval $[0, T]$ with N time steps such that $0 = T_0 < T_1 < \ldots < T_N = T$. The mean drag coefficient in the discrete form is defined as:

$$C_{d,ave} = \frac{1}{N} \sum_{i=1}^{N} C_d(y(T_i), u). \tag{5}$$

The drag coefficient C_d at time T_i depends on the state vector $y(T_i)$ and the control variable u. Since a second order implicit scheme is used in the flow solver for the discretization of unsteady terms, the flow solution at time T_i depends on the flow solution at the times T_{i-1} and T_{i-2}, i.e. $y(T_i) = \phi(y(T_{i-1}), y(T_{i-2}))$. It is clear that in order to apply the reverse mode of AD, the first problem we have to tackle is the reversal of the time loop, which we address in the following section.

3.1 Reversal of the Time Loop

In general, the reversal of time evolution requires the storage of flow solutions at time iterations from T_0 to T_{N-1} during the forward sweep. The stored solutions are then used in solving the adjoint equations in reverse sweep from T_N to T_0. Storing the entire flow history in main memory is commonly known as the *store-all approach* in the AD community. For many practical aerodynamic configurations with millions of grid points and large values of N, the storage costs may become prohibitively expensive. Yet another way of reducing the memory requirements is by pursuing the recompute-all approach. In this method, the flow solutions are recomputed from the initial time T_0 for each time iteration of the reverse sweep. It is very clear that this approach results in minimal memory as there is no storing of the intermediate flow solutions. On the other hand, the CPU time increases drastically as one has to recompute $(N^2 - N)/2$ flow solutions. Thus, it can be argued that this method is practically infeasible from the computational point of view.

A compromise between store-all and recompute-all approaches is the checkpointing strategy. In algorithms based on a checkpointing strategy, the flow solutions are stored only at selective time iterations known as checkpoints. These checkpoints are then used to recompute the intermediate states that have not been stored. In the present example, we chose r ($r \ll N$) checkpoints. We then have $0 = T_0 = T_{C_1} < T_{C_2} < \cdots < T_{C_{r-1}} < T_{C_r} < T_N = T$. Here, T_{C_r} represents the time at r^{th} checkpoint. During the adjoint computation over the subinterval $[T_{C_r}, T_N]$, required flow solutions at intermediate time iterations are recomputed by using the stored solution at T_{C_r} as the initial condition. The above procedure is then repeated over other subintervals $[T_{C_{r-1}}, T_{C_r}]$ until we compute all the adjoints. It may be noted that the checkpoints can be reused as and when they become free. We designate them as intermediate checkpoints. In the present work, we have used the binomial checkpointing strategy, which is implemented in the algorithm `revolve` [11]. This algorithm generates the checkpointing schedules in a binomial fashion so that the number of flow re-computations is proven to be optimal.

3.2 Adjoining the Outer Iterations

After introducing the time reversal scheme for adjoining the unsteady part, we now focus our attention to adjoin the outer iterations at each unsteady time step. If reverse mode AD is applied to the outer iterations in a black-box fashion then the resulting adjoint code will have tremendous memory requirements. This is due to the fact that the black-box approach needs taping the flow solution for all outer iterations in the forward sweep although the reverse sweep requires only the converged flow solution due to the existence of a fixed point in that level [5] (i.e., state vector converges to some solution such that the velocity corrections tend to zero). Therefore, a lot of memory and run time can be saved if we make use of the iterative structure and store only the converged flow solution in each physical timestep. Further, it is highly desirable to have independent convergence criteria for the iterative schemes in forward and reverse sweeps. One way of achieving this objective is by employing the reverse accumulation technique [3, 8], which is also referred as the two-phase method [12]. In this approach, the primal iterative scheme in the forward sweep and the adjoint iterative scheme in the reverse sweep are solved separately, one after the other. In yet another approach known as piggy-backing [10], the primal and adjoint solution iterations are performed simultaneously. In the present work, we pursue the reverse accumulation strategy in adjoining the outer iterations.

Consider the total derivative of a general object function f (e.g. in the present case $C_d\,(y\,(t))$) with respect to the control u at the converged state solution y^* for any time step T_i:

$$\frac{df(y^*, u)}{du} = \frac{\partial f(y^*, u)}{\partial u} + \frac{\partial f(y^*, u)}{\partial y^*}\frac{dy^*}{du}. \tag{6}$$

On the other hand, from the primal fixed point equation $y^* = G(y^*, u)$, we get:

$$\frac{dy^*}{du} = \frac{\partial G(y^*, u)}{\partial u} + \frac{\partial G(y^*, u)}{\partial y^*}\frac{dy^*}{du} = \left(I - \frac{\partial G(y^*, u)}{\partial y^*}\right)^{-1}\frac{\partial G(y^*, u)}{\partial u}. \tag{7}$$

Multiplying on both sides with $\frac{\partial f(y^*, u)}{\partial y^*}$, we obtain

$$\frac{\partial f(y^*, u)}{\partial y^*}\frac{dy^*}{du} = \underbrace{\frac{\partial f(y^*, u)}{\partial y^*}\left(I - \frac{\partial G(y^*, u)}{\partial y^*}\right)^{-1}}_{:=\overline{y}^{*\top}}\frac{\partial G(y^*, u)}{\partial u}. \tag{8}$$

From the definition of $\overline{y}^{*\top}$ in (8) and making use of equation (7), the adjoint fixed point iteration can be written as:

$$\overline{y}^{*\top} = \overline{y}^{*\top}\frac{\partial G(y^*, u)}{\partial y^*} + \frac{\partial f(y^*, u)}{\partial y^*}. \tag{9}$$

The first term on the right hand side of the above equation is the adjoint of a single outer iteration, which can be generated by applying reverse mode of AD to the wrapper subroutine G, which combines all steps done in a single iteration of the SIMPLE scheme. Function f on the other hand can be thought as a post-processing step (i.e. integration of aerodynamic forces around an object under interest). The reduced gradient vectors $(\partial f/\partial y^*)$ and $(\partial f/\partial u)$ come from the adjoint of the post-processor, which is computed only once for each time iteration. Since post-processing is not an iterative process like a flow solver, the adjoint post-processing tool can be easily generated by applying reverse mode of AD in a black-box manner. In the present work, the adjoints of f and G are developed using the source transformation based AD tool Tapenade [13] in reverse mode.

3.3 Parallelization of the Adjoint Solver

For simulations on fine grids with a large number of unsteady time iterations, the primal ELAN solver is executed on a cluster using multi-block grids and MPI parallelization. For example, at the end of each outer iteration, i.e. each ith iteration, the state solution must be exchanged at the block interfaces using ghost layers. Since we differentiate a single outer iteration using the reverse mode, MPI calls must also be adjoined properly. In [14] adjoining MPI for the reverse mode of AD is presented on a general circulation model. Most of the AD tools, including Tapenade cannot differentiate the MPI calls algorithmically. Therefore, MPI calls should be declared as external subroutines prior to differentiation so that Tapenade assumes that the derivative routines are to be supplied by the user. Later, the MPI routines are adjoined manually and provided to the adjoint solver.

The primal solver has two types of communication: a `MPI_Sendrecv` call to exchange information at the block interface and `MPI_Allreduce` call to integrate the aerodynamic coefficients along the wall boundary, which is distributed to several processors. The `MPI_Sendrecv` calls are present inside the outer iterations, whereas `MPI_Allreduce` calls are used in the post-processing subroutines. In the present work, we limit ourself only to the adjoining of these two MPI directives. For the other directives and the details of adjoining MPI communications, the reader might refer to [19].

One can interpret a `MPI_Send(a)` statement and the corresponding receive `MPI_Receive(b)` statement as a simple assignment in the form of $b = a$. The only difference to a simple assignment is that the operation takes place between two processors using MPI communications. The adjoint of the assignment statement is $\overline{a}+ = \overline{b}$; $\overline{b} = 0$. By using the analogy we can conclude that the adjoint of `MPI_Send(a)` is : `MPI_Receive(t)`; $\overline{a}+ = t$. On the other hand the adjoint of `MPI_Receive(b)` is: `MPI_Send(`$\overline{b}$`)`. Combining these rules and applying them to the `MPI_Sendrecv` statement, which exchanges the field variable `PHI` between the processors `I` and `J`, we get the adjoint counterpart as follows:

```
primal:
 CALL MPI_SENDRECV(
&     PHI(index1),count,MPI_DOUBLE_PRECISION,dest,I,
&     PHI(index2),count,MPI_DOUBLE_PRECISION,source,J,
&     MPI_COMM_WORLD,ISTATUS,INFO)

adjoint:
 CALL MPI_SENDRECV_B(
&     PHI_B(index2),count,MPI_DOUBLE_PRECISION,source,J,
&     temp,count,MPI_DOUBLE_PRECISION,dest,I,
&     MPI_COMM_WORLD,ISTATUS,INFO)
 do i=1,count,1
    PHI_B(index1+i-1)=PHI_B(index1+i-1)+temp(i)
    PHI_B(index2+i-1)=0.0
 enddo
```

Here we denote the adjoint variables with suffix "_B"; so that `PHI_B` corresponds to the adjoint of `PHI`. Since this MPI call is a generic one, it can represent the exchange of any field variable u, v, p, etc. It is interesting to note that in the adjoint part, the data flow occurs in the reverse direction from `J` to `I`, which is expected by the nature of the reverse mode.

Another MPI directive we focus on, is the `MPI_Allreduce` call in the post-processing step. With this call, summation of an aerodynamic coefficient (e.g. drag coefficient) over different blocks is realized. An example call with `MPI_SUM` operation is:

```
    CALL MPI_ALLREDUCE(Z,ZS,N,
  & MPI_DOUBLE_PRECISION,MPI_SUM,MPI_COMM_WORLD,ISTATUS)
```

This operation is nothing but a summation statement in the form: $a = b_1 + b_2 + b_3 + .. + b_n$, which can be adjoined easily by: $\overline{b}_1 = \overline{a}, \overline{b}_2 = \overline{a}, ..., \overline{b}_n = \overline{a}$. By using the analogy we can write the adjoint of `MPI_Allreduce` with summation operator as:

```
   DO NN=1,N
      ZS_B(NN)=Z_B(NN)
   END DO
```

It should be noted that this treatment is valid only when `MPI_SUM` operation is used and does not apply to the general `MPI_Allreduce` directive.

4 Validation of the Adjoint Solver

We now present some numerical results to demonstrate the performance of the AD generated unsteady discrete adjoint solver. The test case under consideration is the drag minimization around a rotating cylinder. It is well-known that a turbulent flow around a cylinder causes flow separation, which results in the Karman vortex shedding, increase in drag coefficient, decrease in lift coefficient, etc. However, by rotating the cylinder, the flow separation can be delayed or even avoided, thus suppressing the intensity of the shedding and reducing the drag coefficient significantly. Figure 1 shows the snapshots of the flow and pressure contours around a non-rotating and a rotating cylinder in $2D$ with rate of rotation $u = 1.13$.

The optimization problem associated with this test case can then be regarded as finding the optimal rate of rotation, which results in a minimum drag. Note that the velocity distribution on the cylinder surface is computed by using its rate of rotation. This velocity profile is then used as a boundary condition for the momentum equations in the flow solver. Note that the rate of rotation is the only control parameter in this test problem. However, for practical applications of optimal flow control, the velocity at each grid point on the cylinder surface can be taken as

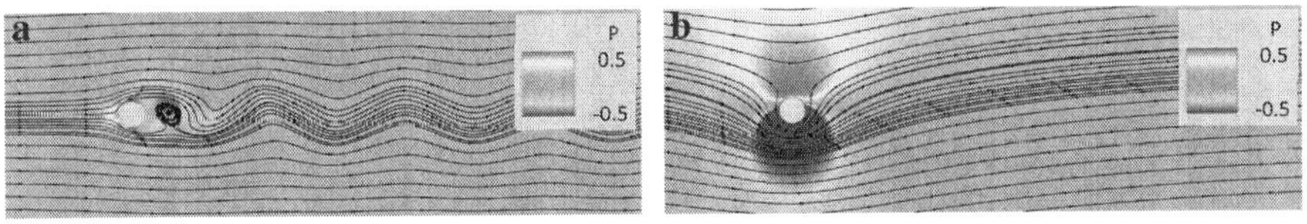

Fig. 1 (**a**) Rate of rotation $u = 0$ and (**b**) Rate of rotation $u = 1.13$, respectively show the snapshots of the flow around a non-rotating and a rotating cylinder

Table 1 A comparison of the sensitivities of the mean drag coefficient with respect to the rate of rotation

Finite volumes	Second order finite differences	Adjoint mode AD code	Forward mode AD code
12,640	−1.59385703413228	−1.59408478020429	−1.59407124877776
24,864	−1.55349944241934	−1.55356122452726	−1.55339068054409

a control parameter. In that case the number of control parameters will increase drastically, and the sensitivities with respect to these parameters can be computed efficiently using the discrete adjoint solver.

Numerical simulations are performed on two grid levels with 12,640 and 24,864 finite volumes respectively on a parallel cluster using eight processors. The Reynolds number is taken as $Re_D = 5{,}000$ while the time step and the rotational rate of the cylinder are chosen as $\Delta t = 0.1$ and $u = 0.1$ respectively. To reduce the storage requirements of the unsteady adjoint code, we have chosen 150 checkpoints. The objective function, which is the mean drag coefficient ($C_{d,ave}$) is defined as

$$C_{d,ave} = \frac{1}{(N - N*)} \sum_{i=N*+1}^{N} C_d\left(y\left(T_i\right), u\right) \tag{10}$$

Numerical simulations show a typical initial transient behavior in C_d up to $T_{N*} = 500$ time steps, which we neglect for our optimization problem. The control is defined from $T_{N*} = 500$ to $T_N = 1{,}500$ time steps. Table 1 shows a comparison of the sensitivities of the mean drag coefficient with respect to the rate of rotation, which is a function of adjoint state vector. It can be observed that the sensitivities based on the adjoint mode AD code are in good agreement with the values obtained from second order accurate finite differences and forward mode AD code. Note that, more accurate sensitivities can be computed by converging the primal and adjoint codes to machine precision. The increase in runtime due to the checkpointing strategy is found to be 1.9240 times compared to the usual store-all approach. It has been observed that the run time of the discrete adjoint code is approximately eight times higher compared to the primal code.

References

1. Caretto, L., Gosman, A., Patankar, S., Spalding, D.: Two calculation procedures for steady, three-dimensional flows with recirculation. In: Proceedings of the Third International Conference on Numerical Methods in Fluid Mechanics, *Lecture Notes in Physics*, vol. 19, pp. 60–68. Springer Berlin / Heidelberg (1973)
2. Carnarius, A., Thiele, F., Özkaya, E., Gauger, N.: Adjoint approaches for optimal flow control. AIAA Paper 2010–5088 (2010)
3. Christianson, B.: Reverse accumulation and attractive fixed points. Optimization Methods and Software **3**, 311–326 (1994)
4. Courty, F., Dervieux, A., Koobus, B., Hascoët, L.: Reverse automatic differentiation for optimum design: from adjoint state assembly to gradient computation. Optimization Methods and Software **18**(5), 615–627 (2003)
5. E. Özkaya, E., Gauger, N.: Automatic transition from simulation to one-shot shape optimization with Navier-Stokes equations. GAMM-Mitteilungen **33**(2), 133–147 (2010). DOI 10.1002/gamm.201010011. URL http://dx.doi.org/10.1002/gamm.201010011
6. Elliot, J., Peraire, J.: Practical 3D aerodynamic design and optimization using unstructured meshes. AIAA Journal **35**(9), 1479–1485 (1997)
7. Ferziger, J.H., Peric, M.: Computational Methods for Fluid Dynamics. Springer, Berlin; Heidelberg (2008)
8. Gauger, N., Walther, A., Moldenhauer, C., Widhalm, M.: Automatic differentiation of an entire design chain for aerodynamic shape optimization. Notes on Numerical Fluid Mechanics and Multidisciplinary Design **96**, 454–461 (2007)
9. Giles, M., Duta, M., Müller, J., Pierce, N.: Algorithm developments for discrete adjoint methods. AIAA Journal **41**(2), 198–205 (2003)
10. Griewank, A., Faure, C.: Reduced functions, gradients and hessians from fixed point iteration for state equations. Numerical Algorithms **30**(2), 113–139 (2002)
11. Griewank, A., Walther, A.: Algorithm 799:revolve: An implementation of checkpointing for the reverse or adjoint mode of computational differentiation. ACM Trans. Math. Software **26**(1), 19–45 (2000)
12. Griewank, A., Walther, A.: Evaluating Derivatives: Principles and Techniques of Algorithmic Differentiation, 2nd edn. No. 105 in Other Titles in Applied Mathematics. SIAM, Philadelphia, PA (2008). URL http://www.ec-securehost.com/SIAM/OT105.html
13. Hascoët, L., Pascual, V.: TAPENADE 2.1 user's guide. Rapport technique 300, INRIA, Sophia Antipolis (2004). URL http://www.inria.fr/rrrt/rt-0300.html
14. Heimbach, P., Hill, C., Giering, R.: An efficient exact adjoint of the parallel MIT general circulation model, generated via automatic differentiation. Future Generation Computer Systems **21**(8), 1356–1371 (2004)
15. Jameson, A.: Aerodynamic design via control theory. J. Sci. Comput. **3**, 233–260 (1988)
16. Jameson, A., Pierce, N., Martinelli, L.: Optimum aerodynamic design using the Navier–Stokes equations. J. Theor. Comp. Fluid Mech. **10**, 213–237 (1998)
17. Nielsen, E., Anderson, W.: Aerodynamic design optimization on unstructured meshes using the Navier-Stokes equations. AIAA Journal **37**(11), 957–964 (1999)
18. Pironneau, O.: On optimum design in fluid mechanics. J. Fluid Mech. **64**, 97–110 (1974)
19. Utke, J., Hascoet, L., Heimbach, P., Hill, C., Hovland, P., Naumann, U.: Toward adjoinable MPI. In: Parallel Distributed Processing, 2009. IPDPS 2009. IEEE International Symposium on, pp. 1–8 (2009). DOI 10.1109/IPDPS.2009.5161165
20. Wilcox, D.: Re-assessment of the scale-determining equation for advanced turbulence models. AIAA Journal **26**(11), 1299–1310 (1988)

21. Xue, L.: Entwicklung eines effizienten parallelen Lösungsalgorithmus zur dreidimensionalen Simulation komplexer turbulenter Strömungen. Ph.D. thesis, Technical University Berlin (1998)
22. Zymaris, A., Papadimitriou, D., Giannakoglou, K., Othmer, C.: Continuous adjoint approach to the Spalart-Allmaras turbulence model for incompressible flows. Computers & Fluids **38**, 1528–1538 (2009)

Applying Automatic Differentiation to the Community Land Model

Azamat Mametjanov, Boyana Norris, Xiaoyan Zeng, Beth Drewniak, Jean Utke, Mihai Anitescu, and Paul Hovland

Abstract Earth system models rely on past observations and knowledge to simulate future climate states. Because of the inherent complexity, a substantial uncertainty exists in model-based predictions. Evaluation and improvement of model codes are one of the priorities of climate science research. Automatic Differentiation enables analysis of sensitivities of predicted outcomes to input parameters by calculating derivatives of modeled functions. The resulting sensitivity knowledge can lead to improved parameter calibration. We present our experiences in applying OpenAD to the Fortran-based crop model code in the Community Land Model (CLM). We identify several issues that need to be addressed in future developments of tangent-linear and adjoint versions of the CLM.

Keywords Automatic differentiation • Forward mode • Climate model

1 Introduction

The Community Earth System Model (CESM) [2], developed by NCAR since 1983 and supported by NSF, NASA, and DOE, is a global climate model for the simulation of Earth's climate system. Composed of five fully coupled submodels of atmosphere, ocean, land, land ice, and sea ice, it provides state-of-the-art simulations for research of Earth's past, present, and future climate states on annual to decadal time scales. The coupled-system approach enables modeling of

A. Mametjanov · B. Norris · X. Zeng · J. Utke · M. Anitescu · P. Hovland (✉)
Mathematics and Computer Science, Argonne National Laboratory, 9700 South Cass Avenue, Argonne, IL 60439, USA
e-mail: hovland@mcs.anl.gov

B. Drewniak
Environmental Science, Argonne National Laboratory, 9700 South Cass Avenue, Argonne, IL 60439, USA

S. Forth et al. (eds.), *Recent Advances in Algorithmic Differentiation*, Lecture Notes in Computational Science and Engineering 87, DOI 10.1007/978-3-642-30023-3_5,

interactions of physical, chemical, and biological processes of atmosphere, ocean, and land subsystems without resorting to flux adjustments at the boundaries of the subsystems.

The CESM has been used in multicentury simulations of various greenhouse gases and aerosols from 1850 to 2100. It has also been used for various "what-if" scenarios of business-as-usual prognoses and prescribed climate policy experiments for acceptable climate conditions in the future up to the year 2100.

The Community Land Model (CLM) is a submodel of CESM for simulations of energy, water, and chemical compound fluxes within the land biogeophysics, hydrology, and biogeochemistry.

Because of the complexities of the global climate state, a significant variability exists in model-based predictions. Therefore, the primary goal of climate modeling is to enable a genuinely predictive capability at variable spatial resolutions and subcontinental regional levels [12].

The increasing availability of computing power enables scientists not only to analyze past climate observations but also to synthesize climate state many years into the future. CESM, for example, is executable not just on leadership-class supercomputers but also on notebook machines. In these settings of unconstrained availability of simulations, one can iteratively run a model in diagnostic mode to tune model parameters and execute prognostic runs with a higher degree of confidence. Nevertheless, because of the large number of parameters and the attendant combinatorial explosion of possible calibrations, uncertainty quantification and sensitivity analysis techniques are needed to estimate the largest variations in model outputs and rank the most sensitive model inputs.

In the optimization of numerical model designs, Automatic Differentiation (AD) [5] provides a method for efficiently computing derivatives of model outputs with respect to inputs. Derivatives can be used to estimate the sensitivity of outputs to changes in some of the inputs. Moreover, accurate derivatives can be obtained at a small multiple of the cost of computing model outputs, which makes AD more efficient than manual parameter perturbation and finite-difference-based calibration. AD has been widely used in applications in the physical, chemical, biological, and social sciences [1, 11]. In addition, AD has been applied successfully to the CLM code for sensitivity analysis of heat fluxes [14]. Our focus is on the biogeochemistry module of the CLM and in particular the carbon-nitrogen interactions within crops. We present our initial findings of differentiating the model code, the commonalities with previous applications, and differences that are specific to the crop model code.

We begin with an overview in Sect. 2 of the CLM model and its crop model subunit. Section 3 provides a brief overview of OpenAD. In Sect. 4, we describe the development of a tangent-linear code with OpenAD. In Sect. 5, we present the results of our experiment in applying OpenAD to the CLM's crop model, including a discussion of our experiences and lessons learned in the differentiation of climate code. Section 6 closes with a brief discussion of future work.

2 Background

The CESM provides a pluggable component infrastructure for Earth system simulations. Each of the five components can be configured in active (fully prognostic), stub (inactive/interface only), or data (intercomponent data cycling) modes, allowing for a variety of simulation cases. There is also a choice of a coupler—either MCT [8] or ESMF [4]—to coordinate the components and pass information between them. During the execution of a CESM case, the active components integrate forward in time, exchanging information with other active and data components and interfacing with stub components.

The land component in active mode models the land surface as a nested subgrid hierarchy, where each grid cell can have a number of different land units, each land unit can have a number of different columns, and each column can have a number of different plant functional types (PFTs). The first subgrid level, land unit, captures the broadest land surface patterns such as glacier, lake, wetland, urban, and vegetated patterns. The second subgrid level, column, has surface patterns similar to those of the enclosing land unit but captures vertical state variability with multiple layers of water and energy fluxes. The third level, PFT, captures chemical differences among broad categories of plants that include grasses, shrubs, and trees.

In order to improve the modeling of carbon and nitrogen cycles, the CLM has been updated with managed PFTs of corn, wheat, and soybean species. Each PFT maintains a state captured in terms of carbon and nitrogen (CN) pools located in leaves, stems, and roots and used for storage or growth. The CN fluxes among PFT structures determine the dynamics of vegetation. A significant contributing factor that affects CN fluxes is the ratio of CN within different structures. A large uncertainty exists regarding the CN ratios, which therefore are the primary targets of calibration in order to improve the overall model accuracy.

3 Automatic Differentiation and OpenAD

Automatic Differentiation [5] is a collection of techniques for evaluating derivatives of functions defined by computer programs. The foundation of AD is the observation that any function implemented by a program can be viewed as a sequence of elementary operations such as arithmetic and trigonometric functions. In other words, a program $\mathscr{P}$ implements the vector-valued function

$$\mathbf{y} = \mathbf{F}(\mathbf{x}) : \mathbb{R}^n \mapsto \mathbb{R}^m \tag{1}$$

as a sequence of p differentiable elemental operations:

$$v_i = \phi(\ldots, v_j, \ldots), i = 1, \ldots, p. \tag{2}$$

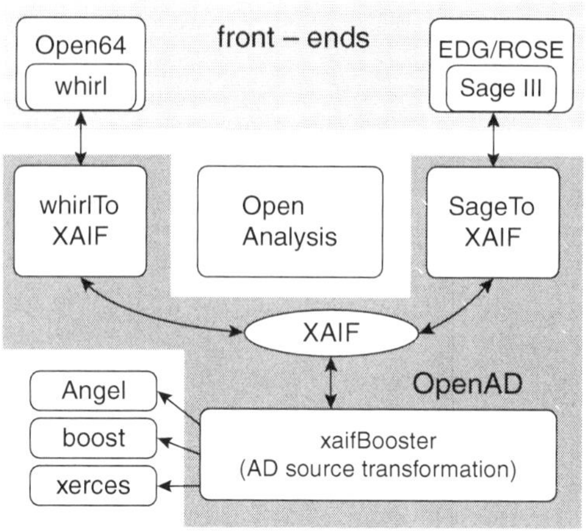

Fig. 1 OpenAD components. Front-ends parse the input source into an IR, which is further translated into XAIF that represents the numerical core of the input. After the AD of the core, the results are unparsed back into source text

The derivatives of elemental operations are composed according to the chain rule in differential calculus.

The key concepts in AD are *independent* variables $u \in \mathbb{R}^a, a \leq n$, and *dependent* variables $v \in \mathbb{R}^b, b \leq m$. Any variable within program $\mathscr{P}$ that depends on (or is *varied* by) values of independent variables and contributes to (or is *useful* for) values of dependent variables is known as *active*. Active variables have value and derivative components.

Because of the associativity of the chain rule, AD has two modes. In the *forward* (or *tangent-linear*) mode, derivative computation follows the original program control flow and accumulates derivative values from independent variables to dependent variables. In the *reverse* (or *adjoint*) mode, derivative computation follows the reverse of the original control flow, accumulating derivatives from dependent to independent variables.

Derivative values of active variables can be computed in at least three ways. First, source-to-source transformations can be used to derive a new program $\mathscr{P}'$ that adds new code to the original program code to propagate derivative values. Second, operator-overloading of elemental operations involving active variables can also be used to propagate the derivatives. Third, a complex-step method [7] can be used to represent active variables as complex numbers, where the real part stores original variable values and the imaginary part propagates derivative values.

OpenAD [15] is a source-to-source, transformation-based AD tool built from components (see Fig. 1). Two front-ends are currently supported: Rose [13] for C/C++ and Open64 [9] for Fortran 90. The intermediate representations (IRs) created by the front-ends are translated by using OpenAnalysis [10] into XML abstract interface format (XAIF) [6], which represents the numerical core of the input source. This representation is transformed to obtain the derivatives, and the result is unparsed back into the front-end's IR for further unparsing into the original source language.

4 AD Development Process

The intended process flow for the Automatic Differentiation of numerical codes is to limit manual intervention to the identification of independent and dependent variables and to let an AD tool generate an efficient code that computes valid derivatives. However, practical implementations of AD are not fully autonomous, and manual development is often necessary to pre- or postprocess the codes or to "stitch" together differentiated and other/external code. Such interventions are cross-cutting, requiring a collaborative effort between domain scientists who developed the original numerical code and AD developers who have expertise in source code analysis and transformation.

In order to reduce the need for manual intervention, it is important to identify patterns of effective (and ineffective) programming practices in the development of numerical codes as a means of making the code that implements the numerical functions more amenable to sensitivity analysis or other analyses requiring derivatives. The emerging patterns can then be targeted and automated by either source code refactoring tools or AD preprocessing tools.

In the current work of model optimization and parameter calibration, our initial goal was to identify whether AD can be performed at all and, if not, to identify obstacles for developing derivative code. To date, we have succeeded in the differentiation of a subunit of the land model code, and we have expended considerable effort into discovering and resolving the obstacles. In the process, we have gained some pattern- and process-related insights, which we report below. As we develop greater expertise in the cross-cutting issues in climate model and intrusive AD analysis domains, we expect greater efficiency and/or automation of AD development. Our goal is to develop and validate AD code for the entire CLM.

4.1 Code Comprehension

The initial step in any AD effort is to understand the original code. Typically, well-maintained codes have documentation in the form of installation guides, user manuals, and HTML documentation generated from source code comments. For AD, one also needs information about source code structures and modules. Dynamic function/procedure call graphs can provide dependency information.

The CLM source code consists of $\approx$70 K lines of code in biogeochemistry, biogeophysics, hydrology, and coupler modules. It is a well-documented Fortran 90 code with a user guide and manual that allow for quick installation and execution. However, most of the documentation is targeted at climate experts, with little information about implementation details or how to modify and extend the model code. Accordingly, we chose the CLM-Crop unit for the initial AD prototype because we had access to the climate scientist (B. Drewniak) who had recently extended the biogeochemistry module of the CLM with a model of managed crop species of corn, wheat, and soybeans [3].

To understand the dependencies between CLM-Crop and other subunits, we constructed a dynamic function call graph. This work entailed porting CESM from PGI compilers to the GNU compiler suite, which provides a built-in dynamic function call profiler `gprof`. Based on the call graph, the first candidates for AD were the nodes that had minimal calls to and from other nodes.

4.2 Preprocessing

Having identified the subroutines for AD, we started preparing the source code for OpenAD transformations. Since the code for differentiation must be visible to the tool, the recommended development pattern is to identify or create a top-level, or *head*, subroutine that invokes all other subroutines subject to AD. The annotations of independent and dependent variables are inserted into the head subroutine. Then, the head and all invoked subroutines are concatenated into a single file, which is transformed by the tool. The advantage of having a head subroutine is that it enables (1) seeding of derivatives of interest before any call to differentiated code and (2) extraction of computed derivatives upon completion of all computation of interest. Both seeding and extraction can be performed in a *driver* subroutine or program that invokes the head subroutine.

One of the frequent patterns that we encountered in the model code is the heavy use of preprocessor directives. They are used to statically slice out portions of code that are not used in a certain model configuration. An example is shown below.

```
psnsun_to_cpool(p)    = psnsun(p) * laisun(p) * 12.011e-6_r8
psnshade_to_cpool(p) = psnsha(p) * laisha(p) * 12.011e-6_r8
#if (defined C13)
  c13_psnsun_to_cpool(p)    = c13_psnsun(p) * laisun(p) * 12.011e-6_r8
  c13_psnshade_to_cpool(p) = c13_psnsha(p) * laisha(p) * 12.011e-6_r8
#endif
```

Here, operations related to *C13* are conditioned on whether that preprocessor flag is set. This kind of programming practice can substantially reduce the amount of code for differentiation, which in turn can produce a more efficient code. However, if the goals of differentiation change (e.g., to include new parameters to calibrate) and include the previously sliced-out code, then the result of the previous AD development effort is not reusable for the new AD goals.

A pattern for improved reusability and maintainability is to use control flow branching to evaluate different sections of code instead of relying on the preprocessor for integrating different semantics. For the example above, the preprocessor directives can be transformed to the following.

```
...
if (is_c13(pft_type(p))) then
  c13_psnsun_to_cpool(p)    = c13_psnsun(p) * laisun(p) * 12.011e-6_r8
  c13_psnshade_to_cpool(p) = c13_psnsha(p) * laisha(p) * 12.011e-6_r8
end if
```

Here, the operations are conditioned on whether the type of PFT p is *C13*. This version of model code promotes reuse by retaining the source code of a different model configuration.

4.3 Transformation

After all the source code has been preprocessed and collected into a file, the code can be passed to OpenAD for transformations. The language-agnostic and modular design of OpenAD allows for incremental transformations as follows:

- *Canonicalize* In order to reduce variability of the input code, it is preprocessed to make it more amenable for differentiation. For example, Fortran intrinsic functions *min* and *max* accept variable number of arguments and do not have closed form for partial derivatives. Calls to these functions are replaced with calls to three-argument library subroutines, which place the result of the first two arguments into the third.
- *Parse (fortran → ir)* The input source code is parsed with the front-end module and converted into its intermediate representation (e.g., Open64's *whirl*).
- *Translate (ir → xaif)* Differentiation of the numerical core of the input program is performed in XAIF. This step filters out various language-dependent features such as replacing dereferences of a user-defined type's element with access to a scalar element.
- *Core transformation (xaif → xaif')* The computational graph of the numerical core is traversed, inserting new elements that compute derivative values.
- *Back-translate (xaif' → ir')* This step adds filtered-out features.
- *Generate (ir' → fortran')* Here, we obtain output source code.
- *Postprocess* Variables that were determined to be active are declared by using the *active* type, and all references are updated with value and derivative component accesses.

4.4 Postprocessing

After a differentiated version of the input code has been obtained, the final stage in the process is to compile and link the output with the rest of the overall code base.

If all the model code is transformed, this step is limited to the invocation of the model's regular build routine. However, if only part of the model code is transformed, then this step requires integration of differentiated (AD) and non-differentiated (external) code. A large part of the reintegration is to convert all uses of activated variables in the external source code to reference the value component(e.g., *my_var* → *my_var%v*). OpenAD automates this conversion by generating a summary source file that declares all activated variables during the

Table 1 Independent and dependent variables for AD-based sensitivity analysis

Inputs	Description	Units
fleafcn	Final leaf CN ratio	gC/gN
frootcn	Final root CN ratio	gC/gN
fstemcn	Final stem CN ratio	gC/gN
leafcn	Leaf CN ratio	gC/gN
livewdcn	Live wood CN ratio	gC/gN
deadwdcn	Dead wood CN ratio	gC/gN
froot_leaf	New fine root C per new leaf C	gC/gC
stem_leaf	New stem C per new leaf C	gC/gC
croot_stem	New coarse root C per new stem C	gC/gC
flivewd	Fraction of new wood that is live	none
fcur	Fraction of allocation that goes to current growth	none
organcn	Organ CN ratio	gC/gN
Outputs		
leafc	Leaf carbon	gC/m^2
stemc	Stem carbon	gC/m^2
organc	Organ carbon	gC/m^2
leafn	Leaf nitrogen	gN/m^2
stemn	Stem nitrogen	gN/m^2
organn	Organ nitrogen	gN/m^2

postprocessing stage. This file is then used by a library script to convert external source code files that reference active variables to dereference the active variables' value component. Finally, an executable of the overall model is built.

In our case, differentiation of the CLM-Crop subunit activated a large number of global state variables in the CLM. Since many of these variables were accessed by external code, the postprocessing stage involved a substantial reintegration effort. Over 60 external source files were modified to properly reference active variable values.

5 Results

In this section, we report the results of the experiment of differentiating the CLM-Crop subunit of the CLM code.

The inputs and outputs chosen for the AD-based sensitivity analysis are summarized in Table 1. As discussed in Sect. 2, the goal of the analysis was to identify the most sensitive parameters for further calibration of model accuracy.

Table 2 briefly summarizes the results of the analysis. For each of the three managed crop types, it reports the derivatives of leaf, stem, and organ carbon and nitrogen with respect to the 12 independent variables. For example, the partial derivative of corn's *leafc* with respect to *fleafcn* is $\frac{\partial leafc}{\partial fleafcn} = 7.0353917$.

Table 2 Derivatives of leaf, stem, and organ C and N with respect to selected CN ratio parameters

	CORN		WHEAT		SOY	
	C	N	C	N	C	N
LEAF						
fleafcn	7.0353917	−93.0305059	5.0544004	−100.2410991	3.2190047	−59.1898603
frootcn	4.6136744	−46.2578484	−10.0406592	−72.8134870	2.7559661	−22.6071324
fstemcn	1.9315305	0.1931531	2.6791610	0.1786107	0.6794228	0.0271769
deadwdcn	0.0000000	0.0000000	0.0000000	0.0000000	0.0000000	0.0000000
STEM						
fleafcn	14.4554170	0.2891083	57.2387074	1.1447741	34.5605004	0.6912100
frootcn	9.4894822	−12.9865931	−116.1697332	−35.8649400	26.2942184	−20.5785076
fstemcn	3.9686602	−25.9019159	30.3402387	−52.7061099	7.2945506	−12.3742598
deadwdcn	0.0000000	0.0000000	0.0000000	0.0000000	0.0000000	0.0000000
ORGAN						
fleafcn	1277.0876953	25.5417539	949.6945091	23.7423627	589.0503999	9.8175067
frootcn	809.0263835	16.1805277	−1938.7040252	−48.4676006	436.6213665	7.2770228
fstemcn	350.6178413	7.0123568	503.3998741	12.5849969	124.3285795	2.0721430
deadwdcn	0.0000000	0.0000000	0.0000000	0.0000000	0.0000000	0.0000000

Table 3 Comparison of selected derivative estimates

	$\frac{\partial leafc}{\partial fleafcn}$	$\frac{\partial stemc}{\partial fleafcn}$	$\frac{\partial organc}{\partial fleafcn}$
OpenAD	0.000021273	0.0023837	0.0062778
Finite differences	0.000020848	0.0023359	0.0061479

Similarly, $\frac{\partial leafn}{\partial fleafcn} = -93.0305059$ and so forth for each intersection of rows and columns. These values represent accumulated derivatives for 1 year, where the model integrates forward in time with half-hour (1800-second) time steps.

Taking a closer look at the table, we can observe that some derivatives are not as large as others, indicating that, comparatively, such parameters are not as important as those with larger derivative values. For example, we can observe that the corn parameter *fstemcn* does not contribute to the variability of *leafc* output as much as *fleafcn* does. Further, we see that some derivatives are zero, indicating that such parameters do not affect the outputs. This information is of clear benefit to model designers because it identifies the most sensitive parameters for model accuracy calibrations. For example, these results indicate that it is best to focus on CN ratios within corn leaves rather than stems, in order to optimize carbon production within corn leaves, stems, and organs. Other values can be interpreted similarly.

We have validated the results using finite differences by perturbing some of the independent variables and calculating the difference between the original and perturbed dependent variable values. Table 3 provides an example of perturbing wheat's *fleafcn* parameter by 1.5% and comparing derivative estimates for one time step obtained by OpenAD and by finite differences. We can observe that the derivatives obtained by the two methods are in agreement with the errors on the order of 0.0001 or better.

6 Conclusion

We presented an initial effort in constructing tangent-linear and adjoint codes for the CLM. We focused on the CLM-Crop subunit that models the growth of managed crops. We determined to which of the model parameters the outputs of interest are most sensitive. This information will be used to improve the subunit and the overall land model code. As part of the experiment, we have acquired substantial knowledge about the model, such as the data structures and dependencies in the code that enable preservation and forward integration of climate state (e.g., deep nesting of global state variables within the hierarchical grid structure). Among the lessons learned is the need for precise tracking of active variables. Activation of a single global variable can lead to numerous changes in the code base. In this context, the utility of automated updates of references to activated global variables—provided by OpenAD—becomes indispensable. Future work in applying AD to the CLM includes differentiation of the overall model and comparison of results obtained using different approaches of forward- and reverse-mode AD, operator-overloaded AD, and complex-step method AD.

Acknowledgements This work was supported by the U.S. Dept. of Energy Office of Biological and Environmental Research under the project of Climate Science for Sustainable Energy Future (CSSEF) and by the U.S. Dept. of Energy Office of Science under Contract No. DE-AC02-06CH11357. We thank our collaborators Rao Kotamarthi (ANL), Peter Thornton (ORNL), and our CSSEF colleagues for helpful discussions about the CLM.

References

1. Community Portal for Automatic Differentiation. http://www.autodiff.org
2. Community Earth System Model. http://www.cesm.ucar.edu
3. Drewniak, B., Song, J., Prell, J., Kotamarthi, V.R., Jacob, R.: Modeling the impacts of agricultural land use and management on u.s. carbon budgets. In prep.
4. Earth System Modeling Framework. http://www.earthsystemmodeling.org
5. Griewank, A., Walther, A.: Evaluating Derivatives: Principles and Techniques of Algorithmic Differentiation, 2nd edn. No. 105 in Other Titles in Applied Mathematics. SIAM, Philadelphia, PA (2008). URL http://www.ec-securehost.com/SIAM/OT105.html
6. Hovland, P.D., Naumann, U., Norris, B.: An XML-based platform for semantic transformation of numerical programs. In: M. Hamza (ed.) Software Engineering and Applications, pp. 530–538. ACTA Press, Anaheim, CA (2002)
7. Martins, J.R.R.A., Sturdza, P., Alonso, J.J.: The complex-step derivative approximation. ACM Transactions on Mathematical Software **29**(3), 245–262 (2003). DOI http://doi.acm.org/10.1145/838250.838251
8. Model Coupling Toolkit. http://www.mcs.anl.gov/mct
9. Open64 compiler. http://www.open64.net
10. OpenAnalysis Web Page. http://www.mcs.anl.gov/research/projects/openanalysis
11. Rall, L.B.: Perspectives on automatic differentiation: Past, present, and future? In: H.M. Bücker, G. Corliss, P. Hovland, U. Naumann, B. Norris (eds.) Automatic Differentiation: Applications, Theory, and Implementations, *Lecture Notes in Computational Science and Engineering*, vol. 50, pp. 1–14. Springer, New York, NY (2005). DOI 10.1007/3-540-28438-9_1
12. Rayner, P., Koffi, E., Scholze, M., Kaminski, T., Dufresne, J.L.: Constraining predictions of the carbon cycle using data. Philosophical Transactions of the Royal Society A **369**(1943), 1955–1966 (2011)
13. ROSE compiler. http://rosecompiler.org
14. Schwinger, J., Kollet, S., Hoppe, C., Elbern, H.: Sensitivity of latent heat fluxes to initial values and parameters of a land-surface model. Vadose Zone Journal **9**(4), 984–1001 (2010)
15. Utke, J., Naumann, U., Fagan, M., Tallent, N., Strout, M., Heimbach, P., Hill, C., Wunsch, C.: OpenAD/F: A modular, open-source tool for automatic differentiation of Fortran codes. ACM Transactions on Mathematical Software **34**(4), 18:1–18:36 (2008). DOI 10.1145/1377596.1377598

Using Automatic Differentiation to Study the Sensitivity of a Crop Model

Claire Lauvernet, Laurent Hascoët, François-Xavier Le Dimet, and Frédéric Baret

Abstract Automatic Differentiation (AD) is often applied to codes that solve partial differential equations, e.g. in geophysical sciences or Computational Fluid Dynamics. In agronomy, the differentiation of crop models has never been performed since these models are more empirical than fully mecanistic, derived from equations. This study shows the feasibility of constructing the adjoint model of a crop model referent in the agronomic community (STICS) with the TAPENADE tool, and the use of this accurate adjoint to perform some sensitivity analysis. This paper reports on the experience from AD users of the environmental domain, in which AD usage is not very widespread.

Keywords Adjoint mode • Agronomic crop model • Sensitivity analysis

---

C. Lauvernet (✉)
Irstea, UR MALY, 3 bis quai Chauveau – CP 220, F-69336, Lyon, France, (previously at INRA Avignon, France)
e-mail: claire.lauvernet@irstea.fr

L. Hascoët
INRIA, Sophia-Antipolis, France
e-mail: Laurent.Hascoet@inria.fr

F.-X. Le Dimet
Université de Grenoble, Grenoble, France
e-mail: Francois-Xavier.Le-Dimet@imag.fr

F. Baret
INRA, Avignon, France
e-mail: baret@avignon.inra.fr

S. Forth et al. (eds.), *Recent Advances in Algorithmic Differentiation*, Lecture Notes in Computational Science and Engineering 87, DOI 10.1007/978-3-642-30023-3_6,

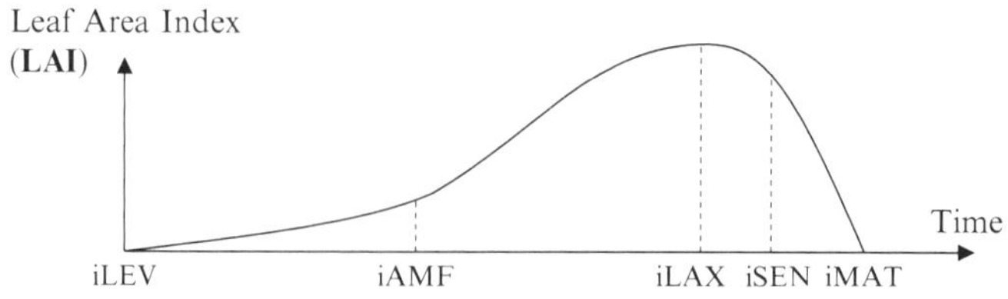

Fig. 1 Simplistic scheme of the stages simulated by the STICS model on dynamics of LAI

1 The Application Domain: The Agronomic Crop Model STICS

STICS [2, 3] is a crop model with a daily time step. Its main aim is to simulate the effects of the physical medium and crop management schedule variations on crop production and environment at the field scale. From the characterization of climate, soil, species and crop management, it computes output variables related to yield in terms of quantity and quality, environment in terms of drainage and nitrate leaching, and to soil characteristics evolution under cropping system.[1]

The two key output variables simulated by STICS that we will need in this paper are the Leaf Area Index (LAI) and the biomass. The LAI is the total one-sided area of leaf tissue per area of ground surface (unitless). This is a canopy parameter that directly quantifies green vegetation biomass. As the leaves are considered to be the main interfaces with the atmosphere for the transfer of mass and energy [16], the LAI indirectly describes properties such as potential of photosynthesis available for primary production, plant respiration, evapotranspiration and carbon flux between the biosphere and the atmosphere, and gives evidence of severely affected areas (fires, parasites ...). Because it is the most observable canopy parameter by remote sensing, the LAI is very commonly used e.g., in crop performance prediction [7], in models of soil-vegetation-atmosphere [15], in crop models [2, 3], in radiative transfer models [20]. Its values can range from 0 for bare soil to 6–7 for a crop during its life cycle, and up to 15 in extreme cases (tropical forests).

STICS simulates the crop growth from sowing to harvest, focusing on the evolution of the LAI at a few selected [2] vegetative stages shown on Fig. 1. These stages involve process thresholds, accounting for some of the differentiation problems described in Sect. 3.2. For a wheat crop, the main phenological stages are known as *ear at* 1 cm, *heading*, *flowering*, and *maturity*. In this work we do not simulate grain yield but only the total biomass. As we focus on the LAI, we only consider the vegetative stages namely: LEV (emergence or budding), AMF (maximum acceleration of leaf area index, equivalent to *ear at* 1 cm), LAX (maximum LAI i.e. end of leaf growth), and SEN (start of net senescence).

[1] http://www.avignon.inra.fr/agroclim_stics_eng

2 Sensitivity Analysis

A model is a more or less realistic or biased simplification of the state variables it simulates. This is especially true for agronomic models, since the functioning of vegetation is not a priori described by exact equations: agronomic models attempt to predict the behavior of the crop by incremental improvements of the simulation code, based on observations made on the field and then published by specialists. Thus, in some parts of the model, this empirical approach is not based on the equations of some underlying physics or chemistry. Sensitivity analysis, which studies the impact of perturbing the control parameters on the model output, gives insights useful to improve or even simplify the model. Sensitivity analysis requires two essential ingredients:

- A model: $F(X, K) = 0$, where X is the state variable (LAI, biomass ...) and K the control variables (parameters, forcing variables ...). F is a differential operator a priori non-linear finite-dimensional, that describes implicitly the evolution of X for a given K. We assume that the system has a unique solution $X(K)$. In this study, what we call the model is exactly the STICS computer program.
- A response function G which combines one or more elements of X into a scalar value, e.g. the final value or the integral over time of an output.

The problem is to evaluate the sensitivity of G with respect to K or in other words the gradient of G with respect to K. With the help of the adjoint model, computing the gradient takes only two steps: run the direct model once for the given K, then solve the adjoint model once [12]. The classical justification is:

$$\nabla G = \frac{dG}{dK}^t = \left(\frac{dG}{dX}.\frac{dX}{dK}\right)^t = \left(\frac{dX}{dK}\right)^t.\left(\frac{dG}{dX}\right)^t$$

where we observe that $\frac{dG}{dX}$ is easily computed from the definition of G alone and the product of $\frac{dX}{dK}^t$ with a vector is achieved by feeding this vector to the adjoint code of STICS, produced by the adjoint mode of Automatic Differentiation.

Sensitivity analysis using an adjoint model is the only way to calculate formally the gradient of the response function at a cost that does not depend on the size of K. It is particularly suitable when the number of entries K is large compared to the size of the response function G [13, 14].

One can also compute the gradient accurately with tangent-linear differentiation, at a cost that is proportional to the size of K. The other sensitivity methods only approximate the gradient: finite difference approximation of the gradient require extensive direct model computations [4]. Stochastic sampling techniques require less mathematical insight as they consist (roughly speaking) in exploring the space of control to determine an overall global sensitivity [10, 18]. Their cost grows rapidly with the dimension of K. These methods have been widely applied to the agronomic models and in particular on STICS [9, 17, 19].

If in many cases, the response function G is a differentiable function of K, it can happen that the model is driven by thresholds e.g., the code uses a lot of branches. Theoretically, a piecewise continuous function is not continuously differentiable, but it has right- and left-derivatives. Differentiation of such a code can only return a sub-gradient. Actually, the methods that do not rely on derivatives (divided differences, stochastic, ...) behave better in these cases, although they remain expensive. In practice, this problem is not considered serious as long as the local sensitivity is valid in a neighborhood of the current K.

3 Automatic Differentiation of STICS

3.1 The TAPENADE Automatic Differentiation Tool

TAPENADE [8] is an Automatic Differentiation (AD) tool based on source transformation. Given a source program written in FORTRAN, TAPENADE builds a new source program that computes some of its derivatives. In "tangent" mode, TAPENADE builds the program that computes directional derivatives. In "adjoint" mode, TAPENADE builds the program that computes the gradient of the output with respect to all input parameters.

Considering the complete set of derivatives of each output with respect to each input, i.e. the Jacobian matrix of the program's function, the tangent mode yields a column of the Jacobian whereas the adjoint mode yields a row of the Jacobian. Therefore in our particular case where the output is a scalar G, one run of the adjoint code will return the complete gradient. In contrast, it takes one run of the tangent mode per input to obtain the same gradient. Although we will experiment with the two modes, the adjoint mode fits our needs better.

However, the adjoint mode evaluates the derivatives in the inverse of the original program's execution order. This is a major difficulty for large programs such as STICS. The AD model copes with this difficulty by a combination of storage of intermediate values and duplicated evaluation of the original program, at a cost in memory and execution time. In TAPENADE, the strategy is mostly based on storage of intermediate values, combined with the storage/recompute tradeoff known as checkpointing, applied automatically at each procedure call.

3.2 STICS Adjoint : The Pains and Sufferings of an AD End-User

The STICS model being written in FORTRAN 77, TAPENADE can in theory build its adjoint. However, there were shortcomings with the early versions of TAPENADE, before 2005. Later versions brought notable improvements but we believe it is worth describing the main problems that we encountered at these early stages.

AD allows for instructions which the symbolic differentiation systems cannot process. It also provides a real gain in computational time. However, a few good programming practices are recommended: the input parameters involved in derivatives must be clearly identified and if possible separate from the other variables. The same holds for the outputs to be differentiated. The precision level of all floating point variables must be coherent, especially for validation purposes: if the chain of computation is not completely "double precision", then the divided difference that is used to validate the analytic derivatives will have a poor accuracy, validation will be dubious and may even fail to detect small errors in the differentiated code.

Validation helped us detect small portability problems in STICS. As divided differences requires to call STICS twice, we discovered that two successive calls to STICS apparently with the same inputs gave different results. In fact the first call was different from all the others, which pointed us to a problem of hidden uninitialized remanent global. Fixing this gave us correct divided differences, and a more portable STICS code.

More specifically to this agronomy application, we had problems with the high number of tests and other conditional jumps in an average run. In more classical situations of Scientific Computing, programs are derived from mathematical equations, typically a set of ODE's or PDE's. This forces some regularity into the code that discretizes and solves these equations: even if branches do occur, they rarely introduce discontinuity and the derivative itself often remains continuous. In our application, the program itself basically is the equation. The model evolves by introducing by hand new subcases and subdivisions, i.e. more tests. If this evolution is not made with differentiation in mind, it may introduce sharp discontinuities that do not harm the original code but make it non-differentiable. It took time to replace faulty branches with a cleaner, differentiable implementation. On the other hand, users agreed that this resulted in a better code.

Still, the number of branches in the STICS model is very large: thresholds, conditions, loops, and other control, all are tests that the adjoint code must remember to run backwards. STICS consumes an unusually large memory for that. Until recently, TAPENADE did not store this control efficiently, using a full INTEGER value to store only a boolean in general.

Checkpointing the time stepping was difficult. Before binomial checkpointing [5] was implemented in TAPENADE, we had to split the main time loop of 400 iterations into two nested loops of 20 iterations each, and place these two loops into two new subroutines to force checkpointing. This tedious manipulations are now spared with the new TAPENADE directives for binomial checkpointing.

More than 5 years after this sensitivity study, both STICS and TAPENADE have evolved. The latest version 6 of STICS is more readily differentiable than before. TAPENADE 3.6 had several bugs fixed and more importantly provides a set of user directives to control checkpointing better. These checkpointing directives are also the answer to the serious performance problem discussed in Sect. 3.3.

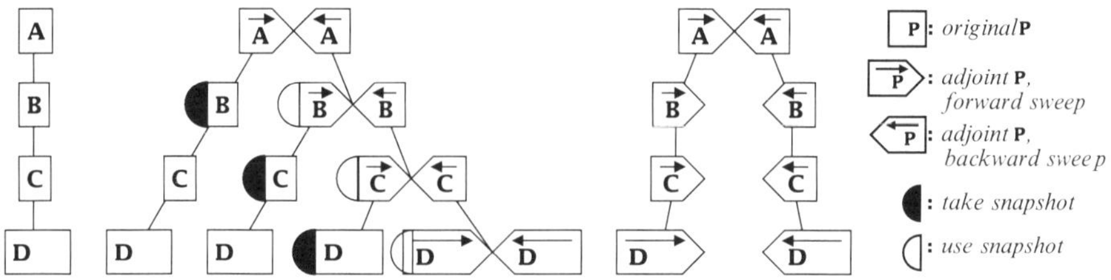

Fig. 2 The cost of checkpointing long chains of nested calls

3.3 Validation of the Adjoint Model

Validation was performed in two steps as usual, and for several directions of perturbation. First, the tangent derivatives were compared with divided differences, and they agreed up to the eighth decimal for an increment of 10^{-8} in the one-sided divided difference. Second, the adjoint derivatives were compared with the tangent derivatives ("dot-product" test [6]) and they agreed up to the 14th decimal. At the time of the study, the run times were:

$$\text{Direct model} : 0.21s \qquad \text{Tangent model} : 0.39s \qquad \text{Adjoint model} : 30.96s$$

The run time of the adjoint code is much higher than the custommary fivefold to tenfold slowdown. The problem was left to the TAPENADE developers to go on with the sensitivity study. Identifying its causes was hard, and pointed to the need for specific profiling tools for adjoint codes. Profiling instructions must be inserted by the AD tool itself, and tools are missing to help interpret the profiling results. Eventually, the problem was found to come from the systematic checkpointing on procedure calls on a chain of four nested procedure calls, each of them doing little else than calling the next nested call, cf. Fig. 2. Checkpointing [6] one call to `P` reduces the peak memory used by the adjoint. This reduction is roughly proportional to the run-time of `P`. On the other hand, it costs one extra run of `P`, plus some memory (a "snapshot") to restore the execution state. Checkpointing nested calls causes an increasing number of extra runs. This is inherent to the approach and beneficial in general, but is a waste for procedures that are nearly empty shells around a deeper call. In our case, the problem was amplified by the size of a very big work array that was restored at each checkpoint. The answer is to deactivate checkpointing on the calls to the "empty shell" procedures. This is known as the "split" mode of adjoint AD [6], and is sketched on the right of Fig. 2. This required development in TAPENADE, plus new directives (`$AD NOCHECKPOINT`) to let the user trigger this split mode on selected procedure calls. Conversely in other cases, it is useful to trigger checkpointing on pieces of a procedure, and TAPENADE new directives (`$AD CHECKPOINT-START`) and (`$AD CHECKPOINT-END`) let the user do that. This results in the following times obtained with TAPENADE 3.6:

Direct model : $0.22s$ Tangent model : $0.52s$ Adjoint model : $0.86s$

4 Results: Sensitivity Analysis of STICS

We decided to compute the gradients of two response functions G: LAI and biomass, and more precisely their integrals over the simulation time from sowing to harvest. These response functions capture well the growth dynamics.

$$G_{LAI} = \sum_{i=1}^{T} LAI(t_i) \qquad G_{biomass} = \sum_{i=1}^{T} biomass(t_i)$$

4.1 Selection of Input Parameters for Sensitivity Analysis of Output Variables

For this feasibility study, the control variables correspond to wheat crops from the Danube's plain in Romania in 2000–2001[2] [1]. The gradient was calculated with respect to the following input parameters[3]: for LAI, we chose the varietal parameters acting on the dynamics of LAI, and *dlaimaxbrut* that strongly characterizes the aerial growth. Parameters were adapted to the ADAM database, including the variety of wheat (Flamura) used here for its particular cold resistance. For biomass, efficiencies at three important phases of the cycle of wheat (juvenile phases, vegetative and grain filling) and *vmax2* were chosen following the experience accumulated by users of the crop model. Table 1 describes the role of these parameters, and their values for this sensitivity study.

4.2 Sensitivity Results of LAI and Biomass

One goal of this sensitivity study was to establish the hierarchy of influent parameters. Therefore Fig. 3 shows the ten influences normalized as percentages,

---

[2] ADAM experiment (Data Assimilation through Agro-Modelling). Project and database at http://kalideos.cnes.fr/spip.php?article68

[3] All the parameters of STICS are described in http://www.avignon.inra.fr/agroclim_stics_eng/notices_d_utilisation

Table 1 Parameter role and values for the ADAM conditions

Parameter	Definition	Value
dlaimaxbrut	Maximum rate of gross leaf surface area production	0.00044
stlevamf	Mumulated development units between the LEV and AMF stages	208.298
stamflax	Mumulated development units between AMF and LAX stages	181.688
jvc	Days of vernalisation (cold days needed to lift)	35
durvieF	Lifespan of a cm of adult leaf	160
adens	Compensation between number of stems and plants density	−0.6
efcroijuv	Maximum growth efficiency during juvenile phase (LEV-AMF)	2.2
efcroiveg	Maximum growth efficiency during vegetative phase (AMF-DRP)	2.2
efcroirepro	Maximum growth effiicency during grain filling phase (DRP-MAT)	4.25
vmax2	Maximum rate of nitrate absorption by the roots	0.05

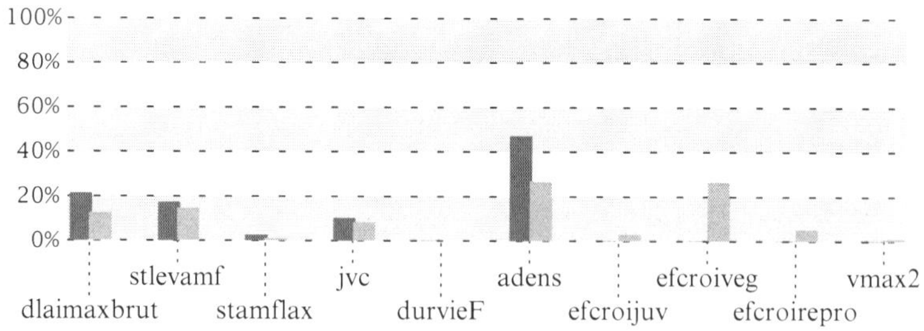

Fig. 3 Relative sensitivity (%) to selected STICS parameters of ouput variables LAI (*left*) and biomass (*right*) computed by the adjoint

totalling 100%. Among the ten selected, the most influential parameters on the LAI are *adens* (47%), *dlaimaxbrut* (21%), *stlevamf* (17%), *jvc* (10%), and finally *stamflax* (2%). *adens* represents the ability of a plant to withstand increasing densities, and since it depends on the species and varieties, its influence may be particularly strong for this type of wheat and less for other crops. For biomass, we observe that the hierarchy is modified by the strong influence of the efficiency *efcroiveg* (maximum growth efficiency during vegetative phase) which is similar to that of *adens* (27%). This means that we can ignore the estimate of *efcroiveg* if we only want to assimilate LAI data, but absolutely not if we need to simulate biomass. *stlevmaf* and *dlaimaxbrut* are of similar importance (14% and 12%). Finally, there is a relatively low sensitivity (5% and 3%) of biomass integrated over the life cycle to the other two parameters of efficiency *efcroirepro* and *efcroijuv*, meaning that the biomass is not so dependant on the juvenile and the grain filling phases but essentially on the vegetative phase. The fact that only the integral over the entire cycle was studied involves a very small influence of the parameters *efcroirepro* and *efcroijuv*, as opposed to *efcroiveg*. These efficiencies with a small influence matter only during short phenological stages: only a sensitivity study restricted to these stages can modify the hierarchy of influent parameters, opening the way to estimation of these low-influence parameters [17]. LAI is actually dependant on four parameters and biomass on five on the ten tested, which will help the user concentrate on these and estimate them better. Uncertainty on the other parameters is of relatively smaller importance.

5 Conclusion and Outlook

This case study illustrates the interest of AD for sensitivity analysis of agronomic models. Coupled with other models, for example radiative transfer model [11], it will allow to assimilate remote sensing data into crop models by using the adjoint to minimize the discrepancy cost function. This work shows the feasibility of applying and developing variationnal methods in agronomy, in the same way as in oceanography or meteorology.

For the agronomic community, the adjoint model of STICS is an interesting tool to perform sensitivity analysis since it requires the calculation only once for each agro-pedo-climatic situation. The most difficult work is the differentiation of the model, which must be done only once, and with the help of AD tools that keep improving. However, the local sensitivity analysis is valid only in a small neighborhood and the hierarchy of sensitivities may vary under different conditions. These results are only a first step. Following work could concentrate on:

1. A "multi-local" sensitivity analysis, keeping the crop management and climate of the base ADAM, but letting the parameters vary in a given range. This would require many runs of the adjoint modes on a representative sample of possible parameter values. This would return a parameter hierarchy with a more general validity.
2. An application of this analysis to other conditions (climate, soil ...) to see whether the hierarchy is preserved in general. Extending to other varieties is also important. Actually, it seems unlikely that this hierarchy is preserved since the change of climate and soil conditions may rapidly hit limiting factors (stress for the plant) and thus modify the parameters influence.
3. A study of the sensitivity at selected phenological stages of the cycle to study the effect of variables temporally valid (especially efficiency) on the general hierarchy.

The adjoint code is able to compute the sensitivities of one response function to all parameters in just one run. There are more parameters to STICS than the 10 we have selected for this sensitivity study. Looking at the influence of all parameters will guide the attention of STICS users on some parameters and modules, according to the users' objectives. Sensitivity study is a preliminary to parameter estimation: many of these agronomic parameters (yield, balance ...) are not directly observable by remote sensing. On the other hand the outputs (biomass) can be measured. The adjoint of the model, by returning the gradient of any discrepancy cost function, is the key to estimate these hidden agronomic parameters from the ones we can measure.

Acknowledgements This study was conducted thanks to a grant provided by CNES within the ADAM project (http://kalideos.cnes.fr/spip.php?article68), during the Ph.D. of the first author at INRA Avignon and the University of Grenoble.

References

1. Baret, F., Vintila, R., Lazar, C., Rochdi, N., Prévot, L., Favard, J., de Boissezon, H., Lauvernet, C., Petcu, E., Petcu, G., Voicu, P., Denux, J., Poenaru, V., Marloie, O., Simota, C., Radnea, C., Turnea, D., Cabot, F., Henry, P.: The adam database and its potential to investigate high temporal sampling acquisition at high spatial resolution for the monitoring of agricultural crops. Romanian Agricultural Research **16**, 69–80 (2001)
2. Brisson, N., Mary, B., Ripoche, D., Jeuffroy, M.H., Ruget, F., Nicoullaud, B., Gate, P., Devienne-Barret, F., Antonioletti, R., Durr, C., Richard, G., Beaudoin, N., Recous, S., Tayot, X., Plenet, D., Cellier, P., Machet, J.M., Meynard, J.M., Delecolle, R.: STICS: a generic model for the simulation of crops and their water and nitrogen balances. I: theory and parameterization applied to wheat and corn. Agronomie **18**(5–6), 311–346 (1998)
3. Brisson, N., Ruget, F., Gate, P., Lorgeou, J., Nicoullaud, B., Tayot, X., Plenet, D., Jeuffroy, M.H., Bouthier, A., Ripoche, D., Mary, B., Justes, E.: STICS: a generic model for simulating crops and their water and nitrogen balances. II: model validation for wheat and maize. Agronomie **22**(1), 69–92 (2002)
4. Castaings, W., Dartus, D., Le Dimet, F.X., Saulnier, G.M.: Sensitivity analysis and parameter estimation for distributed hydrological modeling: potential of variational methods. Hydrol. Earth Syst. Sci. **13**(4), 503–517 (2009)
5. Griewank, A.: Achieving logarithmic growth of temporal and spatial complexity in reverse automatic differentiation. Optimization Methods and Software **1**, 35–54 (1992)
6. Griewank, A., Walther, A.: Evaluating Derivatives: Principles and Techniques of Algorithmic Differentiation, 2nd edn. No. 105 in Other Titles in Applied Mathematics. SIAM, Philadelphia, PA (2008). URL http://www.ec-securehost.com/SIAM/OT105.html
7. Guérif, M., Houlès, V., Makowski, D., Lauvernet, C.: Data assimilation and parameter estimation for precision agriculture using the crop model STICS. In: D. Wallach, D. Makowski, J.W. Jones (eds.) Working with dynamic crop models: evaluating, analyzing, parameterizing and using them, chap. 17, pp. 391–398. Elsevier (2006)
8. Hascoët, L., Pascual, V.: TAPENADE 2.1 user's guide. Rapport technique 300, INRIA, Sophia Antipolis (2004). URL http://www.inria.fr/rrrt/rt-0300.html
9. Houlès, V., Mary, B., Guérif, M., Makowski, D., Justes, E.: Evaluation of the ability of the crop model stics to recommend nitrogen fertilisation rates according to agro-environmental criteria. Agronomie **24**(6), 339–349 (2004)
10. Ionescu-Bujor, M., Cacuci, D.G.: A comparative review of sensitivity and uncertainty analysis of large-scale systems. I: deterministic methods. Nuclear science and engineering **147**(3), 189–203 (2004)
11. Lauvernet, C., Baret, F., Hascoët, L., Buis, S., Le Dimet, F.X.: Multitemporal-patch ensemble inversion of coupled surface-atmosphere radiative transfer models for land surface characterization. Remote Sens. Environ. **112**(3), 851–861 (2008)
12. Le Dimet, F.X., Ngodock, H.E., Navon, I.M.: Sensitivity analysis in variational data assimilation. J. Meteorol. Soc. Japan pp. 145–155 (1997)
13. Le Dimet, F.X., Talagrand, O.: Variational algorithms for analysis and assimilation of meteorological observations: theoretical aspects. Tellus A **38A**(2), 97–110 (1986)
14. Lions, J.L.: Optimal control of systems governed by partial differential equations. Springer-Verlag (1968)

15. Olioso, A., Inoue, Y., Ortega-FARIAS, S., Demarty, J., Wigneron, J., Braud, I., Jacob, F., Lecharpentier, P., Ottl, C., Calvet, J., Brisson, N.: Future directions for advanced evapotranspiration modeling: Assimilation of remote sensing data into crop simulation models and SVAT models. Irrigation and Drainage Systems **19**(3–4), 377–412 (2005)
16. Rosenberg, N.J., Blad, B.L., Verma, S.B.: Microclimate: the biological environment. Wiley-Interscience (1983)
17. Ruget, F., Brisson, N., Delecolle, R., Faivre, R.: Sensitivity analysis of a crop simulation model, STICS, in order to choose the main parameters to be estimated. Agronomie **v. 22(2) p. 133–158** (2002)
18. Saltelli, A., Chan, K., Scott, E.M.: Sensitivity Analysis. Wiley (2000)
19. Varella, H., Guérif, M., Buis, S.: Global sensitivity analysis measures the quality of parameter estimation: The case of soil parameters and a crop model. Environmental Modelling and Software **25**(3), 310–319 (2010)
20. Verhoef, W.: Light scattering by leaf layers with application to canopy reflectance modeling: The SAIL model. Remote Sensing of Environment **16**(2), 125–141 (1984)

Efficient Automatic Differentiation of Matrix Functions

Peder A. Olsen, Steven J. Rennie, and Vaibhava Goel

Abstract Forward and reverse mode automatic differentiation methods for functions that take a vector argument make derivative computation efficient. However, the determinant and inverse of a matrix are not readily expressed in the language of vectors. The derivative of a function $f(\mathbf{X})$ for a $d \times d$ matrix $\mathbf{X}$ is itself a $d \times d$ matrix. The second derivative, or Hessian, is a $d^2 \times d^2$ matrix, and so computing and storing the Hessian can be very costly. In this paper, we present a new calculus for matrix differentiation, and introduce a new matrix operation, the *box product*, to accomplish this. The box product can be used to elegantly and efficiently compute both the first and second order matrix derivatives of *any* function that can be expressed solely in terms of arithmetic, transposition, trace and log determinant operations. The Hessian of such a function can be implicitly represented as a sum of Kronecker, outer, and box products, which allows us to compute the Newton step efficiently. Whereas the direct computation requires $\mathcal{O}(d^4)$ storage and $\mathcal{O}(d^6)$ operations, the indirect representation of the Hessian allows the storage to be reduced to $\mathcal{O}(kd^2)$, where k is the number of times the variable $\mathbf{X}$ occurs in the expression for the derivative. Likewise, the cost of computing the Newton direction is reduced to $\mathcal{O}(kd^5)$ in general, and $\mathcal{O}(d^3)$ for $k = 1$ and $k = 2$.

Keywords Box product • Kronecker product • Sylvester equation • Reverse mode

---

P.A. Olsen (✉) · S.J. Rennie · V. Goel
IBM, TJ Watson Research Center, Yorktown Heights, NY, USA
e-mail: pederao@us.ibm.com; sjrennie@us.ibm.com; vgoel@us.ibm.com

S. Forth et al. (eds.), *Recent Advances in Algorithmic Differentiation*, Lecture Notes in Computational Science and Engineering 87, DOI 10.1007/978-3-642-30023-3_7,

1 Introduction

The computation of the derivatives of scalar functions that take a matrix argument (scalar–matrix functions) has been a topic of several publications, much of which can be found in these books, [5, 6, 10, 12]. There are also two elegant papers by Minka and Fackler that present the derivatives for many scalar–matrix functions, [2, 7]. There has even been a publication in this venue, [3]. These papers contain tables for computing the first derivative (gradient) and the second derivative (Hessian) of scalar–matrix functions $f : \mathbb{R}^{m\times n} \to \mathbb{R}$. The main tool facilitating the organized computation of the second derivative is the Kronecker product. However, the task can be complex and tedious even if the function can be composed from canonical formulas listed in the previous publications. In this paper we introduce a new direct matrix product, the *box product*, that simplifies the differentiation and representation of the derivative of scalar–matrix functions. We show that the box product can be used to compactly represent Hessian matrices, and that the box product reveals structure that can be exploited to efficiently compute the Newton direction.

1.1 Terminology

To simplify the discussion in the following sections we will use the following terminology:

- **Scalar–Scalar Function**: A scalar function that takes a scalar argument and returns a scalar: $f : \mathbb{R} \to \mathbb{R}$. An example involving matrices is $f(x) = \mathbf{a}^\top(\mathbf{Y} + x\mathbf{Z})^{-1}\mathbf{b}$.
- **Scalar–Matrix Function**: A scalar–matrix function is a function that takes a matrix argument and returns a scalar: $f : \mathbb{R}^{m\times n} \to \mathbb{R}$. An example is $f(\mathbf{X}) = \text{trace}((\mathbf{X}\boldsymbol{\Sigma}\mathbf{X}^\top)^{-1})$.
- **Matrix–Matrix Function**: A matrix–matrix function is a function that takes a matrix argument and returns a matrix: $\mathbf{F} : \mathbb{R}^{m_1\times n_1} \to \mathbb{R}^{m_2\times n_2}$. An example is $\mathbf{F}(\mathbf{X}) = \mathbf{A}(\mathbf{X}\boldsymbol{\Sigma}\mathbf{X}^\top)^{-1}\mathbf{B}$.

Some common unary and binary operations that yields matrix–matrix functions from matrices are: matrix multiplication, addition, subtraction, inversion, transposition and scalar-multiplication of matrices. To form scalar–matrix functions we use the trace and determinant operations. The trace can be used to express the matrix inner-product, $f(\mathbf{X}) = \text{vec}^\top(\mathbf{A})\text{vec}(\mathbf{F}(\mathbf{X})) = \text{trace}(\mathbf{A}^\top\mathbf{F}(\mathbf{X}))$, where $\text{vec}(\mathbf{X})$ is standard terminology used to indicate the column vector formed by stacking all the different columns of $\mathbf{X}$ into one big column vector. We can also use the trace to express vector-matrix-vector product, $f(\mathbf{X}) = \mathbf{a}^\top\mathbf{F}(\mathbf{X})\mathbf{b} = \text{trace}(\mathbf{F}(\mathbf{X})\mathbf{b}\mathbf{a}^\top)$. For scalar–matrix functions formed using only these operations, we show how the Kronecker and box products can help understand and organize efficient computation and storage of the first and second order derivatives.

1.2 Matrix Derivatives in Matrix Form

For a scalar–matrix function we define the scalar–matrix derivative to have the same dimensions as the matrix argument:

$$\frac{\partial f}{\partial \mathbf{X}} = \left\{ \frac{\partial f}{\partial x_{ij}} \right\}_{ij}. \tag{1}$$

For a matrix–matrix function, we define the matrix–matrix derivative to be:

$$\frac{\partial \mathbf{F}}{\partial \mathbf{X}} \stackrel{\text{def}}{=} \frac{\partial \text{vec}(\mathbf{F}^\top)}{\partial \text{vec}^\top(\mathbf{X}^\top)} = \begin{pmatrix} \frac{\partial f_{11}}{\partial x_{11}} & \frac{\partial f_{11}}{\partial x_{12}} & \cdots & \frac{\partial f_{11}}{\partial x_{mn}} \\ \frac{\partial f_{12}}{\partial x_{11}} & \frac{\partial f_{12}}{\partial x_{12}} & \cdots & \frac{\partial f_{12}}{\partial x_{mn}} \\ \vdots & \vdots & \ddots & \vdots \\ \frac{\partial f_{kl}}{\partial x_{11}} & \frac{\partial f_{kl}}{\partial x_{12}} & \cdots & \frac{\partial f_{kl}}{\partial x_{mn}} \end{pmatrix}. \tag{2}$$

The matrix–matrix derivative is row-major, whereas the vec operator is column-major. We have made this somewhat unpleasant choice since the standard Kronecker product definition is also row-major. A scalar–matrix function is also a matrix–matrix function whose matrix–matrix derivative is $\text{vec}^\top((\partial f/\partial \mathbf{X})^\top)$, which is different from the scalar–matrix derivative. What form of the derivative is applied should be clear from the context. Note also that the derivative of a scalar–matrix function is a matrix–matrix function, and so the Hessian can be computed by first applying the scalar–matrix derivative followed by the matrix–matrix derivative. The Hessian of a scalar–matrix function can also be written

$$H(f(X)) = \frac{\partial^2 f}{\partial \text{vec}(\mathbf{X}^\top) \partial \text{vec}^\top(\mathbf{X}^\top)}.$$

In Sect. 2 we review properties of the Kronecker product, and in Sect. 3 we introduce the new *box product*. The Kronecker and box products allow us to express the derivative of rational matrix–matrix functions. Next, we review the standard differentiation rules in Sect. 4, and apply these to a simple example scalar–matrix function. Finally, we state some new results related to computing Newton's step for scalar–matrix derivatives.

2 Kronecker Products

For matrices $\mathbf{A} \in \mathbb{R}^{m_1 \times n_1}$ and $\mathbf{B} \in \mathbb{R}^{m_2 \times n_2}$ the Kronecker product $\mathbf{A} \otimes \mathbf{B} \in \mathbb{R}^{(m_1 m_2) \times (n_1 n_2)}$ is defined to be

$$\mathbf{A} \otimes \mathbf{B} = \begin{pmatrix} a_{11}\mathbf{B} & \cdots & a_{1n}\mathbf{B} \\ \vdots & \vdots & \vdots \\ a_{m1}\mathbf{B} & \cdots & a_{mn}\mathbf{B} \end{pmatrix}. \tag{3}$$

It can be verified that this definition is equivalent to $(\mathbf{A} \otimes \mathbf{B})_{(i-1)m_2+j,(k-1)n_2+l} = a_{ik}b_{jl}$, which we simply write as $(\mathbf{A} \otimes \mathbf{B})_{(ij)(kl)} = a_{ik}b_{jl}$, where it is understood that the pairs ij and kl are laid out in row-major order.

Theorem 1 (Kronecker Product Identities). *Define matrices* $\mathbf{A}, \mathbf{A}_1, \mathbf{A}_2 \in \mathbb{R}^{m_1 \times n_1}$, $\mathbf{B}, \mathbf{B}_1, \mathbf{B}_2 \in \mathbb{R}^{m_2 \times n_2}$, $\mathbf{C} \in \mathbb{R}^{n_1 \times o_1}$, $\mathbf{D} \in \mathbb{R}^{n_2 \times o_2}$, $\mathbf{F} \in \mathbb{R}^{m_3 \times n_3}$, *and* $\mathbf{Y} \in \mathbb{R}^{n_2 \times n_1}$. $\mathbf{I}_n \in \mathbb{R}^{n \times n}$ *is used to denote the* $n \times n$ *identity matrix. The following identities hold for the Kronecker product (the matrices in the trace and determinant identities must be square):*

$$(\mathbf{A}_1 + \mathbf{A}_2) \otimes \mathbf{B} = \mathbf{A}_1 \otimes \mathbf{B} + \mathbf{A}_2 \otimes \mathbf{B} \tag{4}$$

$$\mathbf{A} \otimes (\mathbf{B}_1 + \mathbf{B}_2) = \mathbf{A} \otimes \mathbf{B}_1 + \mathbf{A} \otimes \mathbf{B}_2 \tag{5}$$

$$\mathbf{A} \otimes (\mathbf{B} \otimes \mathbf{F}) = (\mathbf{A} \otimes \mathbf{B}) \otimes \mathbf{F} \tag{6}$$

$$(\mathbf{A} \otimes \mathbf{B})(\mathbf{C} \otimes \mathbf{D}) = (\mathbf{AC}) \otimes (\mathbf{BD}) \tag{7}$$

$$(\mathbf{A} \otimes \mathbf{B})^\top = \mathbf{A}^\top \otimes \mathbf{B}^\top \tag{8}$$

$$(\mathbf{A} \otimes \mathbf{B})^{-1} = \mathbf{A}^{-1} \otimes \mathbf{B}^{-1} \tag{9}$$

$$\mathbf{I}_m \otimes \mathbf{I}_n = \mathbf{I}_{mn} \tag{10}$$

$$(\mathbf{A} \otimes \mathbf{B})\mathrm{vec}(\mathbf{Y}) = \mathrm{vec}(\mathbf{BYA}^\top) \tag{11}$$

$$\mathrm{trace}(\mathbf{A} \otimes \mathbf{B}) = \mathrm{trace}(\mathbf{A})\,\mathrm{trace}(\mathbf{B}) \tag{12}$$

$$\det(\mathbf{A} \otimes \mathbf{B}) = (\det(\mathbf{A}))^{m_2} (\det(\mathbf{B}))^{m_1}. \tag{13}$$

3 Box Products

Let us first formally define the box product:

Definition 1 (Box Product). For matrices $\mathbf{A} \in \mathbb{R}^{m_1 \times n_1}$ and $\mathbf{B} \in \mathbb{R}^{m_2 \times n_2}$ we define the box product $\mathbf{A} \boxtimes \mathbf{B} \in \mathbb{R}^{(m_1 m_2) \times (n_1 n_2)}$ to be

$$(\mathbf{A} \boxtimes \mathbf{B})_{(i-1)m_2+j,(k-1)n_1+l} = a_{il}b_{jk} = (\mathbf{A} \boxtimes \mathbf{B})_{(ij)(kl)}. \tag{14}$$

For example, the box product of two 2×2 matrices is

$$\mathbf{A} \boxtimes \mathbf{B} = \begin{pmatrix} a_{11}b_{11} & a_{12}b_{11} & a_{11}b_{12} & a_{12}b_{12} \\ a_{11}b_{21} & a_{12}b_{21} & a_{11}b_{22} & a_{12}b_{22} \\ a_{21}b_{11} & a_{22}b_{11} & a_{21}b_{12} & a_{22}b_{12} \\ a_{21}b_{21} & a_{22}b_{21} & a_{21}b_{22} & a_{22}b_{22} \end{pmatrix}. \tag{15}$$

Theorem 2 (Box Product Identities). *Define matrices* $\mathbf{A}, \mathbf{A}_1, \mathbf{A}_2 \in \mathbb{R}^{m_1 \times n_1}$, $\mathbf{B}, \mathbf{B}_1$, $\mathbf{B}_2 \in \mathbb{R}^{m_2 \times n_2}$, $\mathbf{C} \in \mathbb{R}^{n_2 \times o_2}$, $\mathbf{D} \in \mathbb{R}^{n_1 \times o_1}$, $\mathbf{F} \in \mathbb{R}^{m_3 \times n_3}$, $\mathbf{G} \in \mathbb{R}^{m \times n}$, $\mathbf{H} \in \mathbb{R}^{n \times m}$, $\mathbf{X} \in \mathbb{R}^{n_1 \times m_2}$, $\mathbf{I}_n \in \mathbb{R}^{n \times n}$. *The following identities hold for the box product:*

$$(\mathbf{A}_1 + \mathbf{A}_2) \boxtimes \mathbf{B} = \mathbf{A}_1 \boxtimes \mathbf{B} + \mathbf{A}_2 \boxtimes \mathbf{B} \quad (16)$$
$$\mathbf{A} \boxtimes (\mathbf{B}_1 + \mathbf{B}_2) = \mathbf{A} \boxtimes \mathbf{B}_1 + \mathbf{A} \boxtimes \mathbf{B}_2 \quad (17)$$
$$\mathbf{A} \boxtimes (\mathbf{B} \boxtimes \mathbf{F}) = (\mathbf{A} \boxtimes \mathbf{B}) \boxtimes \mathbf{F} \quad (18)$$
$$(\mathbf{A} \boxtimes \mathbf{B})(\mathbf{C} \boxtimes \mathbf{D}) = (\mathbf{AD}) \otimes (\mathbf{BC}) \quad (19)$$
$$(\mathbf{A} \boxtimes \mathbf{B})^\top = \mathbf{B}^\top \boxtimes \mathbf{A}^\top \quad (20)$$
$$(\mathbf{A} \boxtimes \mathbf{B})^{-1} = \mathbf{B}^{-1} \boxtimes \mathbf{A}^{-1} \quad (21)$$
$$\mathrm{trace}(\mathbf{G} \boxtimes \mathbf{H}) = \mathrm{trace}(\mathbf{GH}) \quad (22)$$
$$(\mathbf{A} \boxtimes \mathbf{B})\mathrm{vec}(\mathbf{X}) = \mathrm{vec}(\mathbf{B}\mathbf{X}^\top \mathbf{A}^\top) \quad (23)$$
$$(\mathbf{A} \boxtimes \mathbf{B})(\mathbf{C} \otimes \mathbf{D}) = (\mathbf{AD}) \boxtimes (\mathbf{BC}) \quad (24)$$
$$(\mathbf{A} \otimes \mathbf{B})(\mathbf{D} \boxtimes \mathbf{C}) = (\mathbf{AD}) \boxtimes (\mathbf{BC}) \quad (25)$$
$$(\mathbf{A} \boxtimes \mathbf{B})(\mathbf{C} \otimes \mathbf{D}) = (\mathbf{A} \otimes \mathbf{B})(\mathbf{D} \boxtimes \mathbf{C}) \quad (26)$$
$$(\mathbf{A} \boxtimes \mathbf{B})(\mathbf{C} \boxtimes \mathbf{D}) = (\mathbf{A} \otimes \mathbf{B})(\mathbf{D} \otimes \mathbf{C}). \quad (27)$$

3.1 Box Products of Identity Matrices

The matrix that permutes $\mathrm{vec}(\mathbf{X})$ into $\mathrm{vec}(\mathbf{X}^\top)$ for $\mathbf{X} \in \mathbb{R}^{m \times n}$ is commonly known as $\mathbf{T}_{m,n}$. This matrix can be expressed as $\mathbf{I}_m \boxtimes \mathbf{I}_n$, a box product of two identity matrices. The box product of two identity matrices is a permutation matrix with many interesting properties.

Theorem 3 (Box Products of Two Identity Matrices). *The Box product of two identity matrices* $\mathbf{I}_m$ *and* $\mathbf{I}_n$ *for* $m, n > 1$ *is a non-trivial permutation matrix* ($\mathbf{I}_n \boxtimes \mathbf{I}_m \neq \mathbf{I}_{mn}$) *satisfying*

$$(\mathbf{I}_m \boxtimes \mathbf{I}_n)^\top = \mathbf{I}_n \boxtimes \mathbf{I}_m \quad (28)$$
$$(\mathbf{I}_m \boxtimes \mathbf{I}_n)^\top (\mathbf{I}_m \boxtimes \mathbf{I}_n) = \mathbf{I}_{mn} \quad (29)$$
$$\det(\mathbf{I}_m \boxtimes \mathbf{I}_n) = (-1)^{mn(m-1)(n-1)/4}. \quad (30)$$

Let $\mathbf{A} \in \mathbb{R}^{m_1 \times n_1}$, $\mathbf{B} \in \mathbb{R}^{m_2 \times n_2}$. *The box product of two identity matrices can be used to switch between box and Kronecker products, or to switch the order of the arguments in the box or Kronecker product.*

$$(\mathbf{A} \boxtimes \mathbf{B})(\mathbf{I}_{n_2} \boxtimes \mathbf{I}_{n_1}) = (\mathbf{A} \otimes \mathbf{B}) \quad (31)$$
$$(\mathbf{A} \otimes \mathbf{B})(\mathbf{I}_{n_1} \boxtimes \mathbf{I}_{n_2}) = (\mathbf{A} \boxtimes \mathbf{B}) \quad (32)$$
$$(\mathbf{I}_{m_2} \boxtimes \mathbf{I}_{m_1})(\mathbf{A} \boxtimes \mathbf{B}) = \mathbf{B} \otimes \mathbf{A} \quad (33)$$
$$(\mathbf{I}_{m_2} \boxtimes \mathbf{I}_{m_1})(\mathbf{A} \otimes \mathbf{B}) = \mathbf{B} \boxtimes \mathbf{A} \quad (34)$$
$$(\mathbf{I}_{n_1} \boxtimes \mathbf{I}_{m_1})\mathrm{vec}(\mathbf{A}^\top) = \mathrm{vec}(\mathbf{A}) \quad (35)$$
$$(\mathbf{I}_{m_1} \boxtimes \mathbf{I}_{n_1})\mathrm{vec}(\mathbf{A}) = \mathrm{vec}(\mathbf{A}^\top) \quad (36)$$

Table 1 Matrix derivative rules for general matrices $\mathbf{X} \in \mathbb{R}^{m\times n}$, with $\mathbf{F}(\mathbf{X}) : \mathbb{R}^{m\times n} \to \mathbb{R}^{k\times l}$

$\mathbf{F}(\mathbf{X})$	$\frac{\partial \mathbf{F}(\mathbf{X})}{\partial \mathbf{X}}$	
$\mathbf{X}$	$\mathbf{I}_{mn}$	(R1)
$\mathbf{X}^\top$	$\mathbf{I}_n \boxtimes \mathbf{I}_m$	(R2)
$\mathbf{AX}$	$\mathbf{A} \otimes \mathbf{I}_n$	(R3)
$\mathbf{XB}$	$\mathbf{I}_m \otimes \mathbf{B}^\top$	(R4)
$\mathbf{AXB}$	$\mathbf{A} \otimes \mathbf{B}^\top$	(R5)
$\mathbf{AX}^\top\mathbf{B}$	$\mathbf{A} \boxtimes \mathbf{B}^\top$	(R6)
$\mathbf{X}^\top\mathbf{X}$	$\mathbf{I}_n \boxtimes \mathbf{X}^\top + \mathbf{X}^\top \otimes \mathbf{I}_n$	(R7)
$\mathbf{F}(\mathbf{X}^\top)$	$\left.\frac{\partial \mathbf{F}(\mathbf{Y})}{\partial \mathbf{Y}}\right\|_{\mathbf{Y}=\mathbf{X}^\top} (\mathbf{I}_n \boxtimes \mathbf{I}_m)$	(R8)
$\mathbf{F}^\top(\mathbf{X})$	$(\mathbf{I}_l \boxtimes \mathbf{I}_k)\frac{\partial \mathbf{F}(\mathbf{X})}{\partial \mathbf{X}}$	(R9)
$\mathbf{AF}(\mathbf{X})\mathbf{B}$	$(\mathbf{A} \otimes \mathbf{B}^\top)\frac{\partial \mathbf{F}(\mathbf{X})}{\partial \mathbf{X}}$	(R10)
$\mathbf{F}(\mathbf{AXB})$	$\left.\frac{\partial \mathbf{F}(\mathbf{Y})}{\partial \mathbf{Y}}\right\|_{\mathbf{Y}=\mathbf{AXB}} (\mathbf{A} \otimes \mathbf{B}^\top)$	(R11)
$\mathbf{F}(\mathbf{G}(\mathbf{X}))$	$\left.\frac{\partial \mathbf{F}(\mathbf{Y})}{\partial \mathbf{Y}}\right\|_{\mathbf{Y}=\mathbf{G}(\mathbf{X})} \frac{\partial \mathbf{G}(\mathbf{X})}{\partial \mathbf{X}}$	(R12)
$\mathbf{H}(\mathbf{X})\mathbf{G}(\mathbf{X})$	$(\mathbf{I}_k \otimes \mathbf{G}^\top(\mathbf{X}))\frac{\partial \mathbf{H}(\mathbf{X})}{\partial \mathbf{X}} + (\mathbf{H}(\mathbf{X}) \otimes \mathbf{I}_l)\frac{\partial \mathbf{G}(\mathbf{X})}{\partial \mathbf{X}}$	(R13)

4 Differentiation Rules

To differentiate matrix–matrix functions there are just a few identities needed: Derivatives for the identity and the transpose, the product and the chain rule and the derivative for a matrix inverse. These, as well as a larger list of differentiation rules for matrix–matrix functions are given in Table 1. Table 2 gives a reference for the derivatives of matrix powers, and Table 3 gives the formulas for the derivatives of the trace and log-determinant of a matrix–matrix function. Together these identities enable us to differentiate any rational matrix–matrix function and any scalar–matrix function formed from arithmetic, transposition, trace and log-determinant operations.

4.1 A Simple Example

To illustrate the difficulty associated with automatically computing the derivative of a scalar–matrix function let us consider the simple example $f(\mathbf{X}) = \text{trace}(\mathbf{X}^\top\mathbf{X})$, where $\mathbf{X} \in \mathbb{R}^{m\times n}$. We can compute the symbolic derivative by hand by noting that $f(\mathbf{X}) = \sum_{i=1}^{m}\sum_{j=1}^{n} x_{ij}^2$, from which it is clear that $\frac{\partial f}{\partial x_{ij}} = 2x_{ij}$ and consequently $\frac{\partial f}{\partial \mathbf{X}} = 2\mathbf{X}$. Let's see what happens if we compute the function and its gradient in *forward mode*. We follow [11], but use matrix terminology. First we compute the

Table 2 Let $\mathbf{X} \in \mathbb{R}^{m \times m}$ be a square matrix, and $k \in \mathbb{N}$ be a positive number. The following is a list of matrix–matrix derivative rules for square matrices

$\mathbf{F}(\mathbf{X})$	$\frac{\partial \mathbf{F}(\mathbf{X})}{\partial \mathbf{X}}$	
$\mathbf{X}$	$\mathbf{I}_{m^2}$	(R14)
$\mathbf{X}^2$	$\mathbf{I}_m \otimes \mathbf{X}^\top + \mathbf{X} \otimes \mathbf{I}_m$	(R15)
$\mathbf{X}^k$	$\sum_{i=0}^{k-1} \mathbf{X}^i \otimes \mathbf{X}^{k-1-i}$	(R16)
$\mathbf{X}^{-1}$	$-\mathbf{X}^{-1} \otimes \mathbf{X}^{-\top}$	(R17)
$\mathbf{X}^{-\top}$	$-\mathbf{X}^{-\top} \boxtimes \mathbf{X}^{-1}$	(R18)
$\mathbf{X}^{-2}$	$-\mathbf{X}^{-1} \otimes (\mathbf{X}^{-2})^\top + \mathbf{X}^{-2} \otimes \mathbf{X}^{-\top}$	(R19)
$\mathbf{X}^{-k}$	$-\sum_{i=-k}^{-1} \mathbf{X}^i \otimes (\mathbf{X}^{-k-1-i})^\top$	(R20)

Table 3 Two identities useful for differentiation of scalar–matrix functions

$\text{vec}\left(\left(\frac{\partial}{\partial \mathbf{X}} \log \det(\mathbf{G}(\mathbf{X}))\right)^\top\right)$	$=$	$\left(\frac{\partial \mathbf{G}}{\partial \mathbf{X}}\right)^\top \text{vec}\left((\mathbf{G}(\mathbf{X}))^{-1}\right)$	(R21)
$\text{vec}\left(\left(\frac{\partial}{\partial \mathbf{X}} \text{trace}(\mathbf{G}(\mathbf{X}))\right)^\top\right)$	$=$	$\left(\frac{\partial \mathbf{G}}{\partial \mathbf{X}}\right)^\top \text{vec}(\mathbf{I})$	(R22)

function value in forward mode:

$$\mathbf{T}_1 = \mathbf{X}^\top \tag{37}$$

$$\mathbf{T}_2 = \mathbf{X} \tag{38}$$

$$\mathbf{T}_3 = \mathbf{T}_1 * \mathbf{T}_2 \tag{39}$$

$$t_4 = \text{trace}(\mathbf{T}_3). \tag{40}$$

Since the variables $\mathbf{T}_1$, $\mathbf{T}_2$ and $\mathbf{T}_3$ are matrices, their derivatives are four dimensional objects that need to be stored. Let us proceed with the forward mode computation for the derivative. Using the identities (R1), (R2), (R13) and (R22) from Tables 1 and 3 we get

$$\frac{\partial \mathbf{T}_1}{\partial \mathbf{X}} = \mathbf{I}_m \boxtimes \mathbf{I}_n \tag{41}$$

$$\frac{\partial \mathbf{T}_2}{\partial \mathbf{X}} = \mathbf{I}_m \otimes \mathbf{I}_n = \mathbf{I}_{mn} \tag{42}$$

$$\frac{\partial \mathbf{T}_3}{\partial \mathbf{X}} = (\mathbf{I}_n \otimes \mathbf{T}_2^\top)\left(\frac{\partial \mathbf{T}_1}{\partial \mathbf{X}}\right) + (\mathbf{T}_1 \otimes \mathbf{I}_n)\left(\frac{\partial \mathbf{T}_2}{\partial \mathbf{X}}\right) \tag{43}$$

$$\text{vec}^\top\left(\left(\frac{\partial t_4}{\partial \mathbf{X}}\right)^\top\right) = \text{vec}^\top(\mathbf{I}_n)\left(\frac{\partial \mathbf{T}_3}{\partial \mathbf{X}}\right). \tag{44}$$

The total storage requirement for the matrices $\frac{\partial \mathbf{T}_1}{\partial \mathbf{X}}$, $\frac{\partial \mathbf{T}_2}{\partial \mathbf{X}}$, $\frac{\partial \mathbf{T}_3}{\partial \mathbf{X}}$ and $\frac{\partial t_4}{\partial \mathbf{X}}$ is $mn + 2m^2n^2 + n^3m$, and computing $\frac{\partial \mathbf{T}_3}{\partial \mathbf{X}}$ requires $2n^4m^2$ multiplications. To implement the derivative computation in forward mode in C++ is a simple matter of operator overloading. Unfortunately, the resulting implementation will neither be memory nor computationally efficient.

As pointed out in [4], reverse mode computation is typically much more efficient, and such is the case here as well. To avoid multiplying the large matrices we pass $\mathrm{vec}(\mathbf{I}_n)$ in reverse to (41)–(44) so that all operations become matrix–vector multiplies.

$$\begin{aligned}
\mathrm{vec}^\top\left(\left(\frac{\partial t_4}{\partial \mathbf{X}}\right)^\top\right) &= \mathrm{vec}^\top(\mathbf{I}_n)\left((\mathbf{I}_n \otimes \mathbf{T}_2^\top)\left(\frac{\partial \mathbf{T}_1}{\partial \mathbf{X}}\right) + (\mathbf{T}_1 \otimes \mathbf{I}_n)\left(\frac{\partial \mathbf{T}_2}{\partial \mathbf{X}}\right)\right) && (45)\\
&= \mathrm{vec}^\top(\mathbf{I}_n\mathbf{I}_n\mathbf{T}_2)\left(\frac{\partial \mathbf{T}_1}{\partial \mathbf{X}}\right) + \mathrm{vec}^\top(\mathbf{T}_1\mathbf{I}_n\mathbf{I}_n)\left(\frac{\partial \mathbf{T}_2}{\partial \mathbf{X}}\right) && (46)\\
&= \mathrm{vec}^\top(\mathbf{T}_2)\left(\frac{\partial \mathbf{T}_1}{\partial \mathbf{X}}\right) + \mathrm{vec}^\top(\mathbf{T}_1)\left(\frac{\partial \mathbf{T}_2}{\partial \mathbf{X}}\right) && (47)\\
&= \mathrm{vec}^\top(\mathbf{T}_2)\mathbf{I}_m \boxtimes \mathbf{I}_n + \mathrm{vec}^\top(\mathbf{T}_1)\mathbf{I}_m \otimes \mathbf{I}_n && (48)\\
&= \mathrm{vec}^\top(\mathbf{T}_2^\top) + \mathrm{vec}^\top(\mathbf{T}_1) && (49)\\
&= \mathrm{vec}^\top(\mathbf{T}_1 + \mathbf{T}_2^\top). && (50)
\end{aligned}$$

In (46) we used the identitites (8) and (11), and in (48) we used the identities (20) and (23). If we look closely at this derivation, we can see that very little computation is needed. The matrix multiplications can be skipped altogether as they all have the identity matrix as one of the arguments. Only the final matrix addition $\mathbf{T}_1 + \mathbf{T}_2^\top$ is necessary. This is a total of $\mathcal{O}(mn)$ arithmetic operations, exactly the same that we achieved when we computed the derivative by hand. The matrix–matrix function required $\mathcal{O}(mn^2)$ arithmetic operations to compute, so the overhead for the derivative computation is very small. We also see that in reverse mode we only need to store the constituent component matrices of the Kronecker and box products – thus significantly reducing the storage requirements. This example can be implemented in C++ using expression templates and operator overloading. The overloaded matrix–matrix functions and operators can be used to record the computational graph, and the scalar–matrix function `trace` (or `det`) can be overloaded to traverse the computation graph in reverse, [13].

This example was easy to work out by hand, and significantly harder using the technique described. For slightly more intricate functions the complexity can grow significantly, and the automatic method becomes the more practical one. For example, the derivative of a function like $f(\mathbf{X}) = \mathrm{trace}((\mathbf{X}\boldsymbol{\Sigma}\mathbf{X}^\top)^{-1})$ can be systematically derived and found to be $\frac{\partial f}{\partial \mathbf{X}} = (\mathbf{X}\boldsymbol{\Sigma}\mathbf{X}^\top)^{-1}\mathbf{X}(\boldsymbol{\Sigma} + \boldsymbol{\Sigma}^\top)(\mathbf{X}\boldsymbol{\Sigma}\mathbf{X}^\top)^{-1}$.

4.2 The Hessian and Newton's Method

For scalar–scalar polynomial functions the derivative is a polynomial of degree one less. We can make similar observations for derivatives of polynomial matrix–matrix functions. For the second degree polynomial $\mathbf{P}(\mathbf{X}) = \mathbf{X}\boldsymbol{\Sigma}\mathbf{X}^\top$ the derivative, $\frac{\partial \mathbf{P}}{\partial \mathbf{X}} = \mathbf{I} \otimes (\boldsymbol{\Sigma}\mathbf{X}^\top) + (\mathbf{X}\boldsymbol{\Sigma}) \boxtimes \mathbf{I}$, has degree one less. For more general polynomials we can observe that the derivative will always have the form

$$\frac{\partial \mathbf{P}}{\partial \mathbf{X}} = \sum_{i=1}^{k_1} \mathbf{A}_i \otimes \mathbf{B}_i + \sum_{i=k_1+1}^{k} \mathbf{A}_i \boxtimes \mathbf{B}_i, \tag{51}$$

where $\mathbf{A}_i$ and $\mathbf{B}_i$ are matrix polynomials in $\mathbf{X}$ with degree($\mathbf{A}_i$) + degree($\mathbf{B}_i$) $\leq$ degree($\mathbf{P}$) $-$ 1, and $\mathbf{A}_i$ and $\mathbf{B}_i$ will in general depend on the constituent parts of $\mathbf{P}$. Furthermore, the number k is less than or equal to the number of instances of $\mathbf{X}$ in the expression for $\mathbf{P}(\mathbf{X})$. This cannot automatically be assumed to be equal to the degree of $\mathbf{P}$ since $\mathbf{P}(\mathbf{X}) = \mathbf{CXC} + \mathbf{DXD}$ cannot be simplified in terms of arithmetic operations, for the case of general matrices $\mathbf{C}$, $\mathbf{D}$. For a general rational matrix function $\mathbf{R}(\mathbf{X})$ we can prove that

$$\frac{\partial \mathbf{R}}{\partial \mathbf{X}} = \sum_{i=1}^{k_1} \mathbf{A}_i \otimes \mathbf{B}_i + \sum_{i=k_1+1}^{k} \mathbf{A}_i \boxtimes \mathbf{B}_i, \tag{52}$$

where $\mathbf{A}_i$, $\mathbf{B}_i$ are rational matrix functions and k is less than or equal to the number of instances of $\mathbf{X}$ in the expression for $\mathbf{R}(\mathbf{X})$. Since the derivative of a scalar–matrix function of the form $f(\mathbf{X}) = \text{trace}(\mathbf{R}(\mathbf{X}))$ or $f(\mathbf{X}) = \log\det(\mathbf{R}(\mathbf{X}))$ is a rational matrix–matrix function it follows that the Hessian of f is of the form (52), where k is the number of times $\mathbf{X}$ occurs in the expression for the scalar–matrix derivative of f.

In general the Hessian of any scalar–matrix function formed by using only arithmetic and transposition matrix operations and the trace and log det operations will lead to Hessians of the form

$$\mathbf{H} = \sum_{i=1}^{k_1} \mathbf{A}_i \otimes \mathbf{B}_i + \sum_{i=k_1+1}^{k_2} \mathbf{A}_i \boxtimes \mathbf{B}_i + \sum_{i=k_2+1}^{k} \text{vec}(\mathbf{A}_i)\text{vec}^\top(\mathbf{B}_i). \tag{53}$$

This is a large class of functions and this result has consequences for optimizing scalar–matrix functions. Due to the special form of the Hessian, it is always possible to compute the Newton step, $(\mathbf{H}(f))^{-1}\nabla f$, efficiently. If, for simplicity, we assume $\mathbf{A}_i, \mathbf{B}_i \in \mathbb{R}^{d\times d}$, then the cost of computing $\mathbf{H}\text{vec}(\mathbf{V})$ for some $\mathbf{V} \in \mathbb{R}^{d\times d}$ is $\mathcal{O}(k_2 d^3 + (k-k_2)d^2)$ operations. If $k < d$ and $\mathbf{H}$ is positive definite we can compute the Newton direction, $\mathbf{H}^{-1}\nabla$, by the conjugate gradient algorithm which uses at most d^2 matrix vector multiplies of the form $\mathbf{H}\text{vec}(\mathbf{V})$, [8]. Thus the total computational cost is $\mathcal{O}(k_2 d^5 + (k - k_2)d^4)$, and the memory cost is $\mathcal{O}(kd^2)$ to store $\mathbf{A}_i, \mathbf{B}_i$.

For $k = 1$ or $k = 2$ there are further savings, and the Newton direction can actually be computed in $\mathcal{O}(d^3)$ arithmetic operations. For $k = 1$ it is clear by (9) and (21). For $k = 2$ the Newton direction is computed by use of the matrix inversion lemma if $k = k_2 + 1$, and otherwise ($k = k_2$) by transforming the Newton direction equation into a Sylvester equation. The Bartels-Stewart algorithm, [1], can then solve the Sylvester equation in $\mathcal{O}(d^3)$ operations. The $k = 2$ case can be solved efficiently because, in general, two matrices can be simultaneously diagonalized. For $k > 2$ we must resort to the conjugate gradient algorithm, unless the matrices happen to be simultaneously diagonalizable.

4.3 An Example Taylor Series

We use these matrix differentiation rules to compute the first two terms in the Taylor series for the log-determinant. Let $f(\mathbf{X}) = \log\det(\mathbf{X})$. Then by rules (R21) and (R18) we have

$$\frac{\partial f}{\partial \mathbf{X}} = \mathbf{X}^{-\top} \qquad \frac{\partial(\mathbf{X}^{-\top})}{\partial \mathbf{X}} = -\mathbf{X}^{-\top} \boxtimes \mathbf{X}^{-1}. \tag{54}$$

The Taylor series around the point $\mathbf{X}_0$ is therefore given by

$$\begin{aligned}\log\det(\mathbf{X}) &= \log\det(\mathbf{X}_0) + \operatorname{trace}((\mathbf{X} - \mathbf{X}_0)^\top \mathbf{X}_0^{-\top}) \\ &\quad - \frac{1}{2!}\operatorname{vec}^\top((\mathbf{X} - \mathbf{X}_0)^\top)\left(\mathbf{X}_0^{-\top} \boxtimes \mathbf{X}_0^{-1}\right)\operatorname{vec}(\mathbf{X} - \mathbf{X}_0)^\top + \mathcal{O}((\mathbf{X} - \mathbf{X}_0)^3) \\ &= \log\det(\mathbf{X}_0) + \operatorname{trace}((\mathbf{X} - \mathbf{X}_0)\mathbf{X}_0^{-1}) \\ &\quad - \frac{1}{2}\operatorname{trace}((\mathbf{X} - \mathbf{X}_0)\mathbf{X}_0^{-1}(\mathbf{X} - \mathbf{X}_0)\mathbf{X}_0^{-1}) + \mathcal{O}((\mathbf{X} - \mathbf{X}_0)^3).\end{aligned} \tag{55}$$

5 Future Work

In this paper we introduced the box product to make the computation of scalar–matrix and matrix–matrix derivatives simple and efficient. We also showed that the box product can be used to reduce the computation and storage requirements of Newton's method. These results have not been presented elsewhere, and were given without proof here. The proofs for these new results, and further properties of the box product, will be given in a future publication, [9]. We also plan to apply these techniques to automatic speech recognition, and to expand the presented optimization theory to non–smooth scalar–matrix functions.

Acknowledgements The authors are indebted to the anonymous reviewers and the editor, Shaun Forth. Their efforts led to significant improvements to the exposition.

References

1. Bartels, R., Stewart, G.: Algorithm 432: Solution of the matrix equation AX+ XB= C [F4]. Communications of the ACM **15**(9), 820–826 (1972)
2. Fackler, P.L.: Notes on matrix calculus. North Carolina State University (2005)
3. Giles, M.B.: Collected matrix derivative results for forward and reverse mode algorithmic differentiation. In: C.H. Bischof, H.M. Bücker, P.D. Hovland, U. Naumann, J. Utke (eds.) Advances in Automatic Differentiation, *Lecture Notes in Computational Science and Engineering*, vol. 64, pp. 35–44. Springer, Berlin (2008). DOI 10.1007/978-3-540-68942-3_4
4. Griewank, A.: On automatic differentiation. In: M. Iri, K. Tanabe (eds.) Mathematical Programming, pp. 83–108. Kluwer Academic Publishers, Dordrecht (1989)
5. Harville, D.A.: Matrix algebra from a statistician's perspective. Springer Verlag (2008)
6. Magnus, J.R., Neudecker, H.: Matrix differential calculus with applications in statistics and econometrics (revised edition). John Wiley & Sons, Ltd. (1999)
7. Minka, T.P.: Old and new matrix algebra useful for statistics. See www.stat.cmu.edu/~minka/papers/matrix.html (2000)
8. Nocedal, J., Wright, S.J.: Numerical Optimization. Springer Series in Operations Research. Springer-Verlag, New York, NY (1999)
9. Olsen, P.A., Rennie, S.J.: The box product, matrix derivatives, and Newton's method (2012). (in preparation)
10. Petersen, K.B., Pedersen, M.S.: The matrix cookbook (2008). URL http://www2.imm.dtu.dk/pubdb/p.php?3274. Version 20081110
11. Rall, L.B., Corliss, G.F.: An introduction to automatic differentiation. In: M. Berz, C.H. Bischof, G.F. Corliss, A. Griewank (eds.) Computational Differentiation: Techniques, Applications, and Tools, pp. 1–17. SIAM, Philadelphia, PA (1996)
12. Searle, S.R.: Matrix algebra useful for statistics, vol. 512. Wiley, New York (1982)
13. Veldhuizen, T.: Expression templates. C++ Report **7**(5), 26–31 (1995)

Native Handling of Message-Passing Communication in Data-Flow Analysis

Valérie Pascual and Laurent Hascoët

Abstract Automatic Differentiation by program transformation uses static data-flow analysis to produce efficient code. This data-flow analysis must be adapted for parallel programs with Message-Passing communication. Starting from a context-sensitive and flow-sensitive data-flow analysis scheme initially devised for sequential codes, we extend this scheme for parallel codes. This extension is independent of the particular analysis and does not require a modification of the code's internal representation, i.e. the flow graph. This extension relies on an accurate matching of communication points, which can't be found automatically in general, and thus new user directives prove useful.

Keywords Data-flow analysis • Activity analysis • Automatic differentiation • Message-passing • MPI

1 Introduction

Static data-flow analysis of programs is necessary for efficient automatic transformation of codes. In the context of Automatic Differentiation (AD), most of the classical data-flow analyses prove useful as well as specific analyses such as activity and TBR analyses [5]. Parallel programs with message-passing pose additional problems to data-flow analysis because they introduce a flow of data that is not induced by the control-flow graph ("flow graph" for short). We propose an extension to data-flow analysis that captures this communication-induced flow of data. This extension applies in the very general framework of flow-sensitive analysis that sweep over the flow graph, possibly using a worklist for efficiency. This extension

V. Pascual (✉) · L. Hascoët
INRIA, Sophia-Antipolis, France
e-mail: Valerie.Pascual@inria.fr; Laurent.Hascoet@inria.fr

S. Forth et al. (eds.), *Recent Advances in Algorithmic Differentiation*, Lecture Notes in Computational Science and Engineering 87, DOI 10.1007/978-3-642-30023-3_8,

makes no particular hypothesis on the specific analysis and only introduces new artificial variables that represent communication channels, together with a generic modification of the flow-sensitive propagation strategy.

2 Context-Sensitive and Flow-Sensitive Data-Flow Analysis

To reach the accuracy that is necessary to generate an efficient transformed program, data-flow analysis should be context-sensitive and flow-sensitive. Context sensitivity operates at the call graph level. In a context sensitive analysis, each procedure uses a context that is built from the information available at its call sites. Even when making the choice of generalization, which means using only one context that summarizes all call sites, this context allows the analysis to find more accurate results inside the called procedure. Flow sensitivity operates at the flow graph level. In a flow-sensitive analysis the propagation of data-flow information follows an order compatible with the flow graph, thus respecting possible execution order.

Data-flow analysis works by propagating information through call graphs and flow graphs. Call graphs may be cyclic in general, due to recursivity. Flow graphs may be cyclic, due to loops and other cyclic control. Completion of the analysis requires reaching a fixed point both on the call graph and on each flow graph. The most effective way to control this fixed point propagation uses worklists [7].

In a naïve implementation a data-flow analysis of a calling procedure would require a recursive data-flow analysis of each called procedure, before the analysis of the calling procedure is completed. This would quickly cause a combinatorial explosion in run-time and in memory. To avoid that, it is wise to introduce a "relative" version of the current analysis that summarizes the effect of each called procedure on the information computed for any calling procedure. For instance in the case of Activity analysis, a variable is active if it depends on an independent input in a differentiable way (it is "varied") and the same time it influences the dependent output in a differentiable way (it is "useful"). This results in two data-flow analyses, both top-down on the call graph: The "varied" analysis goes forward on the flow graph, and the "useful" analysis goes backward on the flow graph. When any of the two reach a procedure call, we don't want to call the analysis recursively on the called procedure. Instead, we use a "differentiable dependency" summarized information that relates each output of the called procedure to each of its inputs on which it depends in a differentiable way. This relative information occupies more space than plain activity, typically the square of the number of visible variables, but it is easily and quickly used in place of the actual analysis of the called procedure. It takes a preliminary data-flow analysis to compute this "dependency", which is this time bottom-up on the call graph. This strategy may have a cost: the summarized relative information may be less accurate than an explicit data-flow analysis of the callee. On the other hand combinatorial behavior is improved, with an initial bottom-up sweep on the call graph to compute the relative information, followed by a top-down sweep to compute the desired information. Each sweep analyses each procedure only once, except for recursive codes.

3 Impact of Message-Passing on Data-Flow Analysis

The above framework for data-flow analysis is originally designed for sequential programs. It does not handle message-passing communication, which introduces a new flow of data unrelated to the flow graph, and that may even apparently go backwards the static flow graph, e.g. in a SPMD context, from a `send` to a `receive` located several lines before. See also in Fig. 1 the data-flow from `MPI_SEND` to `MPI_RECV` that is unrelated to the static flow graph. The propagation algorithm must be extended to capture this additional flow of data. Little research has been done in the domain of static analysis of message-passing programs [3]. Bronewtsky [1] defines parallel control-flow graphs, an extension of flow graphs that is the finite cross-product of the flow graphs of all the processes. This a theoretical framework useful for reasoning about analyses but that does not easily lend itself to implementation in our tools.

In the context of AD, several methods have been tried to solve this problem. `Odyssée` [2] introduced fictitious global communication variables but let flow graph propagation unchanged. This alone cannot capture the effect of communication that goes against the static flow graph order, and may give incorrect results.

A more radical method is to assign the analysis' conservative default value to all variables transmitted through message-passing. This leads to degraded accuracy of data-flow results and a less efficient differentiated code that may contain unnecessary derivative computation, useless differentiated communications, or useless trajectory storage in adjoint mode. This can be compensated partly with user directives understood by the AD tool.

Strout, Kreaseck and Hovland [10] use an "interprocedural control-flow graph" and augment it with communication edges between possible `send/receive` pairs [9]. Heuristics keep the number of communication edges low, based on constant propagation and the `MPI` semantics. This extended data-flow analysis improves the accuracy, e.g. for activity analysis. However these extra edges in the flow graph correspond to no control and have a special behavior: only the variables involved in the communication travel through these edges.

4 Data-Flow Analysis with Flow Graph Local Restart

We believe that introducing new global variables to represent communication channels as in [2] is an element of the solution. A channel is an artificial variable that contains all values currently in transit. However to cope with communication that goes against the flow graph we prefer to modify the data-flow propagation algorithm

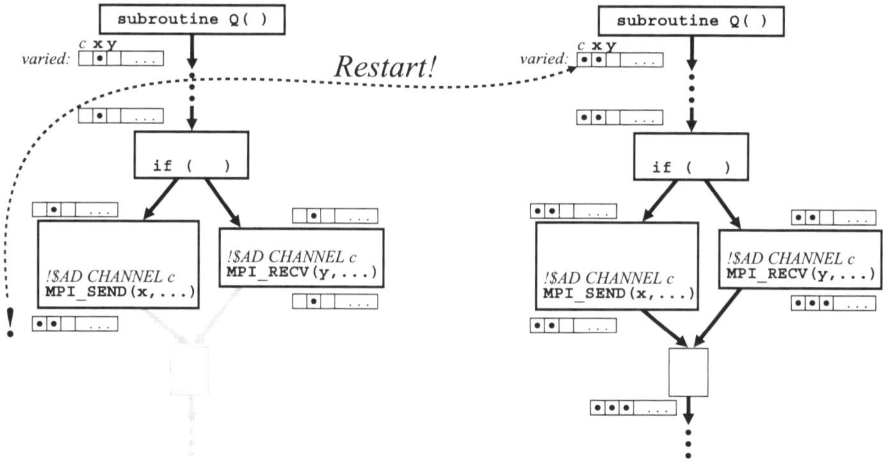

Fig. 1 Flow graph local restart after communication, in the case of the "varied" analysis

rather than modifying the flow graph itself. The arrows of the flow graph really represent an execution order, and adding arrows for communication may blur this useful interpretation. Note that adding flow arrows requires an interprocedural control-flow graph. In either case, modifying the propagation algorithm or modifying the graph it runs on, this can be done in a way mostly independent from the particular analysis.

The run-time context in which a given procedure is executed contains in particular the state of the various channels. During static data-flow analysis the context in which a given procedure is analyzed is an abstraction of this run-time context, only it represents several actual run-time contexts together. Therefore this static analysis context also includes the information on channels. When analysis of a given procedure reaches a communication call that changes the status of a channel, this change must be seen by all processes running in parallel and therefore possibly by all procedures of the code. In particular the static analysis context for the given procedure must change to incorporate the new channel status, and the analysis itself must restart from the beginning of the procedure. However this restart remains local to the given flow graph, as shown by Fig. 1. The effect on the other procedures' analysis will be taken care of by the "relative" version of the analysis. Thus this restart, illustrated by Fig. 1, remains local to the current flow graph: after the `MPI_SEND` is executed with a varied `x`, the artificial variable `c` that represents this particular communication channel becomes varied. The changing "varied" status of `c` restarts the current propagation from the entry of the flow graph. This new sweep, when reaching the `MPI_RECV` that reads the same channel, makes `y` varied in turn. In the frequent case when propagation order of the data-flow analysis is done with a worklist, the restart is achieved by just adding the entry block on top of the worklist, or the exit block in case of a backward propagation. This results in the framework Algorithm 1, common to any forward data-flow analysis. Navigation in the flow graph only needs the EntryBlock, the ExitBlock, plus the successor (succ)

Algorithm 1 Extension of forward analysis to message-passing communication

```
Given entryInfo:
01 ∀ Block b, in(b) := ∅; out(b) := ∅
02 out(EntryBlock) := entryInfo
03 worklist := succ(EntryBlock)
04 while [worklist ≠ {ExitBlock}]
05     b := firstof(worklist)          // i.e. the element with lowest dfst index
06     worklist := worklist\{b}
07     i := ∪_{p∈pred(b)} out(p)
08     o := propagate i through b
09     if [o/channels > out(b)/channels
10             && out(EntryBlock) ⊁ o/channels]
11         out(EntryBlock) := out(EntryBlock) ∪ (o/channels)
12         worklist := worklist ∪ succ(EntryBlock)
13     if [o > out(b)]
14         out(b) := o
15         worklist := worklist ∪ succ(b)
16 exitInfo := ∪_{p∈pred(ExitBlock)} out(p)
```

and predecessor (pred) sets for every block of the flow graph. Blocks are labelled with their *dfst* index, which is such that the index of a block is most often lower than the index of its successors. Actual propagation of the data-flow information through a given block is represented by the analysis-specific "propagate" operation. Operation "o/channels" builds a copy of data-flow information o that concerns only communication channels. Algorithm 1 lines 01–08 and 13–16 is the usual sequential data-flow analysis. Our proposed extension is exactly lines 09–12.

Consider now the call graph level. During the bottom up computation of the "relative" analysis, every individual procedure `Q` is analyzed with an extended algorithm following Algorithm 1, therefore taking care of channels. The relative information that is built thus captures the effect of the procedure on the channels. For instance, the relative "differentiable dependency" information for the procedure `Q` of Fig. 1 will contain in particular that the output values of both y and channel c depend on the input values of `x` and of channel c. During analysis of a procedure `P` that calls `Q`, analysis of the call to `Q` may modify the information attached to the channels accordingly. In other words, analysis of the call to `Q` has an effect similar to analysis of a call to an elementary message-passing procedure. This triggers the local restart mechanism of Algorithm 1 at the level of the flow graph of `P`, and eventually leads to completion of the analysis inside procedure `P`.

5 Performance Discussion

We will discuss the consequences of introducing the Flow Graph local restart on termination and execution time of the analyses. These questions partly depend on the specific data-flow analysis, and each analysis deserves a specific study.

However, we saw that our proposed extension to message-passing is essentially done on the general analysis framework Algorithm 1 so that some general remarks apply to any analysis.

About termination, the argument most frequently used is that the data-flow information, kept in the variables in(b) and out(b) for each block b, belong to a set of possible values that is finite, with a lattice structure wrt the partial order $>$ compatible with the union $\cup$. If one can show that propagation for the particular analysis represented by line 08:

$$\text{o} := \text{ propagate i through b}$$

is such that propagation of a larger i returns a larger o, then termination is granted. This argument is still valid when we introduce the local restart. Every local restart makes out(EntryBlock) grow, so that restarts are in finite number and the process terminates.

The local restart clearly affects the execution time of the analysis. For each propagation through one flow graph, the execution time on a non parallel code depends on its nested loop structure. Classically, one introduces a notion of "depth" of the flow graph which measures the worst-case complexity of the analysis on this graph. On well-structured flow graphs, one can show that this "depth" is actually the depth of the deepest nested loop. On a code with message-passing, extra propagation is needed when the status of a channel variable changes. When the approach chosen to represent communication is to add extra flow edges, the "depth" of the flow graph changes [6]. When these new edges make the graph irreducible, evaluation of the depth even becomes NP-hard in general. Nevertheless, a reasonable rule of thumb is to add the number of communication edges to the nested loop depth to get an idea of the analysis complexity increase. With our approach, which adds no communication edge but rather triggers restarts from the flow graph EntryBlock, the complexity effect is very similar. Local restart can occur once for each individual channel, so that the "depth" is increased by the number of channels. Not surprisingly, an increased number of channels may yield more accurate analysis results, but may increase the analysis time. In practice, this slowdown is quite tolerable. To be totally honest, local restart incurs some redundant propagation compared to [10]: since restart is done from the EntryBlock rather than from the destinations of communication, it goes uselessly through all blocks between the EntryBlock and these destinations. However, this does not change the degree of complexity.

For propagation through the call graph, though, the number of times one given procedure is analyzed does not change with message-passing and still depends only on the structure of recursive calls. The restarts are local to each flow graph, and do not change the behavior at the call graph level. To summarize, the local restart method introduces an extra complexity only into the data-flow analysis of procedures that call message passing communication, directly or indirectly. However, after implementation of the local restart into all the data-flow analyses of the AD tool `Tapenade`, we observe no particular slowdown of the differentiation process.

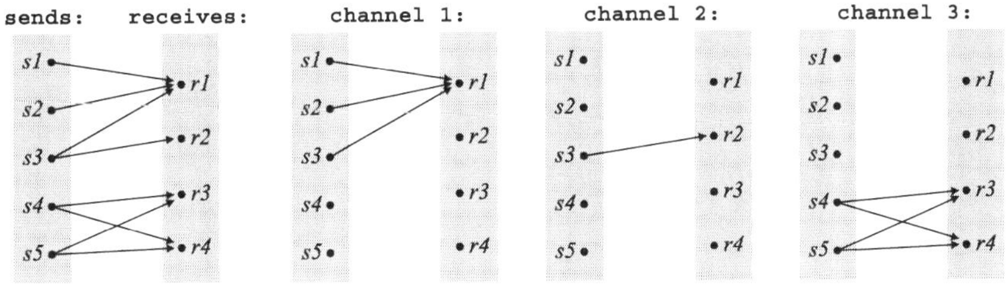

Fig. 2 A minimal biclique edge cover of a communication bipartite graph

6 Choosing a Good Set of Channels

Channel extraction depends on the message-passing communication library, in our case we use the MPI library [4, 8]. Collective communication functions such as broadcast do not need channels as all message-passing communications are done in one function call. We just have to focus on point-to-point communications functions.

We first define a test to match send's with receive's. For MPI point-to-point communication, this matching uses the source and destination, plus when possible the "tag" and "communicator" arguments of the message-passing function calls. If the communicators are identical, if the source and destination processes correspond, and if finally the tags may hold the same integer value, then the send and the receive match, which means that a value may travel from the former to the latter. The quality of this matching, i.e. the lowest possible number of false matches found, clearly depends on the quality of the available static constant propagation. Expressed in terms of channels, a match just means that there must be at least one defined communication channel that is common to the send and the receive.

Unfortunately, this matching depends on many parameters, and these are often computed dynamically in a way that is untractable statically, even with a powerful constant propagation. Static detection of matching send and receives will most often find too many matches, and we'd better resort to the user's knowledge of the code. This is done with a directive that the user can place in the code to designate explicitly the channel(s) affected by any communication call.

This preliminaries done, we end up with a bipartite graph that relates the send's to the matching receive's. We shall use Fig. 2 as an illustration. The question is to find a good set of channels that will exactly represent this communication bipartite graph:

- First, a good set of channels must not introduce artificial communication. On Fig. 2, we see we must not use a single channel to represent communications between $s1, s2, s3$ and $r1, r2$, because this would imply e.g. a spurious communication from $s2$ to $r2$. The rule here is that the bipartite subgraph induced by nodes that share a given channel must be complete.

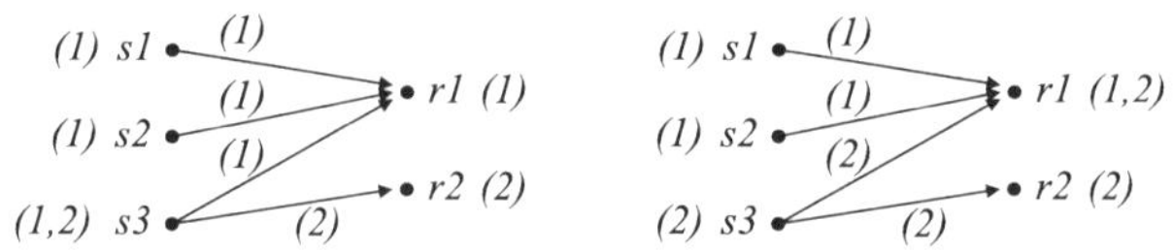

Fig. 3 Two different minimal covers. Channels shown between parentheses

- Second, a good set of channels must be as small as possible. We saw that the number of channels conditions the extra complexity of the analyses. In particular, the trivial choice that assigns one new channel for each edge of the bipartite graph is certainly correct, but too costly in general. On Fig. 2, we could introduce two channels for the two edges $(s1, r1)$ and $(s2, r1)$, but one channel suffices.

This question is already known as the "minimal biclique edge cover", a known NP-complete problem. We have thus a collection of available heuristics to pick from. On Fig. 2, three channels suffice.

Even when all channels were specified by the end-user by means of directives, it is good to run the above minimization problem. The user may have in mind a "logical" set of channels that may be reduced to a smaller set. On Fig. 2, suppose the user defined two channels $c4$ and $c5$, corresponding to send's $s4$ and $s5$ respectively, and that receive's $r3$ and $r4$ can receive from both channels. It turns out that channel minimization will merge $c4$ and $c5$ into a single one, because this captures the same communication pattern.

In general, there is not a unique minimal biclique edge cover. Different solutions, although yielding the same number of channels, may imply a marginally different number of iterations in the analyses. On Fig. 3, we have two minimal covers of a communication bipartite graph. The cover on the left has a send node labelled with two channels. If a forward data-flow analysis reaches this node first, then both channels are affected at once and no other fixpoint iteration will be necessary when later reaching the other send nodes. Conversely, the cover of the right is more efficient for a backward data-flow analysis, as the node with two channels is now a receive node.

There is an unfortunate interaction between this channel mechanism and the choice of generalization during data-flow analyses. If the code is such that native `MPI` calls are encapsulated into wrapper procedures, then attaching the channel to the native `MPI` calls may leave us with only one channel, as there is only one textual `MPI_SEND` present. On the other hand, we probably want to attach different channels to different wrapper calls, as if the wrapper procedures were the primitive communication points. We did not address this problem, which would either require to attach the channel to the wrapper call, or the possibility to opt for specialization instead of generalization for the analysis of each wrapper call, which means that a wrapper procedure will be analyzed once for each of its call sites.

7 Implementation and Outlook

We have implemented a prototype native treatment of MPI communication calls in Tapenade following the ideas of this paper. Implementation amounts to the following:

- Define the basic properties of MPI procedures in Tapenade's standard library.
- Make the tool recognize the MPI calls as message-passing calls, identify them as send, receive, collective ... and distinguish in their arguments those defining the channel and those containing the communicated buffer.
- Implement flow graph local restart into the single parent class of all data-flow analyses.
- Adapt each individual data-flow analysis at the point of propagating data-flow information through one message-passing call

We also updated tangent mode AD to introduce differentiated communication when the communication channel is active. Notice that this also introduces a notion of differentiation for parameters such as "tag", "request", and error "status". For instance, the "tag" of the differentiated communication call must be distinct from the original call's to make sure the receives are done in the correct order. Similar remarks hold between the "request" of nonblocking communication, and also for error "status".

We obtained correct data-flow information on a set of representative small examples, for all data-flow analyses.

We extended validation to a much larger CFD code called AERO, which implements an unsteady, turbulent Navier-Stokes simulation. The code is more that 100,000 lines long, and SPMD parallelization is necessary for most of its applications. Message-passing is done with MPI calls. In addition to point-to-point non-blocking communication MPI_ISEND, MPI_IRECV, and MPI_WAIT, the code uses collective communication MPI_BCAST, MPI_GATHER, and MPI_ALLREDUCE.

Given the current stage of development in Tapenade about message-passing communication, we could only apply tangent differentiation on the code. The resulting derivatives were validated by comparison of the parallel tangent code with divided differences between two runs of the original code, each of them parallel. At the source level, 10 of the 32 calls to MPI were detected active, causing 10 differentiated message-passing calls. On a relatively small test case, average run time per processor of the tangent code was 0.49 s, compared to an original run time per processor of 0.38 s. This increase of 30% is in line with what we observe on sequential codes.

The adjoint mode is still under development. However, we plan to validate soon an adjoint built semi-automatically, using the data-flow information which is already available, and hand-coding the appropriate adjoint communication calls.

We foresee a few extra difficulties for the adjoint mode of AD. As the adjoint differentiation model we have devised [11] exchanges the roles of paired MPI_ISEND or MPI_IRECV on one hand, and MPI_WAIT on the other hand, we need a way of associating those. A solution might be to wrap the MPI_WAIT's into special-purpose MPI_WAIT_SEND's or MPI_WAIT_RECV's containing all the necessary parameters. Another manner would be to run another static data-flow analysis. Matching MPI_ISEND or MPI_IRECV to MPI_WAIT is local to each process, unlike matching MPI_ISEND to MPI_IRECV. Therefore all we need is a local analysis, akin to data-dependence analysis on the "request" parameter. Considering that MPI_ISEND or MPI_IRECV write into their "request" parameter, and that MPI_WAIT reads its "request" then resets it, the two will match when there is a true dependency between them. User-directives may also be of help as a fallback option.

This work was not done with the one-sided communications of MPI-2 in mind. Although its new synchronization primitives may prove difficult to handle, we believe the remote memory of one-sided communications can be treated like a channel.

References

1. Bronevetsky, G.: Communication-sensitive static dataflow for parallel message passing applications pp. 1–12 (2009). DOI http://dx.doi.org/10.1109/CGO.2009.32. URL http://dx.doi.org/10.1109/CGO.2009.32
2. C.Faure, P.Dutto: Extension of odyssée to the mpi library -reverse mode. Rapport de recherche 3774, INRIA, Sophia Antipolis (1999)
3. Gopalakrishnan, G., Kirby, R.M., Siegel, S., Thakur, R., Gropp, W., Lusk, E., de Supinski, B., Schulz, M., Bronevetsky, G.: Formal analysis of mpi based parallel programs: Present and future. Communications of the ACM (2011)
4. Gropp, W., Lusk, E., Skjellum, A.: Using MPI: Portable Parallel Programming with the Message Passing Interface, 2nd edition. MIT Press, Cambridge, MA (1999)
5. Hascoët, L., Naumann, U., Pascual, V.: "To be recorded" analysis in reverse-mode automatic differentiation. Future Generation Computer Systems **21**(8), 1401–1417 (2005). DOI 10.1016/j.future.2004.11.009
6. Kreaseck, B., Strout, M.M., Hovland, P.: Depth analysis of mpi programs. ANL/MCS-P1754-0510 (2010)
7. Muchnick, S.S.: Advanced Compiler Design and Implementation. Morgan Kaufmann (1997)
8. Pacheco, P.S.: Parallel programming with MPI. Morgan Kaufmann Publishers Inc. (1996)
9. Shires, D., Pollock, L., Sprenkle, S.: Program flow graph construction for static analysis of mpi programs. In: Parallel and Distributed Processing Techniques and Applications, pp. 1847–1853 (1999)
10. Strout, M.M., Kreaseck, B., Hovland, P.D.: Data-flow analysis for mpi programs. In: Proceedings of the International Conference on Parallel Processing (ICPP) (2006)
11. Utke, J., Hascoët, L., Heimbach, P., Hill, C., Hovland, P., Naumann, U.: Toward adjoinable MPI. In: Proceedings of the 10th IEEE International Workshop on Parallel and Distributed Scientific and Engineering, PDSEC-09 (2009). http://doi.ieeecomputersociety.org/10.1109/IPDPS.2009.5161165

Increasing Memory Locality by Executing Several Model Instances Simultaneously

Ralf Giering and Michael Voßbeck

Abstract We present a new source-to-source transformation which generates code to compute several model instances simultaneously. Due to the increased memory locality of memory accesses this speeds up the computation on processors using a cache hierarchy to overcome the relative slow memory access. The speedup depends on the model code, the processor, the compiler, and on the number of instances.

Keywords Vector mode • Source-to-source transformation • Ensemble Kalman filter • Genetic optimization

1 Introduction

The majority of processors currently available use a memory hierarchy to overcome the slow memory access compared to CPU speed. Several levels of cache with different sizes and bandwidths buffer the access to memory by transferring cache lines instead of individual data. Requested data can be read from cache much faster than from memory or the next cache level. If the data is missing in the cache (cache misses) it must be fetched from its original memory location or the next cache level. Thus the access to a bunch of data is faster if they are within relatively close storage locations (spatial locality).

Optimizing compilers try to generate efficient object code by increasing the locality of memory access patterns. Various code analyses are applied in order to allow the transformation of the code's internal representation without changing the results. Still, even the most advanced optimizing compilers cannot generate object code that reaches peak performance for all programs.

R. Giering (✉) · M. Voßbeck
FastOpt GmbH, Lerchenstrasse 28a, 22767 Hamburg, Germany
e-mail: Ralf.Giering@FastOpt.com; Michael.Vossbeck@FastOpt.com

S. Forth et al. (eds.), *Recent Advances in Algorithmic Differentiation*, Lecture Notes in Computational Science and Engineering 87, DOI 10.1007/978-3-642-30023-3_9,

However, many applications are based on the repeated execution of a model code for several different inputs. Applying genetic optimization requires the features of many different species to determine the next generation. In weather forecasting an ensemble Kalman filter is processed to estimate the robustness of a forecast and the ensemble is based on several forecasts with different initial states. Usually these model runs are computed sequentially on one platform or in parallel on several processors. Here we suggest running a new transformed code that computes several model instances simultaneously. Due to the increased locality of the transformed code and only single computation of passive parts (see below) this can speed up the overall runtime considerably.

2 Cloning

On a formal level we regard the numerical model M[1] as a mapping between two Euclidian vector spaces $I\!R^n$ and $I\!R^m$:

$$\begin{aligned} M : I\!R^n &\rightarrow I\!R^m \\ x &\mapsto y \end{aligned}$$

As pointed out in Sect. 1, a set of instances, i.e. an N-tuple $\hat{x} := (x_1, \ldots, x_N) \in (R^n)^N$ with some suitable N is processed by evaluating (sequentially or in parallel) N instances of M:

$$y_i = M(x_i) \quad (i = 1, \ldots, N) .$$

In contrast we describe the automatic generation of a transformed numerical code M_{cl}, which we call cloned code, that implements the simultaneous evaluation of M for all elements in the N-tuple $\hat{x}$.

$$\begin{aligned} M_{cl} : (I\!R^n)^N &\rightarrow (I\!R^m)^N \\ \hat{x} &\mapsto \hat{y} \end{aligned}$$

The implementation of this source-to-source transformation requires a setup very similar to that of automatic differentiation. The given code is scanned, parsed, checked semantically, and an internal representation is generated. A backward data flow analysis determines the required variables for a given set of dependent variables. The following forward data flow analysis determines the influenced variables for a given set of independent variables. In contrast to automatic differentiation, influenced means not only in a differentiable way, but in general. This has the important consequence that also integer and logical variables can be active, i.e. are required and influenced [3]. Variables which are not active are called passive.

---

[1] We will use the same symbol for the mapping itself and it's implementation as computer code throughout this text.

All active variables are transformed into cloned variables with an extra dimension holding the number of model instances. This extra dimension must be the innermost dimension in terms of memory layout in order to increase spatial locality. In Fortran that is the leftmost dimension in C/C++ the rightmost.

The core of the transformation mainly consists in generating these cloned variables and for each given routine, a cloned routine which executes the original statements not only for one instance but for several instances simultaneously. In most cases this can simply be done by embedding every assignment into a loop over all instances and replacing each access to an active variable by accessing the current instance of the cloned variable.

For example, the assignment[2]

```
y = 2. * p * sin(x)
```

with active scalar variables x and y and passive variable p is transformed to a loop

```
do ic = 1,cmax
   y_cl(ic) = 2. * p * sin(x_cl(ic))
end do
```

where cmax denotes the number of instances computed simultaneously. If the target code is Fortran-95 it is preferable to generate an array assignment which can be used inside where- and forall-constructs.

```
y_cl(:) = 2. * p * sin(x_cl(:))
```

This is very similar to the vector mode of automatic differentiation.

However, in case the control flow depends on an active variable the loop over all instances must be placed around the corresponding code segment. In the following example the variable x shall be active and thus the condition in the if-clause becomes active:

```
if (x .gt. 0.) then
   y = 2. * p * sin(x)
else
   y = p
endif
```

The clone transformation then yields

```
do ic = 1,cmax
   if (x_cl(ic) .gt. 0.) then
      y_cl(ic) = 2. * p * sin(x_cl(ic))
   else
      y_cl(ic) = p
   endif
end do
```

[2] All code presented conforms to Fortran-95 standard.

and the generated code loses the increased locality since now all statements inside the `if`-construct are computed for each instantiation separately. And even worse, it has a decreased locality because of the added innermost dimension.

Other exceptions must be made for two kinds of array assignments due to Fortran-95 language restrictions. First, a peculiarity arises if the right hand side (RHS) contains a passive array variable. Assuming an active variable `y` and a passive variable `p`, the assignment

```
y(:) = p(:)
```

cannot simply be transformed to

```
y_cl(:,:) = p(:)
```

This is not a legal Fortran-95 assignment because the rank of the left hand side (LHS) is 2 and the rank of the RHS is 1. Instead an explicit loop

```
forall( ic=1:cmax )
   y_cl(ic,:) = p(:)
end forall
```

may be generated or alternatively, one can expand the RHS:

```
y_cl(:,:) = spread( p(:), 1, cmax )
```

Second, a similar problem occurs if the RHS of an array assignment contains an active scalar expression. With `x,y` being active variables the assignment

```
y(:) = x
```

cannot be transformed to

```
y_cl(:,:) = x_cl(:)
```

Again, this is not a legal Fortran-95 assignment because of the conflicting ranks. As above, the solution is to generate an explicit loop

```
forall( ic=1:cmax )
   y_cl(ic,:) = x_cl(ic)
end forall
```

or an array assignment with expanded RHS:

```
y_cl(:,:) = spread( x_cl(:), 2, size(y_cl,2) )
```

Similar exceptions must be made in vector mode of automatic differentiation.

In addition, the transformational intrinsic functions `dot_product`, `merge`, `matmul`, `reshape`, `sum` and all intrinsic functions that have a dimensional argument need special handling.

A subroutine call is transformed into the call of the corresponding cloned subroutine if it is active, otherwise if it computes required variables, the original call is included. In the argument list of the call each variable is either kept or replaced by its cloned counterpart if it is active. A subroutine `sub` with passive dummy argument `k` and active dummy arguments `a,b`

```
subroutine sub( k, a, b )
...
end subroutine sub
```

may be called as follows:

```
call sub( n, x, y )
```

If x,y are active variables and n is a passive variable, it is transformed into a call of the cloned subroutine sub_cl

```
call sub_cl( n, x_cl, y_cl )
```

with x_cl,y_cl being the corresponding cloned variables.

Special care must be taken if a subroutine is called more than once in a different context, i.e. in one case an actual argument is active in the other passive. For example, the same subroutine sub may be called differently

```
call sub( n, x, p )
```

where x is active as above but p is a passive variable. If the generated code should call the same cloned subroutine sub_cl an auxiliary variable p_cl is introduced to be used as an actual argument to sub_cl:

```
p_cl(:) = p
call sub_cl( n, x_cl, p_cl )
p = p_cl(1)
```

In front of the call it gets a copy of the passive variable, and if the variable p is used afterwards it needs to get a copy of one instance of the cloned variable p_cl. Alternatively, one can generate several cloned subroutines, one for each different context. This would avoid auxiliary variables and yield less computations at the cost of longer source code.

We have implemented the clone mode in our tool TAF (Transformation of Algorithms in Fortran)[4]. The transformation is applied by the command line:

taf -toplevel func -input x -output y -clone -vecfor 10 code.f95

where func is the top-level subroutine with independent input variable x and dependent output variable y. The result is a file code_cl.f95 which contains a routine func_cl that computes (in this case) ten instances simultaneously.

3 Applications

To test the performance of the cloned code we applied TAF to a few production codes. Model-specific TAF options have been added to the command line. The generated code was then compiled by the Intel Fortran Compiler, 64 bit, version 12.0 using the option -fast that turns on -ipo, -O3, -no-prec-div, -static, and -xHost (inter-procedural optimization between files, aggressive loop transformations, less precise floating point results). The examples were all run on one core of an Intel Core i5 CPU with 2.8 GHz, which is endowed with three levels of caches (L1:32 KB, L2:256 KB, L3:8 MB). The processor supports Streaming SIMD Extensions version 4.2 (SSE4.2) [7]. Among other accelerations this Single Instructions Multiple Data instruction set allows processing several floating point values simultaneously (two double or four single precision values).

For a number of N instances, we measure the performance of the generated cloned code by comparing its runtime $t(M_{cl})$ with the runtime $t(M)$ of the original model times N and define the speedup s by

$$s := \frac{N \cdot t(M)}{t(M_{cl})}. \tag{1}$$

By definition speedup numbers greater than 1 indicate superior performance of the cloned code. Both run times are recorded by the Fortran-95 intrinsic subroutine `system_clock`. In case the runtime is to short, M (M_{cl} respectively) is evaluated multiple times within a loop and the mean time is recorded.

For an enhanced discussion of the results it is convenient to decompose the overall runtime into the times p and a spent for the computation of passive and active quantities

$$t(M) := p + a.$$

Because the cloned code computes passive quantities only once independent of the number of instances N it's runtime is given by

$$t(M_{cl}) := p + \frac{1}{c_N} \cdot N \cdot a.$$

Here c_N denotes a positive number which accounts for all effects (e.g. increased data locality) that influence the relative efficiency of the cloned code when computing active quantities. Based on this decomposition and Eq. (1) we derive

$$s = \frac{N \cdot (p + a)}{p + \frac{1}{c_N} \cdot N \cdot a}. \tag{2}$$

Note that (2) simplifies to $s = c_N$ for original code without any passive computations ($p = 0$).

BETHY is a land biosphere model that simulates the carbon dioxide flux to the atmosphere. It is part of the Carbon Cycle Data Assimilation System CCDAS [9]. Its inputs are 57 parameters of the biosphere model. The speedup for BETHY reaches a factor of 3 but does not increase much when more than 10 instances are run simultaneously (Fig. 1). We have analyzed the original and the cloned code with the cachegrind/valgrind tool [8]. The results show a sub-linear rise of the overall L1-cache misses with increasing N, which well explains the performance gain.

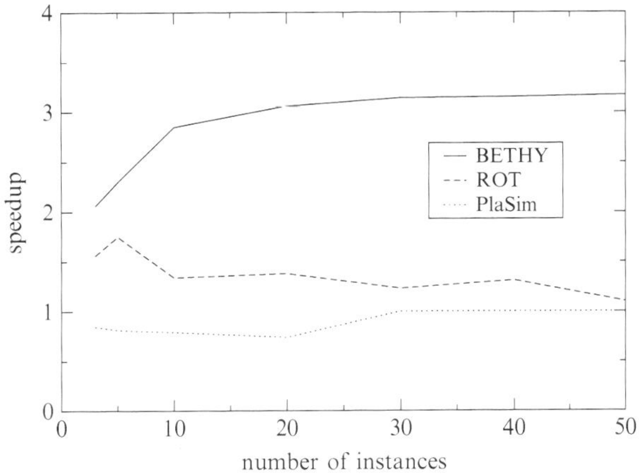

Fig. 1 Speedup of the three model codes BETHY, ROT, PlaSim

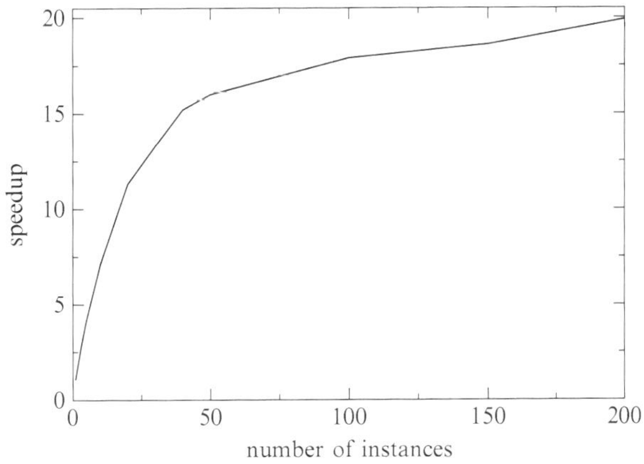

Fig. 2 Speedup NADIM

ROT calculates the stresses and deflections in a rotating disk that has an inner hole and a variable thickness [10]. The speedup gain of the cloned code reaches its peak of 1.7 for the number of 5 instances and has its highest values greater than 1.5 for small numbers of instances between 3 and 7. With more instances the speedup factor slightly decreases but remains above 1.

PlaSim simulates the atmosphere [1, 2]. The cloned version of PlaSim runs always slower than the corresponding number of original model runs. Here many `where`-constructs with array assignments are used in the given code. They already are executed very efficiently. Several of them depend on active variables leading to cloned code with the instantiation-loop around `where`-constructs similar to the cloned `if`-construct shown in Sect. 2. This decreases memory locality and thus performance of the cloned code.

NADIM is a radiative transfer model [5] that simulates the scattering of light by the vegetation. The speedup of the cloned code increases rapidly with the number of instances (Fig. 2). Above about 50 instances the increase flattens significantly.

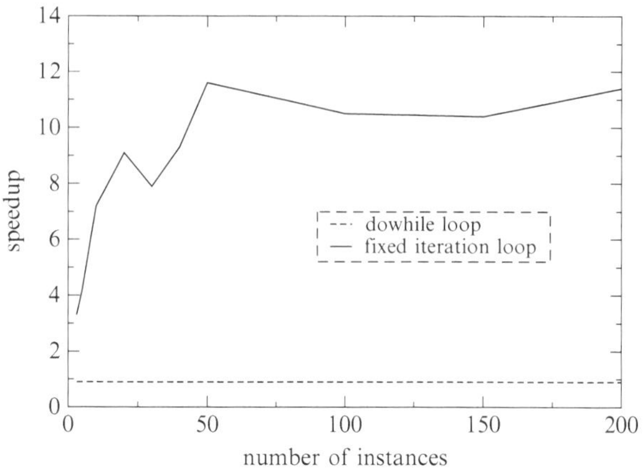

Fig. 3 Speedup NAST2D

The code contains a passive routine which calls intrinsic trigonometric functions indirectly and contributes 85% to the overall runtime. By setting c_N to 1 in Eq. (2) we would derive a speedup between 4.3 and 6.4. Thus, one part of the overall speedup of the cloned code can be explained by computing the passive part only once. The remaining speedup (factor c_N) computed from Eq. (2) and the measured run times varies between 2.8 and 3.5.

NAST2d is a 2-dimensional Navier-Stokes solver [6]. The given source code contains an iterative solver. The corresponding `dowhile`-loop is executed until a stopping criterion is reached. Because of this active condition the whole loop must be embedded into an instantiation-loop degrading the performance of the cloned code (Fig. 3). If the `dowhile`-loop is replaced by a fixed iteration loop (the number of iterations required has been determined before hand) the generated cloned code is much more efficient. The speedup increases with the number of instances up to about 11. An analysis of the two cloned codes shows a significant decrease of L1-cache misses in the version with the fixed iteration loop.

4 Conclusions

We presented a new source-to-source transformation that generates from a given simulation code a new code that computes not only one but several model instances simultaneously. By extending each active variable with an additional dimension the new code has a higher spatial locality. This can speed up the computation of several model instances considerably. Other reasons for a speedup can be that the cloned code

- Avoids unused computations (dead code elimination),
- Computes passive variables only once,
- Reads passive variables from file only once (e.g. forcing fields),
- May use (or increases) the number of vector operations (SIMD, SSE4).

However, if the original code already reaches peak performance on the processor no advantages can be expected. In addition if the control flow depends on the model input the need to put the instantiation loop around the specific control flow section decreases locality for that section and thus may even slow down the computations. Obviously, the memory requirements increase by the number of instances which limits the maximal allowed number.

Many simulation codes require solving a system of equations. If that system is solved iteratively the stopping criterion usually depends on intermediate results and thus on the model inputs. In order to gain speedup for the cloned code, one must avoid the dependence of the stopping criterion on the inputs. This can be done by computing a fixed number of iterations or using a direct solver instead.

The transformation can easily be extended to handle Message Parsing Interface (MPI) calls. Each message would be extended by the number of instances. Overall, computing several model instances simultaneously would require about as many messages as one model simulation. If the performance is communication bound this may lead to an additional speedup.

References

1. Blessing, S., Greatbatch, R., Fraedrich, K., Lunkeit, F.: Interpreting the atmospheric circulation trend during the last half of the 20th century: Application of an adjoint model. J. Climate **21**(18), 4629–4646 (2008)
2. Fraedrich, K., Jansen, H., Kirk, E., Luksch, U., Lunkeit, F.: The planet simulator: Towards a user friendly model. Meteorol. Z **14**, 299–30 (2005)
3. Giering, R., Kaminski, T.: Recipes for adjoint code construction. ACM Transactions on Mathematical Software **24**(4), 437–474 (1998). DOI http://doi.acm.org/10.1145/293686.293695
4. Giering, R., Kaminski, T.: Applying TAF to generate efficient derivative code of Fortran 77-95 programs. In: Proceedings of GAMM 2002, Augsburg, Germany (2002)
5. Gobron, N., Pinty, B., Verstraete, M.M.: A semidiscrete model for the scattering of light by vegetation. Journal of Geophysical Research Atmospheres **102**(D8), 9431–9446 (1997). DOI 10.1029/96JD04013
6. Griebel, M., Dornseifer, T., Neunhoeffer, T.: Numerical Simulation in Fluid Dynamics, a Practical Introduction. SIAM, Philadelphia (1998)
7. Intel Corporation: Intel SSE4 Programming Reference (2007). URL http://softwarecommunity.intel.com/isn/Downloads/Intel%20SSE4%20Programming%20Reference.pdf
8. Nethercote, N., Seward, J.: Valgrind: a framework for heavyweight dynamic binary instrumentation. SIGPLAN Not. **42**, 89–100 (2007). DOI http://doi.acm.org/10.1145/1273442.1250746. URL http://valgrind.org
9. Rayner, P., Knorr, W., Scholze, M., Giering, R., Kaminski, T., Heimann, M., Quere, C.L.: Inferring terrestrial biosphere carbon fluxes from combined inversions of atmospheric transport and process-based terrestrial ecosystem models. In: Proceedings of 6th Carbon dioxide conference at Sendai, pp. 1015–1017 (2001)
10. Timoshenko, S.P., Goodie, J.N.: Theory of Elasticity. McGraw-Hill, New York (1970). 3rd ed.

Adjoint Mode Computation of Subgradients for McCormick Relaxations

Markus Beckers, Viktor Mosenkis, and Uwe Naumann

Abstract In Mitsos et al. (SIAM Journal on Optimization 20(2):573–601, 2009), a method similar to Algorithmic Differentiation (AD) is presented which allows the propagation of, in general nondifferentiable, McCormick relaxations (McCormick, Mathematical Programming 10(2):147–175, 1976; Steihaug, Twelfth Euro AD Workshop, Berlin, 2011) of factorable functions and of the corresponding subgradients in tangent-linear mode. Subgradients are natural extensions of "usual" derivatives which allow the application of derivative-based methods to possibly nondifferentiable convex and concave functions. The software package `libMC` (Mitsos et al. SIAM Journal on Optimization 20(2):573–601, 2009) performs the automatic propagation of the relaxation and of corresponding subgradients based on the principles of tangent-linear mode AD by overloading. Similar ideas have been ported to Fortran yielding `modMC` as part of our ongoing collaboration with the authors of Mitsos et al. (SIAM Journal on Optimization 20(2):573–601, 2009).

In this article an adjoint method for the computation of subgradients for McCormick relaxations is presented. A corresponding implementation by overloading in Fortran is provided in the form of `amodMC`. The calculated subgradients are used in a deterministic global optimization algorithm based on a branch-and-bound method. The superiority of adjoint over tangent-linear mode is illustrated by two examples.

Keywords Non-smooth analysis • McCormick relaxations • Algorithmic differentiation • Adjoint mode • Subgradients

---

M. Beckers (✉)
German Research School for Simulation Sciences and Lehr- und Forschungsgebiet Informatik 12: Software and Tools for Scientific Computing (STCE), RWTH Aachen University, D-52062 Aachen, Germany
e-mail: beckers@stce.rwth-aachen.de

V. Mosenkis · U. Naumann
STCE, RWTH Aachen University, D-52062 Aachen, Germany
e-mail: mosenkis@stce.rwth-aachen.de; naumann@stce.rwth-aachen.de

S. Forth et al. (eds.), *Recent Advances in Algorithmic Differentiation*, Lecture Notes in Computational Science and Engineering 87, DOI 10.1007/978-3-642-30023-3_10,

1 Introduction

We start by recalling essential terminology from the relevant literature. Let therefore $Z \subseteq \mathbb{R}^n$ be an open convex set and let F be a continuous scalar function with domain Z. Local convex underestimations and concave overestimations of F play a central role in the upcoming argument. Such *relaxations* are defined as follows.

Definition 1 (Relaxations of functions). Let $F{:}Z \to \mathbb{R}$ be a scalar function defined on an open convex set Z. A convex (concave) function $F^{cv}(F^{cc})$ for which $F^{cv}(\mathbf{z}) \leq F(\mathbf{z})$ $(F^{cc}(\mathbf{z}) \geq F(\mathbf{z}))$ holds for all $\mathbf{z} = (z_1, \ldots, z_n) \in Z$ is called a *convex (concave) relaxation* of F.

To ensure applicability of AD technology, the function of interest should be decomposable into a sequence of assignments of the results of its *elemental* functions (a subset of the arithmetic operators and intrinsic functions provided by the used programming language; for example, Fortran) to unique intermediate variables. See [4] for details. Such functions are also referred to as *factorable.*

Definition 2 (Factorable functions). A function is called *factorable* if it is defined by a recursive composition of sums, products and a given library of univariate intrinsic functions.

From now on our focus will be on *McCormick relaxations* as introduced in [6].

Definition 3 (McCormick relaxations). Relaxations of factorable functions that are formed by recursive application of rules for the relaxation of univariate composition, binary multiplication, and binary addition to convex and concave relaxations of the univariate intrinsic functions and without the introduction of auxiliary variables are called *McCormick relaxations.*

Figure 1 shows examples of convex and concave relaxations for several elemental functions. Rules for the relaxations of sums, multiplications, and compositions of elemental functions are discussed in [6]. An implementation of the propagation of McCormick relaxations by overloading in C++ is presented in [7].

Since McCormick relaxations are possibly nonsmooth, generalized gradients are used in the context of derivative-based numerical algorithms. Related issues were also discussed in [3]. The relevant *subdifferentials* and *subgradients* are defined, for example, in [5] as follows.

Definition 4. Let $F : Z \to \mathbb{R}$ be a convex function given on an open convex set $Z \subseteq \mathbb{R}^n$ and let $F'(\mathbf{z}, \mathbf{d}) := \lim_{t \to +0} \frac{F(\mathbf{z}+t\mathbf{d})-F(\mathbf{z})}{t}$ denote the directional derivative of F at a point $\mathbf{z}$ in direction $\mathbf{d}$. Then the set

$$\partial F(\mathbf{z}) := \left\{\mathbf{s} \in \mathbb{R}^n \mid \langle \mathbf{s}, \mathbf{d} \rangle \leq F'(\mathbf{z}, \mathbf{d}) \ \forall \mathbf{d} \in \mathbb{R}^n\right\}$$

is called the **subdifferential** of F at $\mathbf{z}$. A vector $\mathbf{s} \in \partial F(\mathbf{z})$ is called a **subgradient** of F at $\mathbf{z}$.

Fig. 1 Relaxations for elemental functions F defined on $Z = [a, b]$

F	F^{cv}	F^{cc}
e^z	e^z	$e^a + \left(\frac{e^b - e^a}{b-a}\right)(z-a)$
$\sqrt{z}$	$\sqrt{a} + \frac{\sqrt{b}-\sqrt{a}}{b-a}(z-a)$	$\sqrt{z}$
$\ln(z)$	$\ln(a) + \frac{\ln(b)-\ln(a)}{b-a}(z-a)$	$\ln(z)$
$\lvert z\rvert$	$\lvert z\rvert$	$\lvert a\rvert + \frac{\lvert b\rvert - \lvert a\rvert}{b-a}(z-a)$
$\frac{1}{z}$ if $0 < a \leq b$ or $a \leq b < 0$	$\frac{1}{z}$ $\frac{1}{a} + \frac{1}{b} - \frac{1}{ab}z$	$\frac{1}{a} + \frac{1}{b} - \frac{1}{ab}z$ $\frac{1}{z}$

If F is concave on Z, then its subdifferential at $\mathbf{z}$ becomes

$$\partial F(\mathbf{z}) := \left\{\mathbf{s} \in \mathbb{R}^n \mid \langle \mathbf{s}, \mathbf{d}\rangle \geq F'(\mathbf{z}, \mathbf{d})\ \forall \mathbf{d} \in \mathbb{R}^n\right\}.$$

The subdifferential of any convex function is nonempty at any point [5]. If the function F is differentiable at $\mathbf{z}$, then its subgradient is unique and identical to the gradient of F at $\mathbf{z}$. In [7] sum, product, and composition rules for the computation of subgradients of McCormick relaxations are introduced. They yield an algorithm for the propagation of subgradients alongside with the respective relaxations. It turns out that this approach is similar to the application of tangent-linear mode AD to McCormick relaxations. The investigation of a corresponding adjoint approach in this article represents the logical next step.

Our interest in adjoint subgradient computation is motivated by a branch-and-bound algorithm for global nonlinear optimization. Subgradients are used there to construct affine underestimators as bounds of the function to be minimized. They are combined with standard interval extensions, thus, aiming for a more effective reduction of the search space. A description of a first version of this algorithm can be found in [2]. Its detailed discussion is beyond the scope of this article. Related approaches include methods based on interval arithmetic enhanced with *slope* computations [9].

Section 2 relates the formalism developed in [7] to standard tangent-linear mode AD. An adjoint method is presented in Sect. 3. First numerical results are reported in Sect. 4. They have been generated with our Fortran implementation of adjoint subgradient propagation for McCormick relaxations, `amodMC`. Conclusions are drawn in Sect. 5 together with a short outlook to our ongoing activities in this area.

2 Tangent-Linear Mode Computation of Subgradients of McCormick Relaxations

Consider the function $f = f(\mathbf{z})$ defined as

$$f \equiv \max: \quad \mathbb{R}^2 \to \mathbb{R}, \quad (z_1, z_2) \mapsto \begin{cases} z_1, & z_1 \geq z_2 \\ z_2, & z_1 < z_2 \end{cases}$$

and used in Proposition 2.7 in [7] for the computation of McCormick relaxations of products. It is differentiable on the set $\{(z_1, z_2) \in \mathbb{R}^2 | z_1 \neq z_2\}$ with gradient

$$\nabla \max(z_1, z_2) = \left(\mathbb{1}_{\{z_1 > z_2\}}, \mathbb{1}_{\{z_1 < z_2\}}\right)^T,$$

where $\mathbb{1}_b = 1$ if the condition b holds and $\mathbb{1}_b = 0$ otherwise. The derivative is undefined within $\{(z_1, z_2) \in \mathbb{R}^2 | z_1 = z_2\}$. However, according to Definition 4, a subgradient can be computed as follows.

$$\frac{f(\mathbf{z} + t\mathbf{d}) - f(\mathbf{z})}{t} = \frac{\max(z_1 + td_1, z_2 + td_2) - \max(z_1, z_2)}{t} = \begin{cases} \frac{td_1}{t} = d_1, & d_1 \geq d_2 \\ \frac{td_2}{t} = d_2, & d_1 < d_2 \end{cases}$$

$$\geq (1\ 0) \cdot \begin{pmatrix} d_1 \\ d_2 \end{pmatrix} = d_1 \ \forall \begin{pmatrix} d_1 \\ d_2 \end{pmatrix} \in \mathbb{R}^2.$$

Hence, the vector

$$s_{\max}(z_1, z_2) = \left(\mathbb{1}_{\{z_1 \geq z_2\}}, \mathbb{1}_{\{z_1 < z_2\}}\right) \tag{1}$$

is an element of the subgradient of max in $\{(z_1, z_2) \in \mathbb{R}^2 | z_1 = z_2\}$.

Similarly an element of the subgradient of min is given by

$$s_{\min}(z_1, z_2) = \left(\mathbb{1}_{\{z_1 \leq z_2\}}, \mathbb{1}_{\{z_1 > z_2\}}\right) \tag{2}$$

It can be shown that the application of (1) and (2) to the propagation of McCormick relaxations yields Theorem 3.3 in [7]. These ideas are implemented in `libMC` and `modMC`.

Another potentially nondifferentiable function with relevance to McCormick relaxations is the midpoint function $f = f(\mathbf{z})$ defined as

$$f = \text{mid}_c : \quad \mathbb{R}^2_{\{z_1 \leq z_2\}} \to \mathbb{R}, \quad (z_1, z_2) \mapsto \begin{cases} z_1, & c < z_1 \\ z_2, & c > z_2 \\ c, & z_1 \leq c \leq z_2. \end{cases}$$

According to Theorem 3.2 in [7] convex and concave relaxations of the composite function $g = F \circ f$, $F : \mathbb{R} \supseteq X \to \mathbb{R}$, $y = F(x)$, can be computed as

$$g^{cv}(\mathbf{z}) = F^{cv}(\text{mid}_{x^{\min}}(f^{cv}(\mathbf{z}), f^{cc}(\mathbf{z}))) \tag{3}$$

and

$$g^{cc}(\mathbf{z}) = F^{cc}(\text{mid}_{x^{\max}}(f^{cv}(\mathbf{z}), f^{cc}(\mathbf{z}))), \tag{4}$$

where $x^{\min} = \text{argmin}_{x \in X} F^{cv}(x)$ and $x^{\max} = \text{argmax}_{x \in X} F^{cc}(x)$, respectively. The tangent-linear extension of the corresponding linearized directed acyclic graph (DAG) [8] is shown in Fig. 2 for $n = 1$.

Obviously, $\text{mid}_c(z_1, z_2)$ is differentiable on $\{c < z_1 \leq z_2\} \cup \{z_1 \leq z_2 < c\} \cup \{z_1 < c < z_2\}$. Let $(z_1, z_2) \in \mathbb{R}^2$ be such that $z_1 < c = z_2$. Then mid_c is locally concave

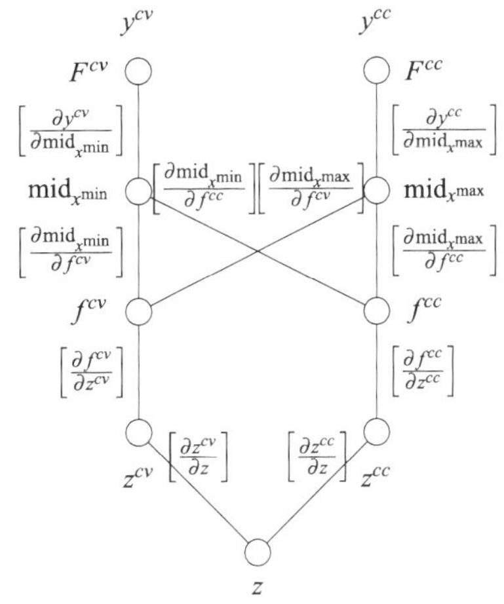

Fig. 2 Tangent-linear extension of the linearized DAG for (3) and (4)

and

$$\lim_{t\to+0}\frac{\text{mid}_c\,(z_1+t\cdot d_1, z_2+t\cdot d_2)-\text{mid}_c(z_1,z_2)}{t}=\lim_{t\to+0}\begin{cases}\frac{c-c}{t}=0, & d_2>0\\ \frac{c+t\cdot d_2-c}{t}=d_2, & d_2\le 0\end{cases}$$

$$=\begin{cases}0, & d_2>0\\ d_2, & d_2\le 0\end{cases}\quad\le 0=(0\ 0)\cdot\begin{pmatrix}d_1\\ d_2\end{pmatrix}\quad\forall \mathbf{d}\in\mathbb{R}^2.$$

In (z_1, z_2) with $z_1 = c < z_2$ the function mid_c is locally convex and

$$\lim_{t\to+0}\frac{\text{mid}_c\,(z_1+t\cdot d_1, z_2+t\cdot d_2)-\text{mid}_c(z_1,z_2)}{t}=\lim_{t\to+0}\begin{cases}\frac{c+t\cdot d_1-c}{t}=d_1, & d_1\ge 0\\ \frac{c-c}{t}=0, & d_1<0\end{cases}$$

$$=\begin{cases}d_1, & d_1\ge 0\\ 0, & d_2< 0\end{cases}\quad\ge 0=(0\ 0)\cdot\begin{pmatrix}d_1\\ d_2\end{pmatrix}\quad\forall \mathbf{d}\in\mathbb{R}^2.$$

In conclusion this shows that

$$s_{\text{mid}_c}(z_1, z_2)=\left(\mathbb{1}_{\{c<z_1\}}, \mathbb{1}_{\{c>z_2\}}\right) \tag{5}$$

is a subgradient of $\text{mid}_c(z_1, z_2)$ in both its convex and concave regions.

According to [1], the interpretation of the chain rule on the tangent-linear extension of the linearized DAG in Fig. 2 yields

$$\begin{aligned}\frac{\partial y^{cv}}{\partial z}&=\frac{\partial z^{cv}}{\partial z}\cdot\frac{\partial f^{cv}}{\partial z^{cv}}\cdot\frac{\partial \text{mid}_{x\min}}{\partial f^{cv}}\cdot\frac{\partial y^{cv}}{\partial \text{mid}_{x\min}}+\frac{\partial z^{cc}}{\partial z}\cdot\frac{\partial f^{cc}}{\partial z^{cc}}\cdot\frac{\partial \text{mid}_{x\min}}{\partial f^{cc}}\cdot\frac{\partial y^{cv}}{\partial \text{mid}_{x\min}}\\ &=1\cdot\left.\frac{\partial f^{cv}(x)}{\partial x}\right|_{x=z^{cv}}\cdot\mathbb{1}_{\{x_{\min}<f^{cv}\}}\left(f^{cv}(z^{cv}), f^{cc}(z^{cc})\right)\cdot\left.\frac{\partial y^{cv}(x)}{\partial x}\right|_{x=x_{\text{mid}}}\\ &\quad+1\cdot\left.\frac{\partial f^{cc}(x)}{\partial x}\right|_{x=z^{cc}}\cdot\mathbb{1}_{\{x_{\min}>f^{cc}\}}\left(f^{cv}(z^{cv}), f^{cc}(z^{cc})\right)\cdot\left.\frac{\partial y^{cv}(x)}{\partial x}\right|_{x=x_{\text{mid}}}\end{aligned}$$

$$= \begin{cases} 0, & f^{cv}(z^{cv}) \leq x^{\min} \leq f^{cc}(z^{cc}) \\ \left.\frac{\partial y^{cv}(x)}{\partial x}\right|_{x=f^{cc}(z^{cc})} \cdot \left.\frac{\partial f^{cc}(x)}{\partial x}\right|_{x=z^{cc}}, & f^{cc}(z^{cc}) < x^{\min} \\ \left.\frac{\partial y^{cv}(x)}{\partial x}\right|_{x=f^{cv}(z^{cv})} \cdot \left.\frac{\partial f^{cv}(x)}{\partial x}\right|_{x=z^{cv}}, & x^{\min} < f^{cv}(z^{cv}), \end{cases} \tag{6}$$

and

$$\frac{\partial y^{cc}}{\partial z} = \frac{\partial z^{cv}}{\partial z} \cdot \frac{\partial f^{cv}}{\partial z^{cv}} \cdot \frac{\partial \mathrm{mid}_{x^{\max}}}{\partial f^{cv}} \cdot \frac{\partial y^{cc}}{\partial \mathrm{mid}_{x^{\max}}} + \frac{\partial z^{cc}}{\partial z} \cdot \frac{\partial f^{cc}}{\partial z^{cc}} \cdot \frac{\partial \mathrm{mid}_{x^{\max}}}{\partial f^{cc}} \cdot \frac{\partial y^{cc}}{\partial \mathrm{mid}_{x^{\max}}}$$

$$= 1 \cdot \left.\frac{\partial f^{cv}(x)}{\partial x}\right|_{x=z^{cv}} \cdot \mathbb{1}_{\{x_{\max} < f^{cv}\}}\left(f^{cv}(z^{cv}), f^{cc}(z^{cc})\right) \cdot \left.\frac{\partial y^{cc}(x)}{\partial x}\right|_{x=x_{\mathrm{mid}}}$$

$$+1 \cdot \left.\frac{\partial f^{cc}(x)}{\partial x}\right|_{x=z^{cc}} \cdot \mathbb{1}_{\{x_{\max} > f^{cc}\}}\left(f^{cv}(z^{cv}), f^{cc}(z^{cc})\right) \cdot \left.\frac{\partial y^{cc}(x)}{\partial x}\right|_{x=x_{\mathrm{mid}}}$$

$$= \begin{cases} 0, & f^{cv}(z^{cv}) \leq x^{\max} \leq f^{cc}(z^{cc}) \\ \left.\frac{\partial y^{cc}(x)}{\partial x}\right|_{x=f^{cc}(z^{cc})} \cdot \left.\frac{\partial f^{cc}(x)}{\partial x}\right|_{x=z^{cc}}, & f^{cc}(z^{cc}) < x^{\max} \\ \left.\frac{\partial y^{cc}(x)}{\partial x}\right|_{x=f^{cv}(z^{cv})} \cdot \left.\frac{\partial f^{cv}(x)}{\partial x}\right|_{x=z^{cv}}, & x^{\max} < f^{cv}(z^{cv}), \end{cases} \tag{7}$$

where $x_{\mathrm{mid}} := \mathrm{mid}_{x^{\min}}\left(f^{cv}(z^{cv}), f^{cc}(z^{cc})\right)$. Generalization of the above for $n > 1$ yields Theorem 3.2 in [7] from the perspective of AD. Hence, overloading (resp., the corresponding semantic source code transformation in AD-enabled compilers/preprocessors) of max, min, and mid_c according to (1), (2), and (5) enables any AD tool to compute correct subgradients of McCormick relaxations.

3 Adjoint Mode Computation of Subgradients of McCormick Relaxations

The propagation of the convex and concave relaxations of the function

$$F: \quad Z \subseteq \mathbb{R}^n \to \mathbb{R}, \quad (z_1, \ldots, z_n) \mapsto y$$

can be considered as the *composition* $g \circ f = \begin{pmatrix} F^{cv} \\ F^{cc} \end{pmatrix}$ of the two functions

$$f: \quad Z \subseteq \mathbb{R}^n \to \mathbb{R}^{2n}, \quad (z_1, \ldots, z_n) \mapsto \left(z_1^{\mathrm{cv}}, z_1^{\mathrm{cc}}, \ldots, z_n^{\mathrm{cv}}, z_n^{\mathrm{cc}}\right) = (z_1, z_1, \ldots, z_n, z_n)$$

and $g = (g^{cv}, g^{cc}) : Z^+ \subseteq \mathbb{R}^{2n} \to \mathbb{R}^2$, $\left(z_1^{\mathrm{cv}}, z_1^{\mathrm{cc}}, \ldots, z_n^{\mathrm{cv}}, z_n^{\mathrm{cc}}\right) \mapsto \left(y^{\mathrm{cv}}, y^{\mathrm{cc}}\right)$, where Z^+ denotes the set $\left\{(z_1, z_1, \ldots, z_n, z_n) \in \mathbb{R}^{2n} | \mathbf{z} = (z_1, \ldots, z_n) \in Z\right\}$. Hence,

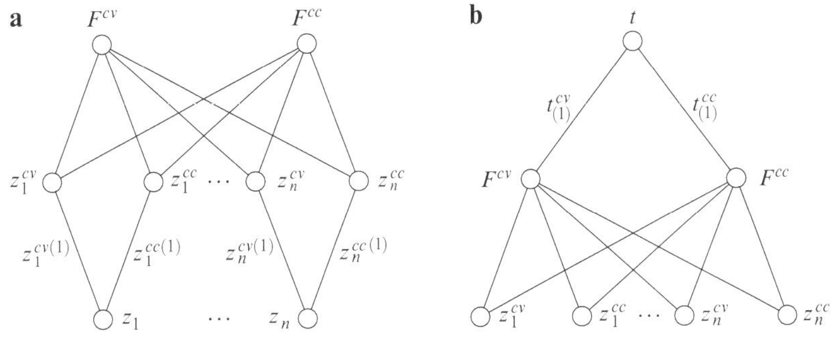

Fig. 3 Tangent-linear (**a**) and adjoint (**b**) extensions of the Jacobian J of (11)

$y^{cv} = F^{cv}(\mathbf{z}) = g^{cv}(f(\mathbf{z})) = g^{cv}(\mathbf{z}^+)$ and $y^{cc} = F^{cc}(\mathbf{z}) = g^{cc}(f(\mathbf{z})) = g^{cc}(\mathbf{z}^+)$, where $\mathbf{z}^+ \in Z^+$ is defined as $\mathbf{z}^+ = (z_1, z_1, \ldots, z_n, z_n)$. The mapping g represents the simultaneous propagation of the convex and concave relaxation based on the duplication f of the input variables of F. See Fig. 3 for illustration. Note that both the convex and concave relaxations of the identity $f(\mathbf{z}) = \mathbf{z}$ are equal to f.

The following theorem enables the transition from tangent-linear mode to adjoint mode for the computation of subgradients of McCormick relaxations.

Theorem 1. *Let $F = g \circ f$, $y = F(\mathbf{z})$, be defined as above, $\mathbf{z} \in Z$, and let*

$$s_{g^{cv}}\left(\mathbf{z}^+\right) = \left(\frac{\partial y^{cv}}{\partial z_1^{cv}}, \frac{\partial y^{cv}}{\partial z_1^{cc}}, \ldots, \frac{\partial y^{cv}}{\partial z_n^{cv}}, \frac{\partial y^{cv}}{\partial z_n^{cc}}\right)$$

denote a subgradient of the convex relaxation g^{cv} at $\mathbf{z}^+$. Similarly, let

$$s_{g^{cc}}\left(\mathbf{z}^+\right) = \left(\frac{\partial y^{cc}}{\partial z_1^{cv}}, \frac{\partial y^{cc}}{\partial z_1^{cc}}, \ldots, \frac{\partial y^{cc}}{\partial z_n^{cv}}, \frac{\partial y^{cc}}{\partial z_n^{cc}}\right)$$

denote a subgradient of the concave relaxation g^{cc} at $\mathbf{z}^+$.

A subgradient of the convex relaxation F^{cv} of F at $\mathbf{z}$ is given by

$$s_{F^{cv}}(\mathbf{z}) := \left(\frac{\partial y^{cv}}{\partial z_1^{cv}} + \frac{\partial y^{cv}}{\partial z_1^{cc}}, \ldots, \frac{\partial y^{cv}}{\partial z_n^{cv}} + \frac{\partial y^{cv}}{\partial z_n^{cc}}\right). \tag{8}$$

Similarly, a subgradient of the concave relaxation F^{cc} of F at $\mathbf{z}$ is given by

$$s_{F^{cc}}(\mathbf{z}) := \left(\frac{\partial y^{cc}}{\partial z_1^{cv}} + \frac{\partial y^{cc}}{\partial z_1^{cc}}, \ldots, \frac{\partial y^{cc}}{\partial z_n^{cv}} + \frac{\partial y^{cc}}{\partial z_n^{cc}}\right). \tag{9}$$

Proof. Obviously, Z^+ is an open convex set and g^{cv} is a convex function for all convex sets $Z \subseteq \mathbb{R}^n$, which implies the existence of the above subgradients. According to Definition 4 it remains to show that

$$\langle s_{F^{cv}}(\mathbf{z}), \mathbf{d}\rangle \leq (F^{cv})'(\mathbf{z}, \mathbf{d}) \quad \forall \mathbf{d} \in \mathbb{R}^n, \tag{10}$$

where $(F^{cv})'(\mathbf{z}, \mathbf{d}) := \lim_{t\to+0} \frac{F^{cv}(\mathbf{z}+t\cdot\mathbf{d})-F^{cv}(\mathbf{z})}{t}$. We observe for arbitrary $\mathbf{d} \in \mathbb{R}^n$

$$\begin{aligned}
\langle s_{F^{cv}}(\mathbf{z}), \mathbf{d}\rangle &= \left(\frac{\partial y^{\mathrm{cv}}}{\partial z_1^{\mathrm{cv}}} + \frac{\partial y^{\mathrm{cv}}}{\partial z_1^{\mathrm{cc}}}, \ldots, \frac{\partial y^{\mathrm{cv}}}{\partial z_n^{\mathrm{cv}}} + \frac{\partial y^{\mathrm{cv}}}{\partial z_n^{\mathrm{cc}}}\right) \cdot \begin{pmatrix} d_1 \\ \vdots \\ d_n \end{pmatrix} \\
&= \left(\frac{\partial y^{\mathrm{cv}}}{\partial z_1^{\mathrm{cv}}}, \frac{\partial y^{\mathrm{cv}}}{\partial z_1^{\mathrm{cc}}}, \ldots, \frac{\partial y^{\mathrm{cv}}}{\partial z_n^{\mathrm{cv}}}, \frac{\partial y^{\mathrm{cv}}}{\partial z_n^{\mathrm{cc}}}\right) \cdot \begin{pmatrix} d_1 \\ d_1 \\ \vdots \\ d_n \\ d_n \end{pmatrix} \\
&= \langle s_{g^{cv}}(\mathbf{z}^+), \mathbf{d}^+\rangle \underset{\text{Def. of subgradient}}{\leq} \lim_{t\to+0} \frac{g^{cv}(\mathbf{z}^+ + t\cdot\mathbf{d}^+) - g^{cv}(\mathbf{z}^+)}{t} \\
&= \lim_{t\to+0} \frac{g^{cv}(f(\mathbf{z}+t\cdot\mathbf{d})) - g^{cv}(f(\mathbf{z}))}{t} = (F^{cv})'(\mathbf{z}, \mathbf{d})
\end{aligned}$$

yielding (10) and thus completing the proof for the convex case. The proof for the concave case considers the convex function $-F^{cc}$. □

In tangent-linear mode projections of the Jacobian

$$J = \begin{pmatrix} \frac{\partial y^{cv}}{\partial z_1^{cv}} & \frac{\partial y^{cv}}{\partial z_1^{cc}} & \cdots & \frac{\partial y^{cv}}{\partial z_n^{cv}} & \frac{\partial y^{cv}}{\partial z_n^{cc}} \\ \\ \frac{\partial y^{cc}}{\partial z_1^{cv}} & \frac{\partial y^{cc}}{\partial z_1^{cc}} & \cdots & \frac{\partial y^{cc}}{\partial z_n^{cv}} & \frac{\partial y^{cc}}{\partial z_n^{cc}} \end{pmatrix} \tag{11}$$

are computed. Tangent-linear and adjoint extensions of the corresponding linearized DAG [8] are shown in Fig. 3. Interpretation of the chain rule according to [1] yields tangent-linear (Fig. 3a) and adjoint (Fig. 3b) modes for subgradients of McCormick relaxations.

Subgradients can be computed in tangent-linear mode by successively *seeding* "relaxed Cartesian basis vectors" of the form

$$\mathbf{z}^{(1)} = (z_1^{cv(1)}, z_1^{cc(1)}, z_2^{cv(1)}, z_2^{cc(1)}, \ldots, z_n^{cv(1)}, z_n^{cc(1)}) = \begin{cases} (1,1,0,0,\ldots,0,0) \\ (0,0,1,1,\ldots,0,0) \\ \vdots \\ (0,0,0,0,\ldots,1,1), \end{cases}$$

see also Theorem 1. Adjoint mode yields projections of the transposed Jacobian J^T. To get (8) and (9) the adjoint $t_{(1)} = (t^{cv}_{(1)}, t^{cc}_{(1)})^T$ needs to be seeded with the Cartesian basis vectors $(1,0)^T$ and $(0,1)^T$ to obtain

$$s^{cv}_{(1)} = \left(\frac{\partial y^{cv}}{\partial z^{cv}_1}, \frac{\partial y^{cv}}{\partial z^{cc}_1}, \ldots, \frac{\partial y^{cv}}{\partial z^{cv}_n}, \frac{\partial y^{cv}}{\partial z^{cc}_n} \right) \tag{12}$$

and

$$s^{cc}_{(1)} = \left(\frac{\partial y^{cc}}{\partial z^{cv}_1}, \frac{\partial y^{cc}}{\partial z^{cc}_1}, \ldots, \frac{\partial y^{cc}}{\partial z^{cv}_n}, \frac{\partial y^{cc}}{\partial z^{cc}_n} \right). \tag{13}$$

Addition of $\frac{\partial y^{cv}}{\partial z^{cv}_i}$ and $\frac{\partial y^{cv}}{\partial z^{cc}_i}$ ($\frac{\partial y^{cc}}{\partial z^{cv}_i}$ and $\frac{\partial y^{cc}}{\partial z^{cc}_i}$) for $i = 1, \ldots, n$ yields the subgradients in (8) (resp., (9)).

4 Numerical Results

Adjoint subgradient propagation for McCormick relaxations has been implemented in Fortran in the previously mentioned `amodMC`.[1] The left table in Fig. 4 compares the run time of subgradient propagation by `amodMC` with its tangent-linear counterpart `modMC` on a standard PC (Intel Core 2 Duo 3 GHz, 4 GB RAM, running Ubuntu 10.04 and `gfortran` at optimization level `-O3`) for the test function $f(x) = \exp\left(\log\left(x_1 + \sum_{i=1}^{n-1} \frac{x_{i+1}}{x_i}\right)\right)$. Note that in general $\exp(\log(x)) \neq x$ in the context of McCormick relaxations.

The run times of a simple branch-and-bound algorithm applied to the minimization of the function $g(x) = \exp\left(-\sum_{j=1}^{n} (x_j - 1)^2\right)$ on the interval [0, 1] for increasing values of n $(m = 4)$ are listed in the right table in Fig. 4. Both modes scale similar to standard AD. The constant overhead of adjoint mode turns out to be more significant due to the more complex differentiation rules that result in more substantial memory traffic per elemental function.

In addition to the above simple test problems, `amodMC` has been applied successfully to the set of real-world case studies in Sect. 5 of [7], including heat conduction, chemical kinetics, and reverse osmosis. While matching the numerical results delivered by `modMC`, no run time benefits were observed due to the relatively small number of independent parameters.

---

[1] See www.stce.rwth-aachen.de/software.

n	tangent-linear	adjoint
1000	0.043	0.011
2000	0.150	0.019
3000	0.314	0.028
4000	0.542	0.037
5000	0.919	0.047

n	tangent-linear	adjoint	Iterations
100	7.85	1.49	4989
200	49.5	6.18	10485
300	156	14	16017
400	360	26	21549
500	713	46	27325

Fig. 4 Run times: subgradient evaluation in adjoint vs. tangent-linear mode (*left*); global optimization based on adjoint vs. tangent-linear mode (*right*)

5 Conclusion and Outlook

Subgradients of McCormick relaxations can be computed both in tangent-linear and adjoint modes of AD. This article rephrases the former in standard AD technology. A corresponding adjoint formulation is proposed. As in standard AD, the computational overhead of subgradient evaluation scales with the number of independent variables in tangent-linear mode whereas in adjoint mode it remains constant.

Ongoing work includes the implementation of a C++ variant (`alibMC`) of `amodMC`. Further efforts are needed to improve the branch-and-bound algorithm, for example, through extension with local search techniques for improved convergence. Both tools will be used within novel algorithms for the solution of bi-level optimization problems based on underdetermined differential-algebraic equation systems. This work is part of a collaborative research project involving colleagues at Forschungszentrum Jülich and at RWTH Aachen University.

Acknowledgements Markus Beckers is supported by the German Research School for Simulation Sciences. Viktor Mosenkis is supported by the German Science Foundation (DFG grant No. 487/2-1).

We would like to acknowledge several fruitful discussions on the subject with Alexander Mitsos from MIT's Mechanical Engineering.

References

1. Bauer, F.: Computational graphs and rounding error. SIAM Journal on Numerical Analysis **11**(1), 87–96 (1974). DOI 10.1137/0711010
2. Corbett, C., Naumann, U.: Demonstration of a branch-and-bound algorithm for global optimization using McCormick relaxations. Tech. Rep. AIB-2011-24, RWTH Aachen (2011). URL http://aib.informatik.rwth-aachen.de/2011/2011-24.pdf
3. Griewank, A.: Piecewise linearization via Algorithmic Differentiation. Twelfth Euro AD Workshop, Berlin (2011)
4. Griewank, A., Walther, A.: Evaluating Derivatives: Principles and Techniques of Algorithmic Differentiation, 2nd edn. No. 105 in Other Titles in Applied Mathematics. SIAM, Philadelphia, PA (2008). URL http://www.ec-securehost.com/SIAM/OT105.html
5. Hiriart-Urruty, J., Lemaréchal, C.: Fundamentals of Convex Analysis. Springer-Verlag (2001)

6. McCormick, G.P.: Computability of global solutions to factorable nonconvex programs: Part I. Convex underestimating problems. Mathematical Programming **10**(2), 147–175 (1976)
7. Mitsos, A., Chachuat, B., Barton, P.I.: McCormick-based relaxations of algorithms. SIAM Journal on Optimization **20**(2), 573–601 (2009)
8. Naumann, U.: The Art of Differentiating Computer Programs. An Introduction to Algorithmic Differentiation. Software, Environments, and Tools. SIAM (2011)
9. Schnurr, M.: The automatic computation of second-order slope tuples for some nonsmooth functions (2007)
10. Steihaug, T.: Factorable programming revisited. Twelfth Euro AD Workshop, Berlin (2011)

Evaluating an Element of the Clarke Generalized Jacobian of a Piecewise Differentiable Function

Kamil A. Khan and Paul I. Barton

Abstract The (Clarke) generalized Jacobian of a locally Lipschitz continuous function is a derivative-like set-valued mapping that contains slope information. Several methods for optimization and equation solving require evaluation of generalized Jacobian elements. However, since the generalized Jacobian does not satisfy calculus rules sharply, this evaluation can be difficult. In this work, a method is presented for evaluating generalized Jacobian elements of a nonsmooth function that is expressed as a finite composition of absolute value functions and continuously differentiable functions. The method makes use of the principles of automatic differentiation and the theory of piecewise differentiable functions, and is guaranteed to be computationally tractable relative to the cost of a function evaluation.

Keywords Forward mode • Generalized gradient • Piecewise differentiable functions • Nonsmooth analysis

1 Introduction

The (Clarke) generalized Jacobian [1] is a set-valued mapping which provides slope information for locally Lipschitz continuous functions that are not everywhere differentiable. Facchinei and Pang present a survey [2, Sect. 7.5] of *semismooth Newton methods* for nonsmooth equation solving that require the evaluation of generalized Jacobian elements. *Bundle methods* [6–8] for optimization of locally

K.A. Khan (✉) · P.I. Barton
Process Systems Engineering Laboratory, Department of Chemical Engineering, Massachusetts Institute of Technology, Cambridge, MA 02139, USA
e-mail: kamil@mit.edu; pib@mit.edu

S. Forth et al. (eds.), *Recent Advances in Algorithmic Differentiation*, Lecture Notes in Computational Science and Engineering 87, DOI 10.1007/978-3-642-30023-3_11,

Lipschitz continuous functions extract descent information from collections of generalized Jacobian elements. These methods require explicit evaluation of matrix-valued generalized Jacobian elements; element-vector products computed with the forward mode of automatic differentiation are not sufficient.

However, as shown in Sect. 2, evaluating a generalized Jacobian element for a composite function can be difficult, since the generalized Jacobian does not obey sharp calculus rules. Nevertheless, several methods exist to evaluate generalized Jacobian elements in special cases. A method by Huang and Ma [5] returns generalized Jacobian elements for scalar-valued functions of the form:

$$f : \mathbb{R}^n \to \mathbb{R} : \mathbf{x} \mapsto \max\left(f_{(1)}(\mathbf{x}), \ldots, f_{(k)}(\mathbf{x})\right), \quad \text{where each } f_{(i)} \in \mathscr{C}^1,$$

and for vector-valued functions whose components are each of this form.

Noting that the generalized Jacobian of a convex function is identical to the *subdifferential* from convex analysis [1], a method by Mitsos, Chachuat and Barton [9] evaluates subgradients for particular convex relaxations of composite $\mathscr{C}^1$ functions.

A method by Nesterov [10] returns generalized Jacobian elements for the class of *lexicographically smooth* functions, which is closed under composition and includes the absolute value function and all differentiable functions. This class includes the familiar max and min functions, since for any $(x, y) \in \mathbb{R}^2$,

$$\max(x, y) = \tfrac{1}{2}(x+y+|x-y|), \qquad \text{and} \qquad \min(x, y) = \tfrac{1}{2}(x+y-|x-y|). \quad (1)$$

For a lexicographically smooth function $\mathbf{f}: X \subset \mathbb{R}^n \to \mathbb{R}^m$, Nesterov demonstrates the existence of some $k \leq n$ so that for all $i \geq k$, the Jacobian of a certain ith-order directional derivative of $\mathbf{f}$ is an element of the generalized Jacobian of $\mathbf{f}$ at $\mathbf{x}$. Griewank notes [3] that using this method, generalized Jacobian elements can be evaluated for any composition of absolute value functions and continuously differentiable functions by evaluating these directional derivatives using the forward mode of automatic differentiation (AD). However, since k is not known a priori in general, this method requires nth-order directional derivatives in the worst case, and so the discussion in [4, Sect. 4.5] implies that its computational cost scales worst-case exponentially with n.

In this article, a method is presented for evaluating a generalized Jacobian element for any finite composition of absolute value functions and continuously differentiable functions. Using the theory of piecewise differentiable functions developed by Scholtes [11], our method uses directional derivatives generated by AD to determine the Jacobian of an essentially active selection function numerically. The method is guaranteed to be computationally tractable in a sense described in Sect. 3.

2 Mathematical Background

In this section, relevant concepts and results from nonsmooth analysis are presented, followed by results concerning piecewise differentiable functions and AD. Notation has been altered from the original sources for consistency.

If a function $f : X \to Y$ satisfies a local property P at every $x \in X$, then f is said to satisfy P (without reference to any particular $x \in X$).

Given an open set $X \subset \mathbb{R}^n$, if a function $\mathbf{f} : X \to \mathbb{R}^m$ is (Fréchet) differentiable at $\mathbf{x} \in X$, then the Jacobian matrix of $\mathbf{f}$ at $\mathbf{x}$ is denoted by $\mathbf{Jf}(\mathbf{x})$.

Given a set $S \subset \mathbb{R}^n$, the *closure* and *interior* of S are denoted by $\mathrm{cl}\,(S)$ and $\mathrm{int}\,(S)$, respectively. The *convex cone* generated by S, $\mathrm{cone}(S)$, is the set of nonnegative combinations of elements of S. If S is finite, then $\mathrm{cone}(S)$ is a *polyhedral cone*. A *conical subdivision* of $\mathbb{R}^n$ is a partition of $\mathbb{R}^n$ into finitely many polyhedral cones with nonempty interior [11].

If a function $\mathbf{f} : X \to \mathbb{R}^m$ on an open set $X \subset \mathbb{R}^n$ is differentiable at $\mathbf{x} \in X$, then it is directionally differentiable at $\mathbf{x}$, with the directional derivative:

$$\mathbf{f}'(\mathbf{x}; \mathbf{d}) = \mathbf{Jf}(\mathbf{x})\,\mathbf{d}, \qquad \forall \mathbf{d} \in \mathbb{R}^n. \tag{2}$$

It follows from the definition of the directional derivative that the absolute value function $\mathtt{abs} : \mathbb{R} \to \mathbb{R} : x \mapsto |x|$ is also directionally differentiable, with

$$\mathtt{abs}'(x; d) = \begin{cases} d & \text{if } x > 0, \text{ or if } x = 0 \text{ and } d \geq 0, \\ -d & \text{if } x < 0, \text{ or if } x = 0 \text{ and } d < 0. \end{cases} \tag{3}$$

The following elements of nonsmooth analysis were developed by Clarke [1].

Definition 1. Given an open set $X \subset \mathbb{R}^n$, some $\mathbf{x} \in X$, and a locally Lipschitz continuous function $\mathbf{f} : X \to \mathbb{R}^m$, let $S \subset X$ be the set on which $\mathbf{f}$ is not differentiable. The *(Clarke) B-subdifferential* $\partial_{\mathrm{B}}\mathbf{f}(\mathbf{x})$ of $\mathbf{f}$ at $\mathbf{x}$ is then defined as

$$\partial_{\mathrm{B}}\mathbf{f}(\mathbf{x}) := \{\mathbf{H} \in \mathbb{R}^{m \times n} : \mathbf{H} = \lim_{i \to \infty} \mathbf{Jf}(\mathbf{x}^{(i)})$$
$$\text{for some sequence } \{\mathbf{x}^{(i)}\}_{i \in \mathbb{N}} \text{ in } X \backslash S \text{ such that } \lim_{i \to \infty} \mathbf{x}^{(i)} = \mathbf{x}\}.$$

The *(Clarke) generalized Jacobian* of $\mathbf{f}$ at $\mathbf{x}$, $\partial\mathbf{f}(\mathbf{x})$, is the convex hull of $\partial_{\mathrm{B}}\mathbf{f}(\mathbf{x})$.

Both $\partial_{\mathrm{B}}\mathbf{f}(\mathbf{x})$ and $\partial\mathbf{f}(\mathbf{x})$ exist, are unique, and are nonempty. If $\mathbf{f}$ is differentiable at $\mathbf{x}$, then $\mathbf{Jf}(\mathbf{x}) \in \partial\mathbf{f}(\mathbf{x})$. If $\mathbf{f}$ is $\mathscr{C}^1$ at $\mathbf{x}$, then $\partial_{\mathrm{B}}\mathbf{f}(\mathbf{x}) = \partial\mathbf{f}(\mathbf{x}) = \{\mathbf{Jf}(\mathbf{x})\}$.

As shown in [1], calculus rules for generalized Jacobians are valid as inclusions, but not necessarily as equations. This creates difficulties in determining generalized Jacobian elements of composite functions, as the following example demonstrates.

Example 1. Consider the functions $f : \mathbb{R} \to \mathbb{R} : x \mapsto x$, $g : \mathbb{R} \to \mathbb{R} : x \mapsto \max(x, 0)$, and $h : \mathbb{R} \to \mathbb{R} : x \mapsto \min(x, 0)$. Then $f = g + h$ on $\mathbb{R}$, $0 \in \partial g(0)$, and $0 \in \partial h(0)$. However, $(0 + 0) \notin \partial f(0) = \{1\}$.

2.1 Piecewise Differentiable Functions

As described in the following definition, piecewise differentiable functions include a broad range of nonsmooth functions, yet preserve many useful properties of $\mathscr{C}^1$ functions. Unless otherwise noted, the definitions and properties presented in this subsection are as stated and proven by Scholtes [11].

Definition 2. Given an open set $X \subset \mathbb{R}^n$, a function $\mathbf{f} : X \to \mathbb{R}^m$ is *piecewise differentiable* ($\mathscr{PC}^1$) at $\mathbf{x} \in X$ if there exists an open neighborhood $N \subset X$ of $\mathbf{x}$ such that $\mathbf{f}$ is continuous on N, and such that there exists a finite collection $\mathscr{F}_{\mathbf{f}}(\mathbf{x})$ of $\mathscr{C}^1$ functions which map N into $\mathbb{R}^m$ and satisfy $\mathbf{f}(\mathbf{y}) \in \{\mathbf{f}^*(\mathbf{y}) : \mathbf{f}^* \in \mathscr{F}_{\mathbf{f}}(\mathbf{x})\}$ for each $\mathbf{y} \in N$. The functions $\mathbf{f}^* \in \mathscr{F}_{\mathbf{f}}(\mathbf{x})$ are called *selection functions* for $\mathbf{f}$ around $\mathbf{x}$, and the collection $\mathscr{F}_{\mathbf{f}}(\mathbf{x})$ is called a *sufficient collection of selection functions* for $\mathbf{f}$ around $\mathbf{x}$.

If there exists a sufficient collection of selection functions for $\mathbf{f}$ that are all linear, then $\mathbf{f}$ is *piecewise linear* ($\mathscr{PL}$).

Any $\mathscr{C}^1$ function is trivially $\mathscr{PC}^1$. The absolute value function abs is $\mathscr{PL}$, since the functions $y \mapsto y$ and $y \mapsto -y$ comprise a sufficient collection of selection functions for abs.

Lemma 1. *Any $\mathscr{PC}^1$ function $\mathbf{f}: X \subset \mathbb{R}^n \to \mathbb{R}^m$ exhibits the following properties:*

1. *$\mathbf{f}$ is locally Lipschitz continuous.*
2. *$\mathbf{f}$ is directionally differentiable, and $\mathbf{f}'(\mathbf{x}; \cdot)$ is $\mathscr{PL}$ for any fixed $\mathbf{x} \in X$.*
3. *Given an open set $Y \subset \mathbb{R}^m$ containing the range of $\mathbf{f}$, and a $\mathscr{PC}^1$ function $\mathbf{g} : Y \to \mathbb{R}^\ell$, the composite function $\mathbf{h} : X \to \mathbb{R}^\ell : \mathbf{h} = \mathbf{g} \circ \mathbf{f}$ is also $\mathscr{PC}^1$. Moreover, the directional derivative of $\mathbf{h}$ satisfies the chain rule $\mathbf{h}'(\mathbf{x}; \mathbf{d}) = \mathbf{g}'(\mathbf{f}(\mathbf{x}); \mathbf{f}'(\mathbf{x}; \mathbf{d}))$ for each $\mathbf{x} \in X$ and each $\mathbf{d} \in \mathbb{R}^n$.*

Definition 3. Given a $\mathscr{PC}^1$ function $\mathbf{f} : X \to \mathbb{R}^m$ at $\mathbf{x} \in X$, a selection function $\mathbf{f}^*$ for $\mathbf{f}$ around $\mathbf{x}$ is *essentially active* if $\mathbf{x} \in \text{cl}\,(\text{int}\,(\{\mathbf{y} \in X : \mathbf{f}(\mathbf{y}) = \mathbf{f}^*(\mathbf{y})\}))$.

Lemma 2. *Given an open set $X \subset \mathbb{R}^n$, a $\mathscr{PC}^1$ function $\mathbf{f} : X \to \mathbb{R}^m$, and some $\mathbf{x} \in X$, $\mathbf{f}$ exhibits the following properties involving essentially active selection functions [11, Propositions 4.1.1, 4.1.3, and A.4.1]:*

1. *There exists a sufficient collection $\mathscr{F}^e_{\mathbf{f}}(\mathbf{x})$ of selection functions for $\mathbf{f}$ around $\mathbf{x}$ which are each essentially active at $\mathbf{x}$.*
2. *For any $\mathbf{d} \in \mathbb{R}^n$, $\mathbf{f}'(\mathbf{x}; \mathbf{d}) \in \{\mathbf{J}\mathbf{f}^*(\mathbf{x})\,\mathbf{d} : \mathbf{f}^* \in \mathscr{F}^e_{\mathbf{f}}(\mathbf{x})\}$.*
3. *The B-subdifferential of $\mathbf{f}$ at $\mathbf{x}$ satisfies $\partial_B \mathbf{f}(\mathbf{x}) = \{\mathbf{J}\mathbf{f}^*(\mathbf{x}) : \mathbf{f}^* \in \mathscr{F}^e_{\mathbf{f}}(\mathbf{x})\} \subset \partial \mathbf{f}(\mathbf{x})$.*

Determining essentially active selection functions for compositions of $\mathscr{PC}^1$ functions is not a simple matter of composing underlying essentially active selection functions, as the following example demonstrates.

Example 2. Consider the functions $f, g, h : \mathbb{R} \to \mathbb{R}$ from Example 1. The mapping $g^* : \mathbb{R} \to \mathbb{R} : x \mapsto 0$ is an essentially active selection function for both g and h at 0, yet $g^* + g^* = g^*$ is not an essentially active selection function for f at 0.

As defined in the subsequent lemma and definition, *conically active selection functions* are introduced in this work to describe the essentially active selection functions for a $\mathscr{PC}^1$ function $\mathbf{f}$ that are necessary to define the directional derivatives of $\mathbf{f}$.

Lemma 3. *Given an open set $X \subset \mathbb{R}^n$, a $\mathscr{PC}^1$ function $\mathbf{f} : X \to \mathbb{R}^m$, and a vector $\mathbf{x} \in X$, there exists a conical subdivision $\Lambda_{\mathbf{f}}(\mathbf{x})$ of $\mathbb{R}^n$ such that for each polyhedral cone $\sigma \in \Lambda_{\mathbf{f}}(\mathbf{x})$, some essentially active selection function $\mathbf{f}_\sigma \in \mathscr{F}^e_{\mathbf{f}}(\mathbf{x})$ satisfies*

$$\mathbf{f}'(\mathbf{x}; \mathbf{d}) = \mathbf{J}\mathbf{f}_\sigma(\mathbf{x})\,\mathbf{d}, \qquad \forall \mathbf{d} \in \sigma. \tag{4}$$

Proof. The result follows from Property 2 in Lemma 1, Property 2 in Lemma 2, and [11, Sect. 2.2.1].

Definition 4. A conical subdivision $\Lambda_{\mathbf{f}}(\mathbf{x})$ as described in Lemma 3 is called an *active conical subdivision* for $\mathbf{f}$ at $\mathbf{x}$. Each cone $\sigma \in \Lambda_{\mathbf{f}}(\mathbf{x})$ is an *active cone* for $\mathbf{f}$ at $\mathbf{x}$. For each active cone σ, an essentially active selection function $\mathbf{f}_\sigma$ satisfying (4) is called a *conically active selection function* for $\mathbf{f}$ at $\mathbf{x}$ corresponding to σ.

An essentially active selection function is not necessarily also conically active.

2.2 Automatic Differentiation

As defined in this section, the class of `abs`-factorable functions is similar to the class $\mathscr{G}$ described in [3], and includes any function that can be expressed as a finite composition of absolute value functions and continuously differentiable functions.

Definition 5. Given an open set $X \subset \mathbb{R}^n$, an `abs`-*factorable function* $\mathbf{f} : X \to \mathbb{R}^m$ is a function for which the following exist and are known:

- An *intermediate function number* $\ell \in \mathbb{N}$,
- A Boolean *dependence operator* $\prec$, such that $(i \prec j) \in \{\mathtt{true}, \mathtt{false}\}$ for each $j \in \{1, 2, \ldots, \ell\}$ and each $i \in \{0, 1, \ldots, j-1\}$,
- An *elemental function* $\varphi_{(j)} : X_{(j)} \to Y_{(j)}$ in the class $\mathscr{C}^1 \cup \{\mathtt{abs}\}$ for each $j \in \{1, \ldots, \ell\}$, where $X_{(j)} \subset \mathbb{R}^{n_j}$ is open, $Y_{(j)} \subset \mathbb{R}^{m_j}$, and $\prod_{\{i:i \prec j\}} Y_{(i)} \subset X_{(j)}$,
- Analytical Jacobians for each continuously differentiable $\varphi_{(j)}$,

and where for any $\mathbf{x} \in X$, $\mathbf{f}(\mathbf{x})$ can be evaluated by the following procedure:

1. Set $\mathbf{v}_{(0)} \leftarrow \mathbf{x}$, and set $j \leftarrow 1$.
2. Set $\mathbf{u}_{(j)} \in X_{(j)}$ to be a column vector consisting of all $\mathbf{v}_{(i)}$s for which $i \prec j$, stacked in order of increasing i. Set $\mathbf{v}_{(j)} \leftarrow \varphi_{(j)}(\mathbf{u}_{(j)})$.
3. If $j = \ell$, then go to Step 4. Otherwise, set $j \leftarrow j + 1$ and return to Step 2.
4. Set $\mathbf{f}(\mathbf{x}) \leftarrow \mathbf{v}_{(\ell)}$.

This procedure is an `abs`-*factored representation* of **f**, and defines **f** completely.

Property 3 in Lemma 1 implies that each `abs`-factorable function is $\mathscr{PC}^1$. The class of `abs`-factorable functions contains a broad range of $\mathscr{PC}^1$ functions encountered in practice, since the class is evidently closed under composition, and since min and max are included according to (1).

The forward mode of AD is defined similarly for `abs`-factorable functions as for $\mathscr{C}^1$-factorable functions.

Definition 6. Given an open set $X \subset \mathbb{R}^n$, an `abs`-factorable function **f** as described in Definition 5, some $\mathbf{x} \in X$ and some $\mathbf{d} \in \mathbb{R}^n$, the *forward mode of AD for* `abs`-*factorable functions* generates a vector $\dot{\mathbf{f}}(\mathbf{x}; \mathbf{d}) \in \mathbb{R}^m$ using the following procedure:

1. Set $\dot{\mathbf{v}}_{(0)} \leftarrow \mathbf{d}$, and set $j \leftarrow 1$.
2. Set $\dot{\mathbf{u}}_{(j)} \in \mathbb{R}^{n_j}$ to be a column vector consisting of all $\dot{\mathbf{v}}_{(i)}$s for which $i \prec j$, stacked in order of increasing i. Evaluate the directional derivative $\varphi_{(j)}{}'(\mathbf{u}_{(j)}; \dot{\mathbf{u}}_{(j)})$ according to (2) if $\varphi_{(j)}$ is $\mathscr{C}^1$, or (3) if $\varphi_{(j)} =$ `abs`. Set $\dot{\mathbf{v}}_{(j)} \leftarrow \varphi_{(j)}{}'(\mathbf{u}_{(j)}; \dot{\mathbf{u}}_{(j)})$.
3. If $j = \ell$, then go to Step 4. Otherwise, set $j \leftarrow j + 1$ and return to Step 2.
4. Set $\dot{\mathbf{f}}(\mathbf{x}; \mathbf{d}) \leftarrow \dot{\mathbf{v}}_{(\ell)}$.

Remark 1. Since the above definitions specify the intermediate variables $\mathbf{v}_{(j)}$, $\dot{\mathbf{v}}_{(j)}$, $\mathbf{u}_{(j)}$, and $\dot{\mathbf{u}}_{(j)}$ uniquely for any particular $\mathbf{x} \in X$ and $\mathbf{d} \in \mathbb{R}^n$, these definitions essentially describe functions $\mathbf{v}_{(j)} : X \to Y_{(j)}$, $\dot{\mathbf{v}}_{(j)} : X \times \mathbb{R}^n \to \mathbb{R}^{m_j}$, $\mathbf{u}_{(j)} : X \to X_{(j)}$, and $\dot{\mathbf{u}}_{(j)} : X \times \mathbb{R}^n \to \mathbb{R}^{n_j}$ that provide the values of these variables for any $(\mathbf{x}, \mathbf{d}) \in X \times \mathbb{R}^n$.

The following result is argued in [3], and depends on Property 3 in Lemma 1.

Theorem 1. *The vector* $\dot{\mathbf{f}}(\mathbf{x}; \mathbf{d})$ *generated by the procedure in Definition 6 is the directional derivative* $\mathbf{f}'(\mathbf{x}; \mathbf{d})$.[1]

---

[1]For functions in the class $\mathscr{G}$ described in [3], any elemental function $\varphi_{(j)}$ other than `abs` must be real analytic. However, the argument yielding Theorem 1 does not make use of higher-order derivatives, and so these elemental functions need only be $\mathscr{C}^1$.

3 Generalized Jacobian Element Evaluation

If any essentially active selection function in $\mathscr{F}_{\mathbf{f}}^{e}(\mathbf{x})$ is known a priori for a $\mathscr{PC}^1$ function $\mathbf{f} : X \subset \mathbb{R}^n \to \mathbb{R}^m$ at $\mathbf{x} \in X$, then an element of $\partial_B \mathbf{f}(\mathbf{x}) \subset \partial \mathbf{f}(\mathbf{x})$ can be evaluated using Property 3 in Lemma 2. Indeed, if $\mathscr{F}_{\mathbf{f}}^{e}(\mathbf{x})$ is known for $\mathbf{f}$ at $\mathbf{x}$, then $\partial \mathbf{f}(\mathbf{x})$ can be described completely.

However, in the spirit of AD, little is typically known about an abs-factorable function a priori beyond its abs-factored representation. Given a point $\mathbf{x}$ in such a function's domain, it is not known in general whether the function is differentiable at given points close to $\mathbf{x}$. Active conical subdivisions for the function are not known in general, yet inferring a generalized Jacobian element according to (4) requires n linearly independent direction vectors in a single active cone. Moreover, Example 2 demonstrates that essentially active selection functions of abs-factorable functions are not known in general.

Nevertheless, as shown in Sect. 2.1, the directional derivatives of $\mathscr{PC}^1$ functions obey a chain rule and are intimately related to the Jacobians of essentially active selection functions. Thus, the following theorem presents a method to evaluate an element of the generalized Jacobian of an abs-factorable function $\mathbf{f}$, by using AD to determine the Jacobian of an essentially active selection function of $\mathbf{f}$ numerically. The subsequent corollary demonstrates that this method is guaranteed to be computationally tractable.

Theorem 2. *Given an open set $X \subset \mathbb{R}^n$, some $\mathbf{x} \in X$, and an* `abs`*-factorable function $\mathbf{f} : X \to \mathbb{R}^m$, suppose a matrix $\mathbf{B} \in \mathbb{R}^{m \times n}$ is constructed by the following procedure:*

1. *For each $k \in \{1, 2, \ldots, n\}$, set a basis vector $\mathbf{q}^{(k)} \in \mathbb{R}^n$ to be the kth unit coordinate vector $\mathbf{e}^{(k)}$ in $\mathbb{R}^n$.*
2. *Use the* `abs`*-factored representation of $\mathbf{f}$ in order to obtain $\mathbf{f}(\mathbf{x})$ and all intermediate variables $\mathbf{v}_{(j)}(\mathbf{x})$. For each $j \in \{1, 2, \ldots, \ell\}$, set a Boolean variable* `IsCD`(j) *to* `false` *if $\varphi_{(j)} =$* `abs` *and $\mathbf{u}_{(j)}(\mathbf{x}) = 0$. Otherwise, set* `IsCD`$(j)$ *to* `true`.
   For each j such that `IsCD`$(j) =$ `false`, *$\varphi_{(j)} =$* `abs`, *and so $\mathbf{u}_{(j)}$ and $\dot{\mathbf{u}}_{(j)}$ are scalar-valued. Hence, these will be denoted in this case as $u_{(j)}$ and $\dot{u}_{(j)}$.*
3. *Use the forward mode of AD to evaluate $\mathbf{f}'(\mathbf{x}; \mathbf{q}^{(k)})$ for each $k \in \{1, \ldots, n\}$. Store the result, along with $\dot{u}_{(j)}(\mathbf{x}; \mathbf{q}^{(k)})$ for each $j \in \{1, \ldots, \ell\}$ such that* `IsCD`$(j) =$ `false`.
4. *Set $j \leftarrow 1$. Carry out the following subprocedure, which iterates over the elemental functions $\varphi_{(1)}, \ldots, \varphi_{(\ell)}$:*
   (a) *If* `IsCD`$(j) =$ `true`, *then go to Step 4c.*
   (b) *Set $k \leftarrow 1$. Carry out the following subprocedure, which iterates over the basis vectors $\mathbf{q}^{(1)}, \ldots, \mathbf{q}^{(n)}$:*
      (i) *Based on the value of $\dot{u}_{(j)}(\mathbf{x}; \mathbf{q}^{(k)})$, one of the following cases will apply:*

- *If* $\dot{u}_{(j)}(\mathbf{x}; \mathbf{q}^{(k)}) \neq 0$*, then go to Step 4(b)ii.*
- *If* $\dot{u}_{(j)}(\mathbf{x}; \mathbf{q}^{(k)}) = 0$ *and* $k < n$*, then set* $k \leftarrow k + 1$ *and return to the start of Step 4(b)i.*
- *If* $\dot{u}_{(j)}(\mathbf{x}; \mathbf{q}^{(k)}) = 0$ *and* $k = n$*, then go to Step 4c.*

(ii) *Store the current* k*-value as* $k^* \leftarrow k$*, and store* $\dot{u}^* \leftarrow \dot{u}_{(j)}(\mathbf{x}; \mathbf{q}^{(k^*)})$.

(iii) *If* $k = n$*, then go to Step 4c. Otherwise, set* $k \leftarrow k + 1$*. One of the following cases will then apply, based on the value of* $\dot{u}_{(j)}(\mathbf{x}; \mathbf{q}^{(k)})$*:*

- *If* $\dot{u}_{(j)}(\mathbf{x}; \mathbf{q}^{(k)})\, \dot{u}^* \geq 0$*, then return to the start of Step 4(b)iii.*
- *If* $\dot{u}_{(j)}(\mathbf{x}; \mathbf{q}^{(k)})\, \dot{u}^* < 0$*, then update* $\mathbf{q}^{(k)}$ *as follows:*

$$\mathbf{q}^{(k)} \leftarrow \mathbf{q}^{(k)} + \left| \frac{\dot{u}_{(j)}(\mathbf{x}; \mathbf{q}^{(k)})}{\dot{u}^*} \right| \mathbf{q}^{(k^*)}. \tag{5}$$

Use the forward mode of AD to evaluate $\mathbf{f}'(\mathbf{x}; \mathbf{q}^{(k)})$*. Store the result, along with* $\dot{u}_{(i)}(\mathbf{x}; \mathbf{q}^{(k)})$ *for each* $i \in \{1, \ldots, \ell\}$ *such that* `IsCD(i) = false`*. Return to the start of Step 4(b)iii.*

(c) *If* $j = \ell$*, then go to Step 5. Otherwise, set* $j \leftarrow j + 1$ *and return to Step 4a.*

5. *Solve the following system of linear equations for the matrix* $\mathbf{B} \in \mathbb{R}^{m \times n}$*:*

$$\mathbf{B}\left[\mathbf{q}^{(1)} \cdots \mathbf{q}^{(n)}\right] = \left[\mathbf{f}'(\mathbf{x}; \mathbf{q}^{(1)}) \cdots \mathbf{f}'(\mathbf{x}; \mathbf{q}^{(n)})\right]. \tag{6}$$

Return $\mathbf{B}$*, and terminate the procedure.*

Then the matrix $\left[\mathbf{q}^{(1)} \cdots \mathbf{q}^{(n)}\right]$ *in (6) is unit upper triangular. The returned matrix* $\mathbf{B}$ *is therefore well defined, and is an element of both* $\partial_{\mathrm{B}}\mathbf{f}(\mathbf{x})$ *and* $\partial\mathbf{f}(\mathbf{x})$*.*

Proof. A detailed proof of this theorem is deferred to a journal article under preparation. An outline of the proof is as follows.

Firstly, a straightforward proof by induction on the number of times (5) has been carried out yields the unit upper triangularity of $\left[\mathbf{q}^{(1)} \cdots \mathbf{q}^{(n)}\right]$ in (6).

Next, to prove that $\mathbf{B} \in \partial_{\mathrm{B}}\mathbf{f}(\mathbf{x})$, it is first proven by strong induction that for each $j \in \{1, \ldots, \ell\}$, after the jth iteration of the subprocedure in Step 4 of Theorem 2, there exists some $\mathbf{v}^*_{(j)} \in \mathscr{F}^e_{\mathbf{v}_{(j)}}(\mathbf{x})$ such that $\mathbf{v}_{(j)}'(\mathbf{x}; \mathbf{d}) = \mathbf{J}\mathbf{v}^*_{(j)}(\mathbf{x})\,\mathbf{d}$ for each $\mathbf{d} \in \mathrm{cone}(\{\mathbf{q}^{(1)}, \ldots, \mathbf{q}^{(n)}\})$, and that this property is unaffected by any further applications of (5). This is accomplished by showing that after the jth iteration of the subprocedure, $\dot{\mathbf{u}}_{(j)}(\mathbf{x}; \mathbf{q}^{(1)}), \ldots, \dot{\mathbf{u}}_{(j)}(\mathbf{x}; \mathbf{q}^{(n)})$ all lie in the same active cone of $\varphi_{(j)}$ at $\mathbf{x}$. The strong inductive assumption, Lemma 1, and Lemma 3 complete the strong inductive step, with $\mathbf{v}^*_{(j)}$ chosen to be an appropriate conically active selection function of $\mathbf{v}_{(j)}$ at $\mathbf{x}$. Since polyhedral cones are closed under nonnegative combinations of their elements, this result is not affected by any further applications of (5).

To complete the proof of the theorem, note that Step 5 of the procedure in Theorem 2 is reached after the ℓth iteration of the subprocedure in Step 4. Thus,

Table 1 Abs-factored representation of f in Example 3

j	Algebraic expression for $\mathbf{v}_{(j)}$	$\mathbf{v}_{(j)}(\mathbf{0})$	IsCD(j)
0	$\mathbf{v}_{(0)} = \mathbf{x}$	$\mathbf{0}$	–
1	$v_{(1)} = -v_{(0),2}$	0	True
2	$v_{(2)} = v_{(0),1} - v_{(1)}$	0	True
3	$v_{(3)} = \|v_{(2)}\|$	0	False
4	$v_{(4)} = \frac{1}{2}\left(v_{(0),1} + v_{(1)} - v_{(3)}\right)$	0	True
5	$v_{(5)} = v_{(0),2} - v_{(0),1}$	0	True
6	$v_{(6)} = v_{(4)} - v_{(5)}$	0	True
7	$v_{(7)} = \|v_{(6)}\|$	0	False
8	$v_{(8)} = \frac{1}{2}\left(v_{(4)} + v_{(5)} + v_{(7)}\right)$	0	True

since $\mathbf{f} = \mathbf{v}_{(\ell)}$, the strong inductive result implies that when the overall procedure is terminated, there is some $\mathbf{f}^* \in \mathscr{F}^e_{\mathbf{f}}(\mathbf{x})$ such that for each $\mathbf{d} \in \mathrm{cone}(\{\mathbf{q}^{(1)}, \ldots, \mathbf{q}^{(n)}\})$, $\mathbf{f}'(\mathbf{x}; \mathbf{d}) = \mathbf{J}\mathbf{f}^*(\mathbf{x})\,\mathbf{d}$. Since $\left[\mathbf{q}^{(1)} \cdots \mathbf{q}^{(n)}\right]$ has been proven to be nonsingular, it follows that $\mathbf{J}\mathbf{f}^*(\mathbf{x})$ is the matrix $\mathbf{B}$ obtained as the solution to (6). Property 3 in Lemma 2 then implies that $\mathbf{B} = \mathbf{J}\mathbf{f}^*(\mathbf{x}) \in \partial_{\mathrm{B}}\mathbf{f}(\mathbf{x}) \subset \partial\mathbf{f}(\mathbf{x})$. □

Corollary 1. *Let p be the number of elements of $\{j \in \{1, \ldots, \ell\} : \varphi_{(j)} = \mathtt{abs}\}$. In the procedure in Theorem 2, the forward mode of AD is applied to* $\mathbf{f}$ *no more than $(np + n - p)$ times. If each $\varphi_{(j)}$ is $\mathscr{C}^1$ at $\mathbf{u}_{(j)}(\mathbf{x})$, the forward mode is applied n times.*

Proof. Since p is no less than the number of elements of the set $\{j \in \{1, \ldots, \ell\} : \mathtt{IsCD}(j) = \mathtt{false}\}$, the result follows by inspection of Theorem 2. □

Remark 2. The computational cost of the method in Theorem 2 is evidently dominated by the cost of performing the forward mode of AD and the cost of solving (6). Since the matrix $\left[\mathbf{q}^{(1)} \cdots \mathbf{q}^{(n)}\right]$ in (6) is upper triangular, the computational cost of solving (6) varies worst-case quadratically with n and linearly with m.

In the following example, the method in Theorem 2 is applied to determine a generalized Jacobian element for an abs-factorable function.

Example 3. Consider the function $f : \mathbb{R}^2 \to \mathbb{R} : (x, y) \mapsto \max(\min(x, -y), y - x)$, and suppose that an element of $\partial f(\mathbf{0})$ is desired. Along the lines of (1), an abs-factored representation of f is given in Table 1.

According to Step 2 of the procedure in Theorem 2, a function evaluation was carried out to determine the values of $f(\mathbf{0}) = v_{(8)}(\mathbf{0})$, all intermediate variables $\mathbf{v}_{(j)}(\mathbf{0})$, and the values of $\mathtt{IsCD}(j)$ for each $j \in \{1, \ldots, 8\}$. These are shown in the rightmost two columns of Table 1. Note that $\mathtt{IsCD}(j) = \mathtt{false}$ only for $j \in \{3, 7\}$, and that the abs-factored representation of f implies that $u_{(3)} \equiv v_{(2)}$ and $u_{(7)} \equiv v_{(6)}$.

According to Step 3 of the procedure, the forward mode of AD was applied to f at $\mathbf{0}$ in the directions $\mathbf{q}^{(1)} = \mathbf{e}^{(1)} = (1, 0)$ and $\mathbf{q}^{(2)} = \mathbf{e}^{(2)} = (0, 1)$. The results are shown in Table 2, along with algebraic instructions for carrying out the forward mode of AD. The directional derivative of abs was evaluated as in (3).

Table 2 Intermediate quantities used to evaluate $\partial f(\mathbf{0})$ in Example 3

j	Algebraic expression for $\dot{\mathbf{v}}_{(j)}$	$\dot{\mathbf{v}}_{(j)}(\mathbf{0};(1,0))$	$\dot{\mathbf{v}}_{(j)}(\mathbf{0};(0,1))$	$\dot{\mathbf{v}}_{(j)}(\mathbf{0};(2,1))$
0	$\dot{\mathbf{v}}_{(0)} = \mathbf{d}$	(1, 0)	(0, 1)	(2, 1)
1	$\dot{v}_{(1)} = -\dot{v}_{(0),2}$	0	−1	−1
2	$\dot{v}_{(2)} = \dot{v}_{(0),1} - \dot{v}_{(1)}$	1	1	3
3	$\dot{v}_{(3)} = \texttt{abs}'(v_{(2)}; \dot{v}_{(2)})$	1	1	3
4	$\dot{v}_{(4)} = \frac{1}{2}\left(\dot{v}_{(0),1} + \dot{v}_{(1)} - \dot{v}_{(3)}\right)$	0	−1	−1
5	$\dot{v}_{(5)} = \dot{v}_{(0),2} - \dot{v}_{(0),1}$	−1	1	−1
6	$\dot{v}_{(6)} = \dot{v}_{(4)} - \dot{v}_{(5)}$	1	−2	0
7	$\dot{v}_{(7)} = \texttt{abs}'(v_{(6)}; \dot{v}_{(6)})$	1	2	0
8	$\dot{v}_{(8)} = \frac{1}{2}\left(\dot{v}_{(4)} + \dot{v}_{(5)} + \dot{v}_{(7)}\right)$	0	1	−1

The subprocedures in Step 4 were then carried out. When $(j,k) = (7,2)$, $\mathbf{q}^{(2)}$ was updated to $(\mathbf{q}^{(2)} + \left|\frac{-2}{1}\right| \mathbf{q}^{(1)}) = (2,1)$. Subsequently, $f'(\mathbf{0};(2,1))$ was evaluated using the forward mode of AD, and is shown in the rightmost column of Table 2. No further basis vector updates took place.

Thus, according to Step 5 of the procedure, the matrix $\mathbf{B}$ was defined so as to solve the linear system:

$$\mathbf{B}\begin{bmatrix}1 & 2\\ 0 & 1\end{bmatrix} = \begin{bmatrix}0 & -1\end{bmatrix}.$$

This system is readily solved to yield $\mathbf{B} = \begin{bmatrix}0 & -1\end{bmatrix}$, which is an element of $\partial f(\mathbf{0})$ according to Theorem 2.

This example is simple enough that the essentially active selection functions of f at $\mathbf{0}$ can be identified to be $(x,y) \mapsto x$, $(x,y) \mapsto -y$, and $(x,y) \mapsto y - x$. Property 3 in Lemma 2 then yields $\partial f(\mathbf{0}) = \text{conv}(\{[1\ 0], [0\ -1], [-1\ 1]\}) \ni \mathbf{B}$, which confirms the obtained result. By contrast, if the basis vectors $\mathbf{q}^{(k)}$ remained unperturbed, this modification to the method would yield $\mathbf{B} = \begin{bmatrix}0 & 1\end{bmatrix}$, which is not an element of $\partial f(\mathbf{0})$.

4 Concluding Remarks

The main theorem of this work provides a computationally tractable method for evaluating generalized Jacobian elements for the broad class of abs-factorable functions. Nevertheless, this procedure can be made more efficient in several ways. The structure of the method allows for significant reductions in computational cost if sparsity is exploited. Moreover, the intermediate AD variables corresponding to $\mathbf{q}^{(k)}$ can be updated using linear algebra instead of remedial applications of AD. This approach would be beneficial for functions whose abs-factored representations include elemental functions with computationally expensive Jacobians, such as the inverse trigonometric functions and the inverse hyperbolic functions.

A detailed proof of the main theorem is deferred to a journal article under preparation, along with a method for generalized Jacobian element evaluation in which a broader range of $\mathscr{PC}^1$ elemental functions is permitted. Such a method would be useful if an `abs`-factored representation of a $\mathscr{PC}^1$ function is inconvenient or impossible to obtain. This generalized method proceeds analogously to the method for `abs`-factorable functions: perturbing basis vectors until the corresponding arguments to each elemental function all lie within the same active cone of the function.

Acknowledgements This work has been funded by the MIT-BP Conversion Program.

References

1. Clarke, F.H.: Optimization and Nonsmooth Analysis. SIAM, Philadelphia, PA (1990)
2. Facchinei, F., Pang, J.S.: Finite-Dimensional Variational Inequalities and Complementarity Problems. Springer-Verlag New York, Inc., New York, NY (2003)
3. Griewank, A.: Automatic directional differentiation of nonsmooth composite functions. In: Recent Developments in Optimization, French-German Conference on Optimization. Dijon (1994)
4. Griewank, A., Walther, A.: Evaluating Derivatives: Principles and Techniques of Algorithmic Differentiation, 2nd edn. No. 105 in Other Titles in Applied Mathematics. SIAM, Philadelphia, PA (2008). URL http://www.ec-securehost.com/SIAM/OT105.html
5. Huang, Z.D., Ma, G.C.: On the computation of an element of Clarke generalized Jacobian for a vector-valued max function. Nonlinear Anal-Theor **72**, 998–1009 (2010)
6. Kiwiel, K.C.: Methods of Descent for Nondifferentiable Optimization. Lecture Notes in Mathematics. Springer-Verlag, Berlin (1985)
7. Lemaréchal, C., Strodiot, J.J., Bihain, A.: On a bundle algorithm for nonsmooth optimization. In: O.L. Mangasarian, R.R. Meyer, S.M. Robinson (eds.) Nonlinear Programming 4. Academic Press, New York, NY (1994)
8. Lukšan, L., Vlček, J.: A bundle-Newton method for nonsmooth unconstrained minimization. Math Program **83**, 373–391 (1998)
9. Mitsos, A., Chachuat, B., Barton, P.I.: McCormick-based relaxations of algorithms. SIAM J Optim **20**, 573–601 (2009)
10. Nesterov, Y.: Lexicographic differentiation of nonsmooth functions. Math Program B **104**, 669–700 (2005)
11. Scholtes, S.: Introduction to piecewise differentiable equations (1994). Habilitation Thesis, Institut für Statistik und Mathematische Wirtschaftstheorie, University of Karlsruhe

The Impact of Dynamic Data Reshaping on Adjoint Code Generation for Weakly-Typed Languages Such as Matlab

Johannes Willkomm, Christian H. Bischof, and H. Martin Bücker

Abstract Productivity-oriented programming languages typically emphasize convenience over syntactic rigor. A well-known example is Matlab, which employs a weak type system to allow the user to assign arbitrary types and shapes to a variable, and it provides various shortcuts in programming that result in implicit data reshapings. Examples are scalar expansion, where a scalar is implicitly expanded to a matrix of the appropriate size filled with copies of the scalar value, the use of row vectors in place of column vectors and vice versa, and the automatic expansion of arrays when indices outside of the previously allocated range are referenced. These features need to be addressed at runtime when generating adjoint code, as Matlab does not provide required information about types, shapes and conversions at compile time. This fact, and the greater scope of reshaping possible, is a main distinguishing feature of Matlab compared to traditional programming languages, some of which, e.g. Fortran 90, also support vector expressions. In this paper, in the context of the AdiMAT source transformation tool for Matlab, we develop techniques generally applicable for adjoint code generation in the face of dynamic data reshapings occurring both on the left- and right-hand side of assignments. Experiments show that in this fashion correct adjoint code can be generated also for very dynamic language scenarios at moderate additional cost.

Keywords Reverse mode • Adjoint code • Dynamic data reshaping • Scalar expansion • Weakly-typed languages • Matlab • Source transformation

---

J. Willkomm (✉) · C.H. Bischof
Scientific Computing Group, TU Darmstadt, Mornewegstrasse 30, 64293 Darmstadt, Germany
e-mail: johannes.willkomm@sc.tu-darmstadt.de; christian.bischof@sc.tu-darmstadt.de

H.M. Bücker
Institute for Scientific Computing, RWTH Aachen University, Seffenter Weg 23, 52074 Aachen, Germany
e-mail: buecker@sc.rwth-aachen.de

S. Forth et al. (eds.), *Recent Advances in Algorithmic Differentiation*, Lecture Notes in Computational Science and Engineering 87, DOI 10.1007/978-3-642-30023-3_12,

1 Introduction

ADiMat [2, 3, 8] is a source transformation tool for the automatic differentiation (AD) of Matlab programs. Matlab is a weakly-typed language, allowing for polymorphism. That is, a variable can assume different types and shapes and the semantics of functions are defined for all possible cases. As a result, the type or shape of a variable usually cannot be inferred statically, a situation different from common languages such as Fortran or C, but a sensible approach for productivity-oriented environments that emphasize expressiveness over syntactic rigor.

This lack of compile-time information does not affect the implementation of the forward mode for Matlab much. As the derivative operations go along with the original statements, usually the same language constructs can be used for the derivative operations as those that were used by the original expression. This is not true for the reverse mode (RM); depending on the shape of the variable we must account for implicit reshapings, when, for example, a matrix is multiplied by a scalar which is implicitly expanded to a matrix of the appropriate size filled with copies of the scalar value. This reshaping called "scalar expansion" is also possible in Fortran 90 and its connection to AD is discussed in [7]. However, Matlab's weak type system allows for a much greater variety of scalar expansion and other implicit reshapings which have to be dealt with in the RM, an example being the need to undo implied changes in the shapes of arrays. The MSAD tool [5] is another source transformation approach to AD for Matlab. However, MSAD, to our knowledge, is restricted to the forward mode on a subset of the language.

In this paper, we address the RM issues arising from the dynamic data reshapings occurring in a weakly-typed language such as Matlab. In Sect. 2 we develop generally applicable approaches for scalar expansion occurring on the right-hand side (RHS) of an assignment and examine the impact of these strategies in Sect. 3. In Sect. 4 we address data reshapings occurring on the left-hand side (LHS) and assess their impact in Sect. 5.

2 Implicit Scalar Expansion and Array Reshape

Variables occurring in a statement can be scalars, vectors, matrices, or tensors. To shorten the exposition of the code examples, we assume that an assignment will be the first one to that variable in the code. So, we do not include code to `push` or `pop` the value of the variable overwritten nor do we zero out the adjoint corresponding to the overwritten variable. Consider the following example for addition, where a_x denotes the adjoint variable corresponding to a variable x in the original program:

Original code	Adjoint code
`Y = A + B;`	`a_A = a_A + a_Y;` `a_B = a_B + a_Y;`

If A and B have the same shape, i.e., no implicit reshaping takes place, and an adjoint variable has the same shape as the corresponding original program variable, all the shapes in the adjoint code fit together. The same is true for all other component-wise operators, namely $+$, $-$, $.*$, $./$, and $.\hat{}$, when we replace a_Y on the RHSs of the adjoint statements by the adjoint expression induced by each of these operators.

2.1 Binary Component-Wise Operators

However, it is also allowed that one of the operands of component-wise binary operators is a scalar, while the other is non-scalar. Here, the scalar variable is implicitly expanded to the shape of the other operand. If, in the previous example, the variable B is scalar, the following operation is implicitly carried out in this case:

$$\mathrm{Y}_i \leftarrow \mathrm{A}_i + \mathrm{B}, \quad \forall i. \tag{1}$$

The single index i runs over all components of the arrays Y and A, regardless of their shape. The corresponding adjoint computations w.r.t. B are:

$$\overline{\mathrm{B}} \leftarrow \overline{\mathrm{B}} + \overline{\mathrm{Y}}_i, \quad \forall i, \tag{2}$$

where adjoint variables are denoted by an overbar. We use **sum** and the single wildcard index expression x(:) returning all components of x in a column vector:

Original code:

```
Y = A + B;
```

Adjoint code:

```
a_A = a_A + a_Y;
a_B = a_B + sum(a_Y(:));
```

This approach also applies to Fortran 90 [7]. However, in Fortran 90 the dimensions of all variables are declared and hence the decision whether a summation is necessary to update the adjoint of an implicitly expanded scalar variable can be taken at transformation time. In Matlab, we do not have that information and have to defer this decision to runtime. To this end, we introduce a runtime function **adjred**:

Original code:

```
Y = A + B;
```

Adjoint code:

```
a_A = a_A + adjred(A, a_Y);
a_B = a_B + adjred(B, a_Y);
```

If A is scalar, `adjred`(A, a_Y) returns `sum`(a_Y(:)), otherwise it returns a_Y unchanged. This way the correct adjoint will be computed in either of the possible cases. The same approach works with the other component-wise binary operators

−, .∗, ./, and .^, if we place the corresponding adjoint expression in the second argument of **adjred**. As an example, consider component-wise multiplication:

Original code	Adjoint code
`Y = A .* B;`	`a_A = a_A + adjred(A, a_Y .* B);` `a_B = a_B + adjred(B, A .* a_Y);`

2.2 *General Expressions*

When adjoining general expressions, entire subexpressions may occur as the first argument to **adjred**. If the RHS of the original statement reads A + B .∗ C, the adjoint code is the following, where the value of the subexpression B .∗ C is recomputed:

```
a_A = a_A + adjred(A, a_Y);
a_B = a_B + adjred(B, adjred(B .* C, a_Y) .* C);
a_C = a_C + adjred(C, B .* adjred(B .* C, a_Y));
```

The resulting performance degradation can be avoided by *outlining* nested expressions in a preprocessing step, i.e. splitting them into binary parts, assigning subexpressions to temporary variables. Then only plain or indexed variables will occur as the first argument to **adjred**.

2.3 *Matrix Multiplication*

As we apply outlining to our code, we only consider the binary case Y ← AB. According to the basic adjoint rules [4], the corresponding adjoint operations are:

$$\overline{A} \leftarrow \overline{A} + \overline{Y}B^T \quad \text{and} \quad \overline{B} \leftarrow \overline{B} + A^T\overline{Y}. \tag{3}$$

However, it is possible in Matlab that any of the operands is a scalar and then an implicit scalar expansion takes place, i.e., the operator ∗ has the same function as the component-wise variant .∗. To properly handle matrix multiplication we introduce two runtime functions, **adjmultl**, and **adjmultr**, employed depending on the position of the adjoint in a multiplication.

Original code	Adjoint code
`Y = A * B;`	`a_A = a_A + adjmultl(A, a_Y, B);` `a_B = a_B + adjmultr(B, A, a_Y);`

If the first argument A is a scalar, `adjmultl`(A, a_Y, B) returns the sum of all components of a_Y .* B, and otherwise a_Y * B.', which corresponds to (3). The function **adjmultr** is defined analogously, except that it returns A.' * a_Y, when B is not scalar. To handle the general case of more than two matrices, a third function **adjmultm** is required when the adjoint occurs in the middle of a multiplication. Matrix division by a scalar or non-singular matrix divisor is treated similarly [4].

2.4 One-to-Many Index Assignment and Implicit Array Reshape

Scalar expansion also occurs in the assignment of a scalar to multiple components of an array, using an indexed expression on the LHS. Here, we also have to sum the components of the adjoint expression for the LHS before adding them to the adjoint of the RHS. In addition, however, the shape of the RHS expression may implicitly change. This must be undone when computing the adjoint expression of the LHS. We use the function `adjreshape`(val,adj) returning `reshape`(adj, `size`(val)) if val is not scalar and adj otherwise. The adjoint of the LHS of an indexed expression is then constructed as the composition of calls to **adjreshape** and **adjred**:

Original code	Adjoint code
`Y(I) = A;`	`a_A = a_A + adjred(A, adjreshape(A, a_Y(I)));`

To illustrate, let Y be a 3×4 matrix, and I a logical array of the same size and shape. If 10 items in I are non-zero, the variable A may legally contain either a scalar or an array of any shape with 10 items, for example a 2×5 matrix. In the adjoint computations, the scalar case for A is correctly handled by the **adjred** mechanism. Otherwise, in the resulting adjoint computation, a_A = a_A + a_Y(I);, the plus operator requires that a_A and a_Y(I) have the same shape. In this example, a_Y(I) is a column vector of 10 items, and a_A is 2×5 matrix. The correct adjoint computation is a_A = a_A+ `reshape`(Y(I), `size`(A));, which undoes the implied shape change.

3 Performance of the Adjred Mechanism

As an example consider the Rosenbrock function:

$$f(\mathbf{x}) = \sum_{i=1}^{n-1} \left[100 \left(x_{i+1} - x_i^2 \right)^2 + (1 - x_i)^2 \right]. \tag{4}$$

Its vectorized implementation is shown in Listing 1 and the corresponding adjoint code, showing the impact of outlining as discussed in Sect. 2.2, in Listing 2.

Listing 1 Vectorized implementation of the Rosenbrock function (4).

```
function y = rosenbrock(x)
n = length(x);
y = sum(100.*(x(2:n)-x(1:n-1).^2).^2+(1-x(1:n-1)).^2);
```

Listing 2 Adjoint code of Listing 1.

```
function [a_x nr_y] = a_rosenbrock(x, a_y)
  n = length(x);
  ca7 = 1 - x(1 : n-1);
  ca6 = ca7 .^ 2;
  ca5 = x(1 : n-1) .^ 2;
  ca4 = x(2 : n) - ca5;
  ca3 = ca4 .^ 2;
  ca2 = 100 .* ca3;
  ca1 = ca2 + ca6;
  y = sum(ca1);
  nr_y = y;
  [a_ca7 a_ca6 a_ca5 a_ca4 a_ca3 a_ca2 a_ca1 a_x] = ...
      a_zeros(ca7, ca6, ca5, ca4, ca3, ca2, ca1, x);
  a_ca1 = a_ca1 + a_sum(a_y, ca1);
  a_ca2 = a_ca2 + adjred(ca2, a_ca1);
  a_ca6 = a_ca6 + adjred(ca6, a_ca1);
  a_ca3 = a_ca3 + adjred(ca3, 100 .* a_ca2);
  a_ca4 = a_ca4 + adjred(ca4, 2 .* ca4 .* a_ca3);
  a_x(2 : n) = a_x(2 : n) + adjred(x(2 : n), a_ca4);
  a_ca5 = a_ca5 + adjred(ca5, -a_ca4);
  a_x(1 : n-1) = a_x(1 : n-1) + adjred(x(1 : n-1), 2 .* x(1 ...
      : n-1) .* a_ca5);
  a_ca7 = a_ca7 + adjred(ca7, 2 .* ca7 .* a_ca6);
  a_x(1 : n-1) = a_x(1 : n-1) + adjred(x(1 : n-1), -a_ca7);
end
```

The function **a_sum** is the adjoint function corresponding to the Matlab built-in **sum**. Note that in this code example, all calls to **adjred** are superfluous. No implicit shape expansion occurs in this code, regardless of the length of the vector x.

To assess the overhead of **adjred**, we generate two adjoint versions of the code, one with the calls to **adjred** and one without. Setting x to a random vector of increasing length n, we run the adjoint code 20 times and report average runtimes.

The results are given in Fig. 1, showing the absolute times in seconds and the ratio of AD-generated code to function evaluation for the variants: The original function is referred to as f, the adjoint function using **adjred** as $\partial f_{\mathrm{red}}$ and the one without is referred to as $\partial f_{\mathrm{nored}}$. Except for very small ($n < 10$) cases, the penalty incurred due to the **adjred** calls is between 12% and 29%, and this is, in our view, a small and reasonable price to pay for correct adjoint code.

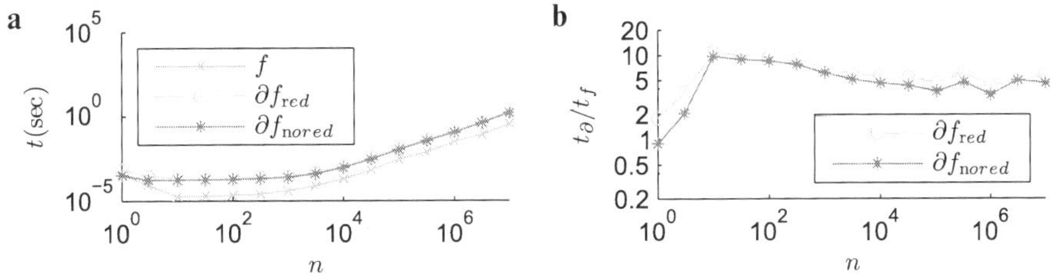

Fig. 1 Absolute runtimes (**a**) and AD overhead ratio (**b**) of the two different adjoint versions

4 Implicit Array Enlargement

In the previous sections we made the simplifying assumption that the variables being assigned need not be stored, i.e., no **push**, **pop**, nor adjoint zeroing statements were necessary, and considered implicit reshapings that occurred on the RHS of an assignment. We also assumed that the size of the variable on the LHS is not changed by the assignment. However, such an enlargement can implicitly occur. To illustrate, consider the statement X(I) = A, where I can be an integer or a vector or array of integers. When some indices in I are outside of the range currently allocated for X, then Matlab will automatically enlarge X to the size required to accommodate all indices. The same occurs when I is a logical array of size larger than X. If we ignore this issue and generate code as described in Sect. 2.4, we push the value of the LHS variable before the assignment, restore it in the adjoint code, and zero its associated adjoint:

Forwardsweep

```
push(X(I));
X(I) = A;
```

Adjoint code

```
X(I) = pop;
a_A = a_A + adjred(A, adjreshape(A, a_X(I)));
a_X(I) = a_zeros(X(I));
```

This adjoint code works if the shape of X remains unchanged, but fails when X was enlarged due to the assignment: The `push` call fails, because it tries to read the LHS expression before the assignment, which causes an out-of-range exception. In addition, the generic **pop** operation does not work, as it has to restore X to its previous size. Lastly, the zeroing of the adjoint of X after the adjoint incrementation also has to resize the adjoint variable to the size X had before the assignment.

In order to handle enlarging indexed assignments, we introduce three new runtime functions as alternatives to **push**, **pop** and **a_zeros**. Their use in the adjoint code is shown in the following example.

Forwardsweep	Adjoint code

`push_index(X, I);`
`X(I) = A;`
	`X = pop_index(X, I);`
	`a_A = a_A + adjred(A, adjreshape(A, a_X(I)));`
	`a_X = a_zeros_index(a_X, X, I);`

push_index and **pop_index** are shown in Listing 3, Listing 4, respectively.

Listing 3 The function **push_index**.

```
function push_index(var, ...
    varargin)
  try
    push(var(varargin{:}));
    push(0);
  catch
    push(var);
    push(1);
  end
```

Listing 4 The function **pop_index**.

```
function r = pop_index(r, ...
    varargin)
  [val whatHappened] = pop;
  switch whatHappened
   case 0
    r(varargin{:}) = val;
   case 1
    r = val;
  end
```

In **push_index**, we detect an enlarging assignment by trying to read and push X(I). If that attempt fails, the catch branch saves X in its entirety on the stack. To distinguish between the two cases, a marker value of type double with a value of either 0 or 1 is also pushed on the stack. Depending on this marker, the suitable value is then restored in **pop_index**.

a_zeros_index takes the arguments adj, var, and an arbitrary number of arguments holding the index expressions, called varargin. When no enlargement has happened, adj and var are of the same size and the function sets adj(varargin {:}) = `a_zeros`(var(varargin{:})) and returns adj. Otherwise, we have to shrink adj to the size of var, without destroying the values in adj. We do this by creating two logical arrays of the size of adj. In the first, selwrit, we set the components that have been assigned to, to one, and in the other, selold, we set those that were present before the enlargement to one. First we zero the components indexed by selwrit, then we select the components indexed by selold and reshape them to the shape of var.

5 Performance of the push_index/pop_index Mechanism

To illustrate the impact of LHS index expansion, we now consider a different implementation of the Rosenbrock function shown in Listing 5, called **rosenbrock_prealloc** which is written using a **for** loop. The code computes the terms of the sum shown in (4) explicitly and stores them in a temporary array t. The final result is then computed as the sum of t.

Listing 5 Naïve implementation of the Rosenbrock function (4).

```
1 function y = rosenbrock_prealloc(x)
2   n = length(x);
3   t = zeros(n-1,1);
4   for i=1:n-1
5     t(i) = (1-x(i)).^2 + 100.*(x(i+1)-x(i).^2).^2;
6   end
7   y = sum(t);
```

We also consider another variant, called **rosenbrock_noprealloc**, which is identical to **rosenbrock_prealloc**, except that the pre-allocation of the array t in line 3 of Listing 5 is omitted. Hence, the array t is grown by one item in each iteration of the loop. Again, this is not recommended, because the enlargement of t in each iteration will cause the copying of $\sum_{i=1}^{n-1}(i-1) \asymp O(n^2)$ data items (cf. [6]). From the discussion in the previous section it is clear that **rosenbrock_noprealloc** can only be differentiated correctly in the RM with **push_index** and **pop_index**. The ADiMat option allowArrayGrowth triggers the replacement of the normal **push** and **pop** functions by the "index" variants described in the previous section.

We conduct four tests: Test 1 uses the function **rosenbrock**, tests 2 and 3 use the function **rosenbrock_prealloc**, and test 4 uses the function **rosenbrock_noprealloc**. In tests 1 and 2 the option allowArrayGrowth is disabled, while in tests 3 and 4 it is enabled. So in particular, a comparison of tests 3 with test 2 will show the additional overhead of the **push_index** and **pop_index** operations compared to normal **push** and **pop** operations. We run the functions and their adjoints with random vectors of increasing length n as arguments. When the runtime for one n becomes larger than 600 s, that test series is aborted.

We use a stack which writes the data to a file on disk asynchronously. In the case of **rosenbrock_noprealloc** the amount of data written increases with $O(n^2)$ as the whole current vector t is pushed on the stack in each iteration. The amount of data becomes so large that we can not run the tests with $n > 10^{4.5}$. At this problem size, the runtime of **rosenbrock_noprealloc** is 42 ms, but the stack is 3.7 GB.

The outlining canonicalization splits the large expression on the RHS of the single statement in the loop body into multiple assignments to temporary variables. Each of these will cause a pair of **push** and **pop** operations, and only the last statement, assigning to t(i), will cause a **push_index** and **pop_index** operation when the option allowArrayGrowth is enabled.

The resulting runtimes are shown in Fig. 2. Here, f denotes the original function and ∂f the adjoint function, while the index refers to the number of the test. There is no f_3 in the figure as it is identical to f_2. We also show the ratios t_∂/t_f in Fig. 2.

The results show that the runtime of the functions **rosenbrock** and **rosenbrock_prealloc** remains largely constant for input vector lengths up to about 10^3. Afterwards the runtimes begin to increase slightly. This shows that most of the time is spent interpreting the code and not on the actual computations. In the case of **rosenbrock_noprealloc** the runtimes are markedly higher, reflecting the quadratic number of copy operations due to the incremental array enlargement.

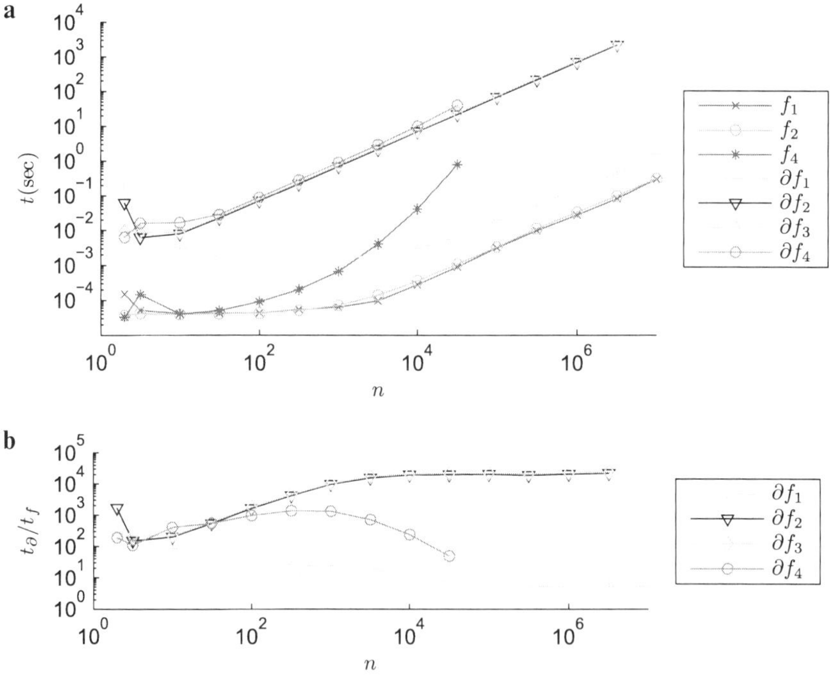

Fig. 2 Absolute runtimes (**a**) and AD overhead ratio (**b**) for the different implementations

The runtime of **a_rosenbrock** is largely constant up to $n = 10^4$. This reflects the longer interpretation time due to the larger amount of code in that function. In the other three cases the runtime of the adjoint function increases linearly with n from very small values of n onwards. This reflects the large amount of code inside the adjoint **for** loop being interpreted $n - 1$ times.

The difference between the two versions of **a_rosenbrock_prealloc** is rather small, but **a_rosenbrock_prealloc** is about 19% faster with allowArrayGrowth off for sizes of $n = 100$ up to $n = 10^4$ and then drops to about 3% for $n \approx 10^{6.5}$. This shows that, for this example, the added overhead of the option allowArrayGrowth is tolerable. Remember that in **a_rosenbrock_prealloc**, no array enlargements actually happen. The runtimes for **a_rosenbrock_noprealloc** are noticeably larger than those of **a_rosenbrock_prealloc**, but the main problem caused by the incremental array enlargement is not the computation time but the storage space, which prevents larger examples from being concluded in the time allowed.

Finally, we consider the ratios t_∂ / t_f: The results for **rosenbrock** in Fig. 2b show that we can achieve runtime ratios smaller than 10, when the code is vectorized and is run with sufficiently large problem sizes. This is comparable to runtime ratios achievable with AD tools for C++ and Fortran. On the other hand, when the number of statements in the original function depends on the problem size, as is the case for **rosenbrock** in tests 2 and 3, then the RM results in a very large overhead ratio of about 10^3 and even 10^4. Surprisingly, the ratios for **rosenbrock_noprealloc** are again rather good, "only" about 100. This is, however, due to the fact that **rosenbrock_noprealloc** is already very slow due to the large amount of data movement triggered by the array enlargement.

6 Conclusion

In this paper, we considered the impact of implicit data reshapings due to scalar expansion and during assignment operations and incremental array allocation on reverse-mode generation for automatic differentiation of Matlab. The weak type system of Matlab prevents a static analysis of these cases, thus necessitating a runtime strategy. For the case of scalar expansion, the function **adjred** was introduced to properly sum up adjoint contributions in the case of a scalar operand in a component-wise binary operation, and we addressed matrix multiplication similarly. The implicit reshaping during an assignment operation has to be undone in the adjoint code. To this end, we introduced the function **adjreshape** which, in conjunction with **adjred**, ensures that indexed assignments in Matlab are handled properly throughout. Implicit data conversions occurring on the LHS of indexed array assignments also required a more general strategy for the pushing and popping of LHS variables. Experimental results with various implementations of the Rosenbrock function showed that the extra overhead incurred to safely differentiate Matlab code employing these features is moderate.

We believe these results to be relevant not only for Matlab, but also for other programming languages, that, in order to improve programming productivity, employ a weak type system and polymorphic programming paradigm. Our conclusion is that in most cases, a generally safe reverse mode strategy can be implemented at moderate additional cost. In particular, for programming environments where convenience and productivity, not execution speed, is the primary motivation, AD can safely be employed. Where a safe default strategy is more expensive, directives would then be a sensible approach to provide the user with some control over performance, while ensuring safe AD differentiation in all other cases.

References

1. Bischof, C.H., Bücker, H.M., Hovland, P.D., Naumann, U., Utke, J. (eds.): Advances in Automatic Differentiation, *Lecture Notes in Computational Science and Engineering*, vol. 64. Springer, Berlin (2008). DOI 10.1007/978-3-540-68942-3
2. Bischof, C.H., Bücker, H.M., Lang, B., Rasch, A., Vehreschild, A.: Combining source transformation and operator overloading techniques to compute derivatives for MATLAB programs. In: Proceedings of the Second IEEE International Workshop on Source Code Analysis and Manipulation (SCAM 2002), pp. 65–72. IEEE Computer Society, Los Alamitos, CA, USA (2002). DOI 10.1109/SCAM.2002.1134106

3. Bücker, H.M., Petera, M., Vehreschild, A.: Code optimization techniques in source transformations for interpreted languages. In: Bischof et al. [1], pp. 223–233. DOI 10.1007/978-3-540-68942-3_20
4. Giles, M.B.: Collected matrix derivative results for forward and reverse mode algorithmic differentiation. In: Bischof et al. [1], pp. 35–44. DOI 10.1007/978-3-540-68942-3_4
5. Kharche, R.V., Forth, S.A.: Source transformation for MATLAB automatic differentiation. In: V.N. Alexandrov, G.D. van Albada, P.M.A. Sloot, J. Dongarra (eds.) Computational Science – ICCS 2006, *Lecture Notes in Computer Science*, vol. 3994, pp. 558–565. Springer, Heidelberg (2006). DOI 10.1007/11758549_77
6. MathWorks: Code vectorization guide (2009). URL http://www.mathworks.com/support/tech-notes/1100/1109.html
7. Pascual, V., Hascoët, L.: Extension of TAPENADE toward Fortran 95. In: H.M. Bücker, G. Corliss, P. Hovland, U. Naumann, B. Norris (eds.) Automatic Differentiation: Applications, Theory, and Implementations, *Lecture Notes in Computational Science and Engineering*, vol. 50, pp. 171–179. Springer, New York, NY (2005). DOI 10.1007/3-540-28438-9_15
8. Vehreschild, A.: Automatisches Differenzieren für MATLAB. Dissertation, Department of Computer Science, RWTH Aachen University (2009). URL http://darwin.bth.rwth-aachen.de/opus3/volltexte/2009/2680/

On the Efficient Computation of Sparsity Patterns for Hessians

Andrea Walther

Abstract The exploitation of sparsity forms an important ingredient for the efficient solution of large-scale problems. For this purpose, this paper discusses two algorithms to detect the sparsity pattern of Hessians: An approach for the computation of exact sparsity patterns and a second one for the overestimation of sparsity patterns. For both algorithms, corresponding complexity results are stated. Subsequently, new data structures and set operations are presented yielding a new complexity result together with an alternative implementation of the exact approach. For several test problems, the obtained runtimes confirm the new theoretical result, i.e., a significant reduction in the runtime needed by the exact approach. A comparison with the runtime required for the overestimation of the sparsity pattern is included together with a corresponding discussion. Finally, possible directions for future research are stated.

Keywords Sparsity patterns for Hessians • Nonlinear interaction domains • Nonlinear frontiers • Conservative second-order dependencies

1 Introduction

For numerous applications, the computation of sparse Hessian matrices is required. Prominent examples are optimization tasks of the form

$$\min_{x\in\mathbb{R}^N} f(x) \qquad \text{or} \qquad \min_{x\in\mathbb{R}^N} f(x), \quad \text{s.t.} \quad c(x) = 0$$

with $f:\mathbb{R}^N \mapsto \mathbb{R}$ and $c:\mathbb{R}^N \mapsto \mathbb{R}^M$, ignoring inequality constraints for simplicity. In these cases, optimization algorithms may benefit considerably from the provision

A. Walther (✉)
Institut für Mathematik, Universität Paderborn, Paderborn, Germany
e-mail: andrea.walther@uni-paderborn.de

S. Forth et al. (eds.), *Recent Advances in Algorithmic Differentiation*, Lecture Notes in Computational Science and Engineering 87, DOI 10.1007/978-3-642-30023-3_13,

of the Hessian $\nabla^2 f(x)$ in the unconstrained case and the Hessian of the Lagrangian function

$$\mathscr{L} : \mathbb{R}^{N+M} \mapsto \mathbb{R}, \quad \mathscr{L}(x,\lambda) = f(x) + \lambda^T c(x),$$

in the constrained case. For almost all large scale problems, these derivatives matrices are sparse, a fact that can be exploited to solve the optimization tasks very efficiently with solvers like Ipopt [9]. When using algorithmic differentiation (AD) for the computation of any sparse Hessian H, one usually follows the following procedure:

1. Determine the sparsity pattern P_H of H. Ideally, this step is performed only once.
2. Obtain a seed matrix S that defines a column partition of H using graph coloring methods for the graph induced by P_H. Once more, this step is performed ideally only once.
3. Compute at each iterate the compressed Hessian matrix $B = HS$ and use these derivative values within the optimizer.

The present paper concentrates on the first step, i.e., the efficient provision of a sparsity pattern of H. In Sect. 2, algorithms for the exact determination of the sparsity pattern P_H and for the computation of an overestimated sparsity pattern OP_H are presented including a first complexity analysis. Section 3 discusses data structures and corresponding set operations for an efficient implementation of the two algorithms yielding a new complexity result. The resulting computing times are presented and discussed in Sect. 4. Finally, conclusions and an outlook are contained in Sect. 5.

2 Algorithms for Exact and Overestimated Sparsity Patterns

Throughout it is assumed, that the function $y = f(x)$ to be differentiated is evaluated as shown in Table 1, where $\varphi_i(v_j)_{j\prec i} = \varphi_i(v_j)$ or $\varphi_i(v_j)_{j\prec i} = \varphi_i(v_j, v_k)$ with $j < i$ and $j, k < i$, respectively. Hence, the *precedence relation* $j \prec i$ denotes that v_i depends directly on v_j. To compute sparsity patterns of Hessians, *index domains*

$$\mathscr{X}_k \equiv \{j \le n \,:\, j - n \prec^* k\} \quad \text{for} \quad 1 - n \le k \le l$$

with $\prec^*$ as transitive closure of $\prec$ as already defined in [3, Sect. 7.1] for all intermediate variables v_k are required. Furthermore, one may employ the *nonlinear interaction domains* (NID)

$$\left\{ j \le n \,:\, \frac{\partial^2 y}{\partial x_i \partial x_j} \ne 0 \right\} \subseteq \mathscr{N}_i \,.$$

Table 1 The first loop of Algorithm I copies $x_1, \ldots, x_n$ into the internal variables $v_{1-n}, \ldots, v_0$. The function is evaluated in the second loop. Finally the value of y is extracted from the v_l. Each *elemental function* φ_i may have one or two arguments

Algorithm I: Function evaluation

for $i = 1, \ldots, n$
  $v_{i-n} = x_i$
for $i = 1, \ldots, l$
  $v_i = \varphi_i(v_j)_{j \prec i}$
$y = v_l$

Table 2 Algorithms for the exact sparsity pattern P_H and the overestimated sparsity pattern OP_H

Algorithm II: NIDs

for $i = 1, \ldots, n$
  $\mathcal{X}_{i-n} \leftarrow \{i\}, \quad \mathcal{N}_i \leftarrow \emptyset$
for $i = 1, \ldots, l$
  $\mathcal{X}_i \leftarrow \bigcup_{j \prec i} \mathcal{X}_j$
  if φ_i **nonlinear then**
    if $v_i = \varphi_i(v_j)$ **then**
      $\forall r \in \mathcal{X}_i : \mathcal{N}_r \leftarrow \mathcal{N}_r \cup \mathcal{X}_i$
    if $v_i = \varphi_i(v_j, v_k)$ **then**
      if v_i **linear in** v_j **then**
        $\forall r \in \mathcal{X}_j : \mathcal{N}_r \leftarrow \mathcal{N}_r \cup \mathcal{X}_k$
      else
        $\forall r \in \mathcal{X}_j : \mathcal{N}_r \leftarrow \mathcal{N}_r \cup \mathcal{X}_i$
      if v_i **linear in** v_k **then**
        $\forall r \in \mathcal{X}_k : \mathcal{N}_r \leftarrow \mathcal{N}_r \cup \mathcal{X}_j$
      else
        $\forall r \in \mathcal{X}_k : \mathcal{N}_r \leftarrow \mathcal{N}_r \cup \mathcal{X}_i$

Algorithm III: NLFs and CSODs

for $i = 1, \ldots, n$
  $\mathcal{X}_{i-n} \leftarrow \{i\}, \quad nlf_i \leftarrow \emptyset$
for $i = 1, \ldots, l$
  $\mathcal{X}_i \leftarrow \bigcup_{j \prec i} \mathcal{X}_j$
  if φ_i **nonlinear then**
    $nlf_i = \{i\}$
  else
    $nlf_i \leftarrow \bigcup_{j \prec i} nlf_j$
$csod = \bigcup_{j \in nlf_l} \mathcal{X}_j \times \mathcal{X}_j$

for all independent variables as defined in [10]. This previous paper proposed Algorithm II as shown on the left-hand side of Table 2 to propagate the index domains and the linear interaction domains forward through the function evaluation yielding the exact sparsity pattern of the corresponding Hessian as the entries of the $\mathcal{N}_i$, $1 \le i \le n$. Analyzing the occurring set operations, one obtains the following result:

Theorem 1 (Complexity Estimate for Computing P_H I). *Let $OPS(P_H)$ be the number of elemental operations, as, e.g., memory accesses, needed by Algorithm II* *to generate all $\mathcal{N}_i$, $1 \le i \le n$. Then, one has*

$$OPS\big(P_H\big) \le 6(1+\hat{n})\sum_{i=1}^{l}\bar{n}_i \lesssim l\,\mathcal{O}(\hat{n}^2),$$

where l is the number of elemental functions evaluated to compute the function value, $\bar{n}_i = |\mathcal{X}_i|$, and $\hat{n} = \max_{1\le i\le n}|\mathcal{N}_i|$.

Proof. See [10]. □

Recently, Varnik, Naumann, and coauthors proposed the computation of an overestimation of the sparsity pattern, which is based on so-called *nonlinear frontiers* (*nlf*s) and *conservative second-order dependencies* (*csod*s) [8]. They also introduced the propagation of these index sets through the function evaluation as shown on the right-hand side of Table 2. Analyzing the occurring set operations, one can show the following result:

Theorem 2 (Complexity Estimate for Computing OP_H). *Let $OPS(OP_H)$ be the number of operations needed by Algorithm III to generate the overestimation of the sparsity pattern given by the index tuples contained in csod. Then, one has*

$$OPS\big(OP_H\big) \le |nlf_l|\,\mathcal{O}(\hat{n}^2) + l\,\mathcal{O}(\hat{n}+\tilde{n}),$$

where l is the number of elemental functions evaluated to compute the function value, $\tilde{n} = max_{1\le i\le l}|nlf_i|$, $\hat{n} = \max_{1\le i\le n}|\mathcal{N}_i|$, and $\mathcal{N}_i$ the ith nonlinear interaction domain as defined above.

Proof. See [7]. □

The main difference in these complexity estimates is that the upper bound in Theorem 1 has the number l of elemental functions as constant in front of the quadratic term whereas in Theorem 2 this constant reduces to $|nlf_l|$ and hence in almost all cases to a much smaller number. This observation is verified by some numerical examples in [7, 8]. For the present paper, we used the test Problem 5.1 of [4] for a varying number N of independents as equality constrained optimization problem. The runtimes obtained with the current implementation of Algorithm II in ADOL-C version 2.2 [11] and an implementation of Algorithm III as described in [7] are illustrated by Fig. 1. One clearly sees the quadratic behavior of Algorithm II and the in essence linear behavior of Algorithm III. However, when implementing Algorithm III using the propagation of the index domains $\mathcal{X}_i$ as already available in ADOL-C, one obtains an even higher runtime as required by Algorithm II as can be seen also in Fig. 1. This indicates that the data structures and set operations play a crucial role for the runtime behavior of the two algorithms. For this purpose, the next section discusses different options for the implementation and the resulting consequences for the runtime and memory requirements, respectively.

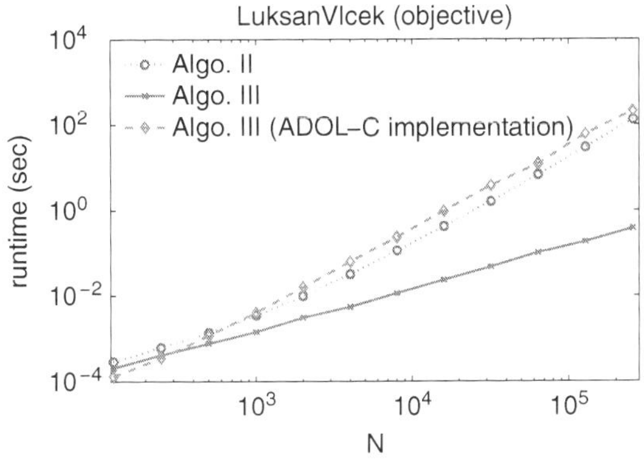

Fig. 1 Runtimes for the computation of P_H as implemented in ADOL-C, version 2.2.1, of OP_H with an implementation according to [7], and of OP_H using the implementation of the index domains as in ADOL-C, version 2.2.1 for the test [4, Problem 5.1]

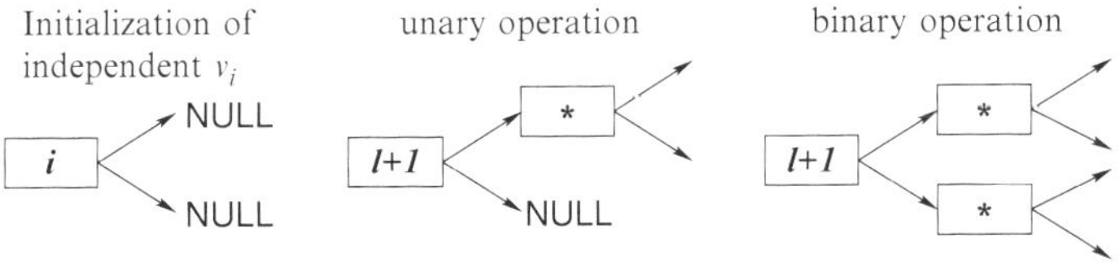

Fig. 2 Initialization and handling of data structures for the propagation of the index domains

3 New Data Structures and Corresponding Set Operations

ADOL-C, version 2.2, uses an adapted data structure and corresponding operations on this data structure to implement the set operations for the propagation of the index domains $\mathcal{X}_i$ required by Algorithm II as true set operations. That is, the entries are ordered and no entry occurs more than once. Alternatively, Varnik used a completely different representation [7], which is based on a graph structure that employs pointers for a simple union of different sets.

As a related strategy, the new ADOL-C version introduces an alternative data structure that consists of one entry and two pointers to this data structure. For each operation $i \in \{1, \ldots, l\}$ occurring in the function evaluation, a variable ind_dom[i] of this data structure is allocated, where the entry either stores the index of an independent variable if appropriate or gets a marker as illustrated in Fig. 2, where the special value $l + 1$ is employed to signal that this node is not a leaf. Obviously, other values are possible here. Using this special data structure, for all unary or binary operations $v_i = \varphi_i(v_j)_{j \prec i}$ the union of the index domains given by

$$\bigcup_{j \prec i} \mathcal{X}_j = \begin{cases} \mathcal{X}_j & \text{if } \varphi_i(v_j) \text{ is unary} \\ \mathcal{X}_j \cup \mathcal{X}_k & \text{if } \varphi_i(v_j, v_k) \text{ is binary} \end{cases}$$

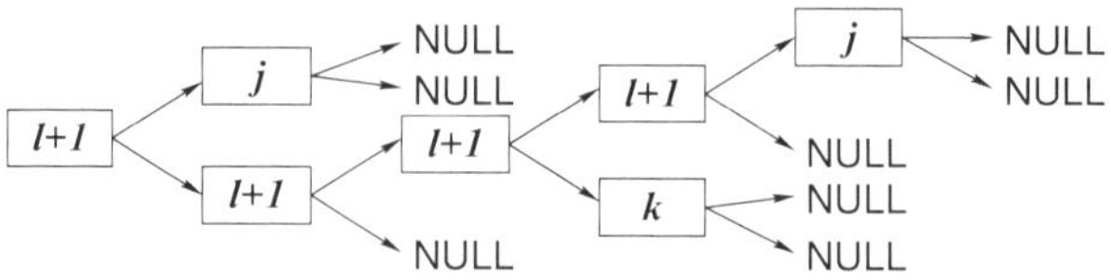

Fig. 3 One possible representation of $\mathscr{X}_i$. The entry k occurs only once, the entry j twice

can be performed with the complexity $\mathscr{O}(1)$ as illustrated on the very right in Fig. 2. However, using this simple representation of the sets, it may happen that an entry occurs multiple times in the tree ind_dom[i] representing the index domain $\mathscr{X}_i$ as sketched in Fig. 3. Hence, on one hand one obtains a remarkable reduction in the runtime complexity, on the other hand the memory requirement increases. This increase in memory can not be predicted in a general way since it depends very much on the specific evaluation procedure. However, for all examples considered so far, this additional memory requirement does not influence the runtime behavior noticeable.

Obviously, for Algorithm III one can use the same strategy to implement the nonlinear frontiers nlf_i, $1 \leq i \leq l$. Hence during the propagation of the index sets, all set operations are performed with a temporal complexity of $\mathscr{O}(1)$. Only once, the more complicated set operation $csod = \bigcup_{j \in nlf_l} \mathscr{X}_j \times \mathscr{X}_j$ yielding the overestimated sparsity pattern OP_H has to be performed. This explains the complexity result of Theorem 2 and the linear runtime observed in Fig. 1.

During the propagation of the index sets, Algorithm II requires not only simple unions of sets but the more complicated operations $\forall r \in \mathscr{X}_i : \mathscr{N}_r \leftarrow \mathscr{N}_r \cup \mathscr{X}_k$. Hence, it is not straightforward to extend the approach used for the implementation of Algorithm III also for the propagation of the nonlinear index domains $\mathscr{N}_i$, $1 \leq i \leq n$. Therefore, the new ADOL-C version uses for the implementation of the nonlinear index domains also an alternative data structure consisting of one entry and two pointers. That is, for each nonlinear interaction domain $\mathscr{N}_i$ there exists a variable nonl_dom[i]. For the appropriate propagation of these nonlinear interaction domains, Table 3 illustrates the implementation of the central set operation of Algorithm III. As can be seen, one has to perform one complete traversal of the tree given by ind_dom[j]. Additionally, $\mathscr{O}(|$ ind_dom[j] $|)$ operations for the corresponding unions of sets are required since one union can be performed in $\mathscr{O}(1)$. Here | ind_dom[j] | denotes the number of entries in ind_dom[j] that are smaller than $l+1$. Hence, using this implementation for the propagation of the index domains and the nonlinear interaction domains, one obtains the following runtime complexity result:

Theorem 3 (Complexity Estimate for Computing P_H II). *Let $OPS(P_H^*)$ be the number of elemental operations needed by Algorithm II to generate all $\mathscr{N}_i$, $1 \leq i \leq n$ when using trees for the representation of $\mathscr{X}_i$ and $\mathscr{N}_i$ as described above. Then,*

Table 3 Implementation of $\forall r \in \mathscr{X}_j : \mathscr{N}_r \leftarrow \mathscr{N}_r \cup \mathscr{X}_k$

```
Algorithm IV: traverse(ind_dom[j], ind_dom[k], nonl_dom)
```

```
if ind_dom[j].left ≠ NULL
traverse(ind_dom[j].left, ind_dom[k], nonl_dom)
if ind_dom[j].right ≠ NULL
    traverse(ind_dom[j].right, ind_dom[k], nonl_dom)
else
    if ind_dom[j].entry < l+1
        nonl_dom[ind_dom[j].entry].left = nonl_dom[ind_dom[j].entry]
        nonl_dom[ind_dom[j].entry].right = ind_dom[k]
        nonl_dom[ind_dom[j].entry].entry = l+1
```

$$OPS(P_H^*) \leq \mathscr{O}(l + \sum_{j \prec i, i \in I_{nl}} \bar{n}_j),$$

where l is the number of elemental functions evaluated to compute the function value, $\bar{n}_j$ = | ind_dom[j] |, i.e., the number of entries of $\mathscr{X}_j$ when using the tree representation, and

$$I_{nl} = \{i \in \{1, \ldots, l\} \mid \varphi_i \text{ is nonlinear}\}.$$

Since the Hessian matrix is assumed to be sparse, it follows that $|\mathscr{N}_i|$ is small for all $i \in \{1, \ldots, n\}$. Therefore, also $|\mathscr{X}_j|$ is small for all $j \prec i, i \in I_{nl}$ yielding a small $\bar{n} = \max_{j \prec i, i \in I_{nl}} |\mathscr{X}_j|$. Hence, if no multiple entries were allowed in the tree representation, one would obtain

$$\text{OPS}(P_H^*) \leq (1 + \bar{n})\mathscr{O}(l), \tag{1}$$

i.e., a linear complexity. Obviously, the multiple occurrence of one entry that is allowed in the simple data structure for the trees representing the index domains and the nonlinear interaction domains might destroy this good complexity result. More complex data structures and set operations that avoid multiple entries and achieve also linear complexity are proposed for example in [6]. Hence, those data structures would ensure the complexity estimate (1). However, as illustrated by the numerical examples in the next section, one observes a linear behavior of the runtime complexity for a wide range of test cases even with the simple data structure described above. Therefore, it is not clear whether the runtime would really benefit from the approaches presented in [6].

4 Numerical Examples

To verify the theoretical results of the previous section, results for the following five scalable test cases are considered: LuksanVlcek as test problem 5.1 of [4], MittelmannBndryCntrlDiri and MittelmannBndryCntrlDiri3Dsin as provided in the example directory of the Ipopt package and described in [5], aug2d from the CUTEr test set [2], and binary as the binary example from [7]. The first four of these test problems define a target function $f : \mathbb{R}^N \mapsto \mathbb{R}$ and constraints $c : \mathbb{R}^N \mapsto \mathbb{R}^M$. Therefore, runtimes for the computation of the (overestimated) sparsity pattern of $\nabla^2 f(x)$ and of $\nabla^2 \mathscr{L}(x)$ will be given. The last one, i.e., the binary example, only defines a target function $f : \mathbb{R}^N \mapsto \mathbb{R}$. Hence, only runtimes for computing the (overestimated) sparsity pattern of $\nabla^2 f(x)$ are considered.

As can be seen in Fig. 4, the runtimes obtained with the new data structures are much smaller than the runtimes obtained with the old version of ADOL-C. Furthermore, using the new data structures one clearly observes the linear temporal complexity. One finds that for the Hessian of the objective only, the computing time for the overestimated sparsity pattern is similar to or less than the computing times required for the exact sparsity pattern using the new data structures. When analyzing the runtimes for the Lagrangian, the situation changes. That is, in these cases, the computing time for the overestimated sparsity pattern is similar to or larger than the computing time required for the exact sparsity pattern using the new data structures. The reason for this runtime behavior might lie in the fact that the potentially overestimated sparsity pattern computed with Algorithm III agrees with the exact one for the objective functions of these test problems. Hence, the lesser computing time reflects the simpler set propagations of Algorithm III. For the Lagrangian function, one obtains a severe overestimation of the sparsity pattern with more than twice the number of nonzeros for the test cases MittelmannBndryCntrlDiri, MittelmannBndryCntrlDiri3Dsin, and aug2d resulting also in a larger computing time. A similar effect can be observed for the binary test example, where the target function is defined by

$$f_4(x_1, x_2, x_3, x_4) = x_1 * x_2 + x_3 * x_4$$

$$f_{16}(x_1, \ldots, x_{16}) = f_4(x_1, \ldots, x_4) * f_4(x_5, \ldots, x_8) + f_4(x_9, \ldots, x_{12}) * f_4(x_{13}, \ldots, x_{16})$$

and correspondingly for larger $N = 4^i, i = 3, 4, \ldots$, as can be seen in Fig. 5. Also in this case the number of nonzeros in the overestimated sparsity pattern is twice as much as the exact number of nonzeros. This binary example is chosen as a test case because the ratio of nonlinear operations compared to the linear operations in the function evaluation is rather high. Therefore, the propagation of sparsity pattern is especially challenging. This fact is also illustrated by the observed runtime since in this somehow extreme case the positive effects of the new data structure on the overall runtime are considerably less in comparison to the first four standard test cases from the equality constrained optimization context.

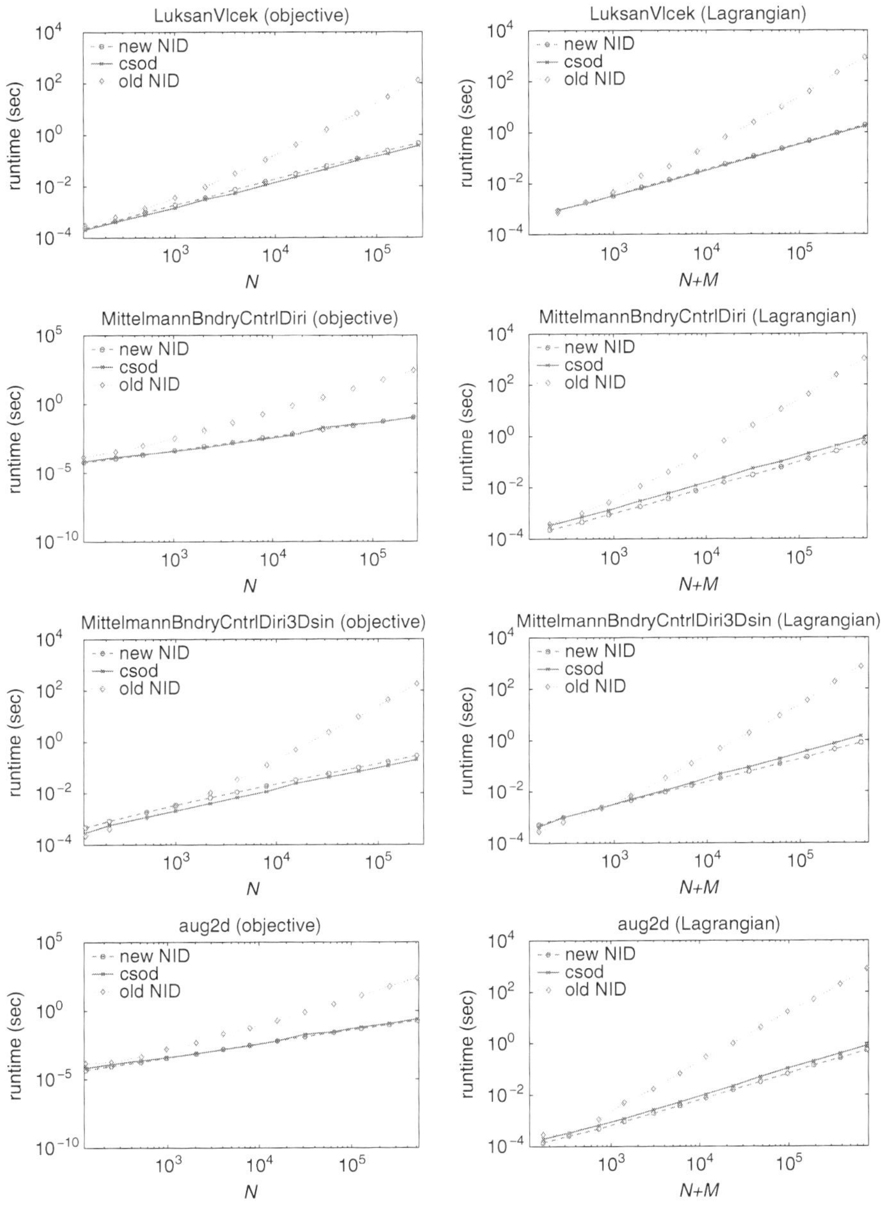

Fig. 4 Runtimes for the computation of P_H as implemented in ADOL-C, version 2.2.1 (old NID), of OP_H using an implementation according to [7] (*csod*), and of P_H using the implementation as described in Sect. 3 (new NID)

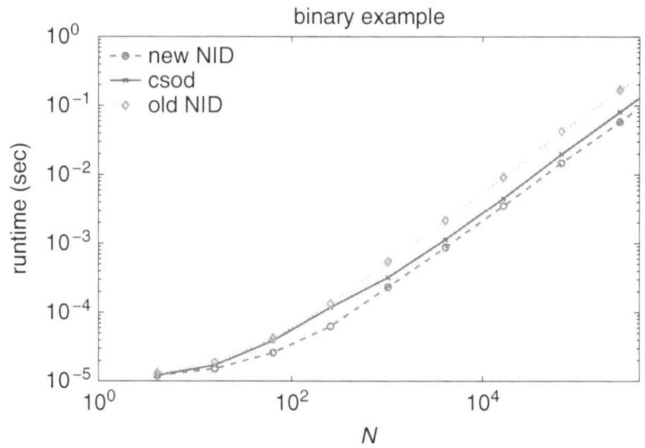

Fig. 5 Runtimes for the computation of P_H as implemented in ADOL-C, version 2.2.1 (old NID), of OP_H using an implementation according to [7] (*csod*), and of P_H using the implementation as described in Sect. 3 (new NID) for the binary example

Table 4 Number of colors required by the graph coloring

	Objective		Lagrangian	
	Exact	Inexact	Exact	Inexact
LuksanVlcek	3	3	5	6
MittelmannBndryCntrlDiri	1	1	8	15–18
MittelmannBndryCntrlDiri3Dsin	14–17	14–17	14–17	25–36
aug2d	1	1	5,15,31, ...	10
binary	2	2		

It is remarkable that the different sparsity patterns also have a non-obvious influence on the number of colors required for the graph coloring that is performed in the second step of the overall procedure as mentioned in the first section. For the sparsity patterns obtained with Algorithm II and III, Table 4 shows the number of colors required for the five test cases considered here when using the default options in ColPack [1]. Hence, there is certainly room for further investigations to clarify the influence of the structure of the actual sparsity pattern on the number of colors required by the subsequent coloring.

5 Conclusions and Outlook

This paper discusses several options for the propagation of index sets to compute a (possibly overestimated) sparsity pattern of Hessian matrices. A data structure that allows considerable simple unions of sets but yields an increase in the memory requirement is presented. An appropriate algorithm for the exact computation of sparsity pattern is presented and analyzed with respect to the temporal complexity.

Numerical tests verify the theoretical results. The resulting consequences of exact and overestimated sparsity patterns for the subsequent graph coloring forms an open questions and might be the subject of interesting future research.

References

1. Gebremedhin, A., Nguyen, D., Patwary, M., Pothen, A.: Colpack: Software for graph coloring and related problems in scientific computing. Tech. rep., Purdue University (2011)
2. Gould, N., Orban, D., Toint, P.: CUTEr and SifDec: a constrained and unconstrained testing environment, revisited. ACM Trans. Math. Softw. **29**(4), 373–394 (2003)
3. Griewank, A., Walther, A.: Evaluating Derivatives: Principles and Techniques of Algorithmic Differentiation, 2nd edn. No. 105 in Other Titles in Applied Mathematics. SIAM, Philadelphia, PA (2008). URL http://www.ec-securehost.com/SIAM/OT105.html
4. Luksan, L., Vlcek, J.: Sparse and partially separable test problems for unconstrained and equality constrained optimization. ICS AS CR V-767, Academy of Sciences of the Czech Republic (1998)
5. Maurer, H., Mittelmann, H.: Optimization techniques for solving elliptic control problems with control and state constraints. II: Distributed control. Comput. Optim. Appl. **18**(2), 141–160 (2001)
6. Tarjan, R.: Data structures and network algorithms, *CBMS-NSF Regional Conference Series in Applied Mathematics*, vol. 44. Society for Industrial and Applied Mathematics (SIAM), Philadelphia, PA (1983)
7. Varnik, E.: Exploitation of structural sparsity in algorithmic differentiation. Ph.D. thesis, RWTH Aachen (2011)
8. Varnik, E., Razik, L., Mosenkis, V., Naumann, U.: Fast conservative estimation of Hessian sparsity. In: Abstracts of Fifth SIAM Workshop of Combinatorial Scientific Computing, no. 2011-09 in Aachener Informatik Berichte, pp. 18–21. RWTH Aachen (2011)
9. Wächter, A., Biegler, L.: On the Implementation of a Primal-Dual Interior Point Filter Line Search Algorithm for Large-Scale Nonlinear Programming. Math. Program. **106**(1), 25–57 (2006)
10. Walther, A.: Computing sparse Hessians with automatic differentiation. ACM Transaction on Mathematical Software **34**(1), 3:1–3:15 (2008). URL http://doi.acm.org/10.1145/1322436.1322439
11. Walther, A., Griewank, A.: Getting started with ADOL-C. In: U. Naumann, O. Schenk (eds.) Combinatorial Scientific Computing. Chapman-Hall (2012). see also http://www.coin-or.org/projects/ADOL-C.xml

Exploiting Sparsity in Automatic Differentiation on Multicore Architectures

Benjamin Letschert, Kshitij Kulshreshtha, Andrea Walther, Duc Nguyen, Assefaw Gebremedhin, and Alex Pothen

Abstract We discuss the design, implementation and performance of algorithms suitable for the efficient computation of *sparse* Jacobian and Hessian matrices using Automatic Differentiation via operator overloading on multicore architectures. The procedure for exploiting sparsity (for runtime and memory efficiency) in serial computation involves a number of steps. Using nonlinear optimization problems as test cases, we show that the algorithms involved in the various steps can be adapted to multithreaded computations.

Keywords Sparsity • Graph coloring • Multicore computing • ADOL-C • ColPack

1 Introduction

Research and development around Automatic Differentiation (AD) over the last several decades has enabled much progress in algorithms and software tools, but it has largely focused on differentiating functions implemented as serial codes. With the increasing ubiquity of parallel computing platforms, especially desktop multicore machines, there is a greater need than ever before for developing AD capabilities for parallel codes. The subject of this work is on AD capabilities for multithreaded functions, and the focus is on techniques for exploiting the *sparsity* available in large-scale Jacobian and Hessian matrices.

B. Letschert (✉) · K. Kulshreshtha · A. Walther
Institut für Mathematik, Universität Paderborn, Paderborn, Germany
e-mail: Benny.Letschert@web.de; kshitij@math.upb.de; andrea.walther@uni-paderborn.de

D. Nguyen · A. Gebremedhin · A. Pothen
Department of Computer Science, Purdue University, West Lafayette, IN, USA
e-mail: nguyend@purdue.edu; agebreme@purdue.edu; apothen@purdue.edu

S. Forth et al. (eds.), *Recent Advances in Algorithmic Differentiation*, Lecture Notes in Computational Science and Engineering 87, DOI 10.1007/978-3-642-30023-3_14,

Derivative calculation via AD for parallel codes has been considered in several previous studies, but the focus has largely been on the source transformation approach [1–4, 11]. This is mainly because having a compiler at hand during the source transformation makes it relatively easy to detect parallelization function calls (as in MPI) or parallelization directives (as in OpenMP). Detecting parallel sections of code for an operator overloading tool is much harder since the corresponding parallelization function calls or directives are difficult or even impossible to detect at runtime. For that reason, the operator overloading tool ADOL-C [13] uses its own wrapper functions for handling functions that are parallelized with MPI. For parallel function evaluations using OpenMP, ADOL-C uses the concept of nested taping [8, 9] to take advantage of the parallelization provided by the simulation for the derivative calculation as well. In this paper we extend this approach to exploit sparsity in parallel.

By exploiting sparsity is meant avoiding computing with zeros in order to reduce (often drastically) runtime and memory costs. We aim at exploiting sparsity in *both* Jacobian and Hessian computations. In the serial setting, there exists an established scheme for efficient computation of sparse Jacobians and Hessians. The scheme involves four major steps: automatic sparsity pattern detection, seed matrix determination via graph coloring, compressed-matrix computation, and recovery. We extend this scheme to the case of multithreaded computations, where both the function evaluation and the derivative computation are done in parallel. The AD-specific algorithms we use are implemented in ADOL-C. The coloring and recovery algorithms are independently developed and implemented via ColPack [6], which in turn is coupled with ADOL-C. We show the performance of the various algorithms on a multicore machine using PDE-constrained optimization problems as test cases.

2 Parallel Derivative Computation in ADOL-C

Throughout this paper we assume that the user provides an OpenMP parallel program as sketched in Fig. 1. That is, after an initialization phase, calculations are performed on several threads, with a possible finalization phase performed by a dedicated single thread (say thread 1). The current "mode" of operation of ADOL-C when differentiating such OpenMP parallel codes is illustrated in Fig. 2. Here, the *tracing* part represents essentially the parallel function evaluation provided by the user. For computing the derivatives also in parallel, the user has to change in the function evaluation all double-variables to adouble-variables, include the headers adolc.h and adolc_openmp.h, and insert the pragma omp parallel firstprivate(ADOLC_OpenMP_Handler) before the trace generation in the initialization phase. Then, ADOL-C performs a parallel derivative calculation using the OpenMP strategy provided by the user as sketched in Fig. 2. Hence, once the variables are declared in each thread, the traces are written on each thread

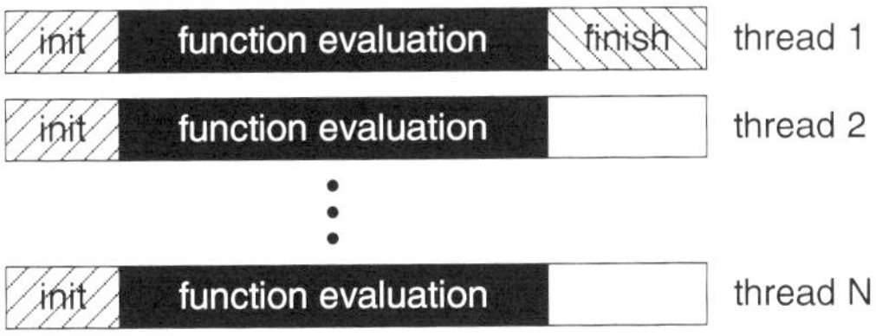

Fig. 1 Function evaluation of an OpenMP parallel code

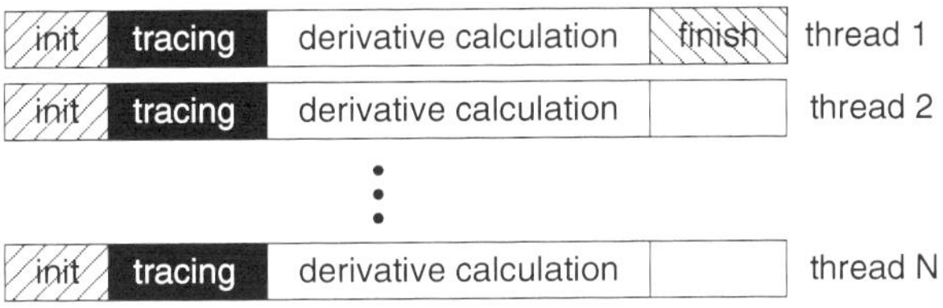

Fig. 2 Derivative calculation with ADOL-C for an OpenMP parallel code

separately during the tracing phase. Subsequently, each thread has its own internal function representation. This allows for the computation of the required derivative information on each thread *separately* as described in [8].

3 Parallel Sparse Derivative Computation

In this work, we extend this functionality of ADOL-C such that *sparse* Jacobians and Hessians can be computed efficiently in a parallel setting. Figure 3 illustrates the approach we take for parallel, sparsity-exploiting derivative computation. As in Fig. 2 derivatives on each thread are computed separately, but this time, the per-thread computation is comprised of several steps: automatic sparsity pattern detection, seed matrix generation and derivative calculation.

3.1 Sparsity Pattern Detection

In the case of a Jacobian matrix, we propagate in parallel on each thread the so-called index domains

$$\mathscr{X}_k \equiv \{j \leq n \,:\, j - n \prec^* k\} \quad \text{for} \quad 1 - n \leq k \leq l$$

determining the sparsity pattern corresponding to the part of the function on that thread. Here, n denotes the number of independent variables, l denotes the number of intermediate variables, and $\prec^*$ denotes precedence relation in the decomposition of function evaluation into elementary components. Since it is not possible to exchange data between the various threads when using OpenMP for parallelization,

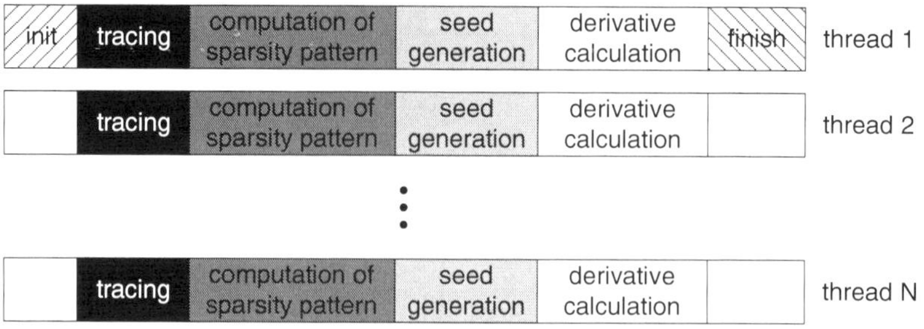

Fig. 3 Derivative calculation with ADOL-C for an OpenMP parallel code when exploiting sparsity

the layout of the data structure storing these partial sparsity patterns has to allow a possibly required reunion of the sparsity pattern, for example during the finalization phase performed by thread 1. However, since the user provides the parallelization strategy, this reunion can not be provided in a general way.

To determine the sparsity pattern of the Hessian of a function $y = f(x)$ of n independent variables, in addition to the index domains, so-called nonlinear interaction domains

$$\left\{ j \leq n \ : \ \frac{\partial^2 y}{\partial x_i \partial x_j} \not\equiv 0 \right\} \subseteq \mathcal{N}_i \ , \quad \text{for} \quad 1 \leq i \leq n$$

are propagated on each thread. Once more, each thread computes only the part of the sparsity pattern originating from the internal function representation available on the specific thread. Therefore, in the Hessian case also, the data structure storing the partial sparsity patterns of the Hessian must allow a possibly required reunion to compute the overall sparsity pattern. Again, this reunion relies on the parallelization strategy chosen by the user.

3.2 Seed Matrix Determination

A key need in compression-based computation of an $m \times n$ Jacobian or an $n \times n$ Hessian matrix A with a known sparsity pattern is determining an $n \times p$ *seed* matrix S of minimal p that would be used in computing the compressed representation $B \equiv AS$. The seed matrix S in our context encodes a *partitioning* of the n columns of A into p groups. It is a zero-one matrix, where entry (j, k) is one if the jth column of the matrix A belongs to group k in the partitioning and zero otherwise. The columns in each group are pair-wise structurally "independent" in some sense. For example, in the case of a Jacobian, the columns in a group are structurally orthogonal to each other. As has been shown in several previous studies (see [5] for a survey), a seed matrix can be obtained using a *coloring* of an appropriate graph representation of the sparsity pattern of the matrix A. In this work we rely on the coloring models and functionalities available in (or derived from) the package ColPack [6].

In ColPack, a Jacobian (nonsymmetric) matrix is represented using a *bipartite graph* and a Hessian (symmetric) matrix is represented using an *adjacency graph*. With such representations in place, we obtain a seed matrix suitable for computing a Jacobian J using a *distance-2 coloring* of the column vertices of the bipartite graph of J. Similarly, we obtain a seed matrix suitable for computing a Hessian H using a *star coloring* of the adjacency graph of H [7]. These colorings yield seed matrices suitable for direct recovery, as opposed to recovery via substitution, of entries of the original matrix A from the compressed representation B.

Just as the sparsity pattern detection was done on each thread focusing on the part of the function evaluation on that thread, the colorings are also done on the "local" graphs corresponding to each thread. For the results reported in this paper, we use parallelized versions of the distance-2 and star coloring functionalities of ColPack.

3.3 Derivative Calculation

Once a seed matrix per thread is determined, the compressed derivative matrix (Jacobian or Hessian) is obtained using an appropriate mode of AD. The entries of the original derivative matrix are then recovered from the compressed representation. For recovery purposes, we rely on ColPack. In Fig. 3 the block "derivative calculation" lumps together the compressed derivative matrix computation and recovery steps.

4 Experimental Results

We discuss the test cases used in our experiments in Sect. 4.1 and present the results obtained in Sect. 4.2.

4.1 Test Cases

We consider optimization problems of the form

$$\min_{x \in \mathbb{R}^n} f(x), \quad \text{such that} \quad c(x) = 0, \tag{1}$$

with an objective function $f : \mathbb{R}^n \mapsto \mathbb{R}$ and a constraint function $c : \mathbb{R}^n \mapsto \mathbb{R}^m$, ignoring inequality constraints for simplicity. Many state-of-the-art optimizers, such as Ipopt [12], require at least first derivative information, i.e., the gradient $\nabla f(x) \in \mathbb{R}^n$ of the target function and the Jacobian $\nabla c(x) \in \mathbb{R}^{m \times n}$. Furthermore, they benefit considerably in terms of performance from the provision of exact second order derivatives, i.e., the Hessian $\nabla^2 \mathscr{L}$ of the Lagrangian function

$$\mathscr{L} : \mathbb{R}^{n+m} \mapsto \mathbb{R}, \quad \mathscr{L}(x, \lambda) = f(x) + \lambda^T c(x).$$

Optimization tasks where the equality constraints represent a state description as discretization of a partial differential equation (PDE) form an important class of optimization problems having the structure shown in (1). Here, sparsity in the derivative matrices occurs inherently and the structure of the sparsity pattern is not obvious when a nontrivial discretization strategy is used.

In [10] several scalable test cases for optimization tasks with constraints given as PDEs are introduced. The state in these test cases is always described by an elliptic PDE, but there are different ways in which the state can be modified, i.e., controlled. For four of the test problems, serial implementations in C++ are provided in the example directory of the Ipopt package. From those, we chose the `MittelmannDistCntrlDiri` and the `MittelmannDistCntrlNeumA` test cases for our experiments. These represent optimization tasks for a distributed control with different boundary conditions for the underlying elliptic PDE. Inspecting the implementation of these test problems, one finds that the evaluation of the constraints does not exploit the computation of common subexpressions. Therefore, when taking the structure of the optimization problem (1) into account, a straightforward parallelization based on OpenMP distributes the single target function and the evaluation of the m constraints equally on the available threads. The numerical results presented in Sect. 4.2 rely on this parallelization strategy.

Problem sizes. The results obtained for the `MittelmannDistCntrlDiri` and `MittelmannDistCntrlNeumA` showed similar general trends. Therefore, we present results here only for the former. We consider three problem sizes $\tilde{n} \in \{600, 800, 1,000\}$, where $\tilde{n}$ denotes the number of inner grid nodes per dimension. The number of constraints (number of rows in the Jacobian ∇c) is thus $m = \tilde{n}^2$. Due to the distributed control on the inner grid nodes and the Dirichlet conditions at the boundary nodes, the number of variables in the corresponding target function (number of columns in the Jacobian ∇c) is $n = \tilde{n}^2 + (\tilde{n} + 2)^2$. Further, the Hessian $\nabla^2 \mathscr{L}$ of the Lagrangian function is of dimension $(n + m) \times (n + m)$. The number of nonzeros in each $m \times n$ Jacobian is $6 \cdot \tilde{n}^2$. Here, five of the nonzero entries per row stem from the discretization of the Laplacian operator occurring in the elliptic PDE, and the sixth entry comes from the distributed control. Similarly, the number of nonzeros in each $(n+m) \times (n+m)$ Hessian is $8 \cdot \tilde{n}^2$. The two additional nonzeros in the Hessian case come from the target function involving a sum of squares and a regularization of the control in the inner computational domain. Table 1 provides a summary of the sizes of the three test problems considered in the experiments.

Table 1 Summary of problem sizes used in the experiments

$\tilde{n}$	m	n	$n+m$	nnz (∇c)	nnz ($\nabla^2 \mathscr{L}$)
600	360,000	722,404	1,082,404	2,160,000	2,880,000
800	640,000	1,283,204	1,923,204	3,840,000	5,120,000
1,000	1,000,000	2,004,004	3,004,004	6,000,000	8,000,000

4.2 Runtime Results

The experiments are conducted on an Intel, Fujitso-Siemens, model RX600S5 system. The system has four Intel X7542, 2.67 GHz, processors each of which has six cores; the system thus supports the use of a maximum of 24 cores (threads). The node memory is 128 GByte DDR3 1066, and the operating system is Linux (CentOS). All codes are compiled with gcc version 4.4.5 with -O2 optimization enabled.

Figure 4 shows runtime results of the computation of the Jacobian of the constraint function for the three problem sizes summarized in Table 1 and various numbers of threads. Figure 5 shows analogous results for the computation of the Hessian of the Lagrangian function. The plots in Fig. 4 (and Fig. 5) show a breakdown of the total time for the sparse Jacobian (and Hessian) computation into four constituent parts: tracing, sparsity pattern detection, seed generation, and derivative computation. The results in both figures show the times needed for the "distributed" (across threads) Jacobian and Hessian computation, excluding the time needed to "assemble" the results. We excluded the assembly times as they are nearly negligibly small and would have obscured the trends depicted in the figure. (The assembly time is less than 0.03 s for $\tilde{n} = 600$ and less than 0.09 s for $\tilde{n} = 1,000$ for the Jacobian case, and less than 0.17 s for the Hessian case for both sizes.)

Note that the vertical axis in Fig. 4 is in linear scale, while the same axis in Fig. 5 is in log scale, since the relative difference in the time spent in the four phases in the Hessian case is too big. Note also the magnitude of the difference between the runtimes in the Jacobian and Hessian cases: the runtimes in the various phases of the Jacobian computation (Fig. 4) are in the order of seconds, while the times in some of the phases in the Hessian case (Fig. 5) are in the order of thousands of seconds. We highlight below a few observations on the trends seen in Figs. 4 and 5.

- *Tracing*: In the Jacobian case, this phase scales poorly with number of threads. A likely reason for this phenomenon is that the phase is memory-intensive. In the Hessian case, tracing accounts for only a small fraction of the overall time that its scalability becomes less important.
- *Sparsity pattern detection*: The routine we implemented for this phase involves many invocations of the `malloc()` function, which essentially is serialized in an OpenMP threaded computation. To better reflect the algorithmic nature of the routine, in the plots we report results after subtracting the time spent on the `mallocs`. In the Jacobian case, the phase did not scale with number of threads,

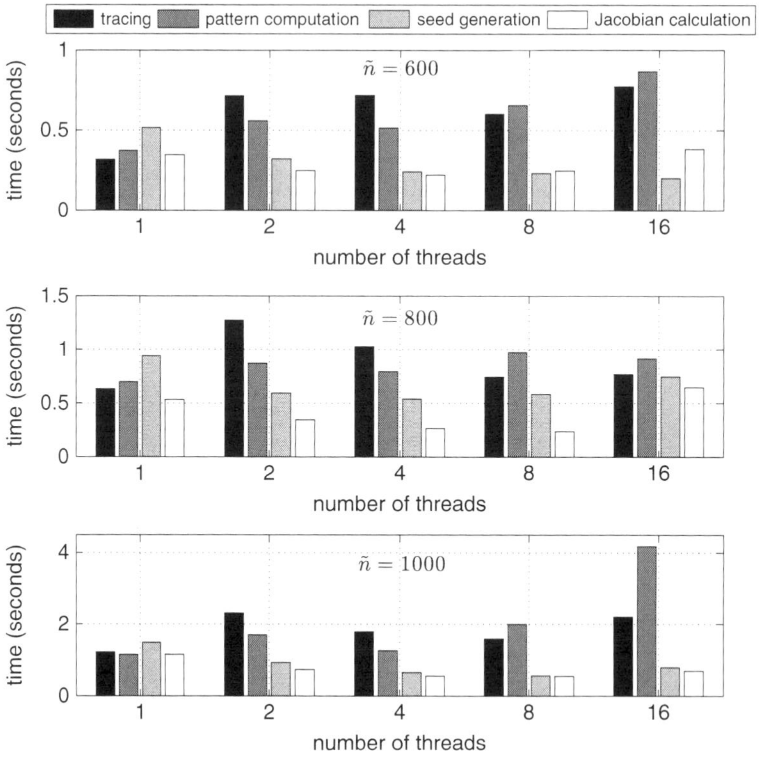

Fig. 4 Timing results for multithreaded computation of the Jacobian ∇c when sparsity is exploited. Three problem sizes are considered: $\tilde{n} = 600$ (*top*), $\tilde{n} = 800$ (*middle*), and $\tilde{n} = 1,000$ (*bottom*)

whereas in the Hessian case it scales fairly well. A plausible reason for the poorer scalability in the Jacobian case is again that the runtime for that step (which is about 1 s) is too short to be impacted by the use of more threads.

- *Seed generation*: For this phase, we depict the time spent on coloring (but not graph construction) and seed matrix construction. It can be seen that this phase scales relatively well. Further, the number of colors used by the coloring heuristics turned out to be optimal (or nearly optimal). In particular, in the Jacobian case, for each problem size, seven colors were used to distance-2 color the local bipartite graphs consisting of n column vertices and m/N row vertices on each thread, where N denotes the number of threads. Since each Jacobian has six nonzeros per row this coloring is optimal. In the Hessian case, again for each problem size, 6 colors were used to star color the local adjacency graphs (consisting of $n + m$ vertices) on each thread.
- *Derivative computation*: This phase scales modestly in both the Jacobian and Hessian cases.

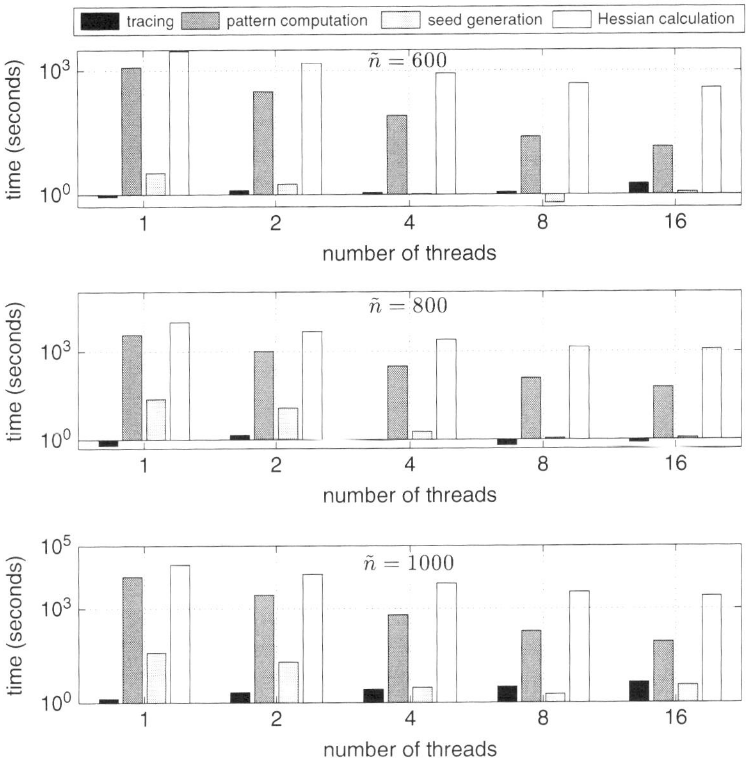

Fig. 5 Timing results for multithreaded computation of the Hessian $\nabla^2 \mathscr{L}$ when sparsity is exploited. Three problem sizes considered: $\tilde{n} = 600$ (*top*), $\tilde{n} = 800$ (*middle*), and $\tilde{n} = 1,000$ (*bottom*)

- *Comparison with dense computation*: The relatively short runtime of the coloring algorithms along with the drastic dimension reduction (compression) the colorings provide enables enormous overall runtime and space saving compared to a computation that does not exploit sparsity. The runtimes for the dense computation of the Jacobian for $\tilde{n} = 600$, for example, are at least three to four orders of magnitude slower requiring hours instead of seconds even in parallel (we therefore omitted the results in the reported plots). For the larger problem sizes, the Jacobian (or Hessian) could not be computed at all due to excessive memory requirement to accommodate the matrix dimensions (see Table 1).

5 Conclusion

We demonstrated the feasibility of exploiting sparsity in Jacobian and Hessian computation using Automatic Differentiation via operator overloading on multi-threaded parallel computing platforms. We showed experimental results on a modest number of threads. Some of the phases in the sparse computation framework scaled reasonably well, while others scaled poorly. In future work, we will explore ways in which scalability can be improved. In particular, more investigation is needed to improve the scalability of the sparsity pattern detection algorithm used for Jacobian computation (Fig. 4) and the tracing phase in both the Jacobian and Hessian case. Another direction for future work is the development of a parallel optimizer that could take advantage of the distributed function and derivative evaluation.

Acknowledgements We thank the anonymous referees for their helpful comments. The experiments were performed on a computing facility hosted by the Paderborn Center for Parallel Computing (PC^2). The research is supported in part by the U.S. Department of Energy through the CSCAPES Institute grant DE-FC02-08ER25864 and by the U.S. National Science Foundation through grant CCF-0830645.

References

1. Bücker, H.M., Rasch, A., Vehreschild, A.: Automatic generation of parallel code for Hessian computations. In: M.S. Mueller, B.M. Chapman, B.R. de Supinski, A.D. Malony, M. Voss (eds.) OpenMP Shared Memory Parallel Programming, Proceedings of the International Workshops IWOMP 2005 and IWOMP 2006, Eugene, OR, USA, June 1–4, 2005, and Reims, France, June 12–15, 2006, *Lecture Notes in Computer Science*, vol. 4315, pp. 372–381. Springer, Berlin / Heidelberg (2008). DOI 10.1007/978-3-540-68555-5_30
2. Bücker, H.M., Rasch, A., Wolf, A.: A class of OpenMP applications involving nested parallelism. In: Proceedings of the 19th ACM Symposium on Applied Computing, Nicosia, Cyprus, March 14–17, 2004, vol. 1, pp. 220–224. ACM Press, New York (2004). DOI 10.1145/967900.967948. URL http://doi.acm.org/10.1145/967900.967948
3. Conforti, D., Luca, L.D., Grandinetti, L., Musmanno, R.: A parallel implementation of automatic differentiation for partially separable functions using PVM. Parallel Computing **22**, 643–656 (1996)
4. Fischer, H.: Automatic differentiation: Parallel computation of function, gradient and Hessian matrix. Parallel Computing **13**, 101–110 (1990)
5. Gebremedhin, A.H., Manne, F., Pothen, A.: What color is your Jacobian? Graph coloring for computing derivatives. SIAM Review **47**(4), 629–705 (2005)
6. Gebremedhin, A.H., Nguyen, D., Patwary, M., Pothen, A.: ColPack: Software for graph coloring and related problems in scientific computing. Tech. rep., Purdue University (2011)
7. Gebremedhin, A.H., Tarafdar, A., Manne, F., Pothen, A.: New acyclic and star coloring algorithms with applications to Hessian computation. SIAM J. Sci. Comput. **29**, 1042–1072 (2007)
8. Kowarz, A.: Advanced concepts for Automatic Differentiation based on operator overloading (1998). PhD Thesis, TU Dresden
9. Kowarz, A., Walther, A.: Parallel derivative computation using ADOL-C. In: W. Nagel, R. Hoffmann, A. Koch (eds.) Proceedings of PASA 2008, Lecture Notes in Informatics, Vol. 124, pp. 83–92. Gesellschaft fr Informatik (2008)
10. Maurer, H., Mittelmann, H.: Optimization techniques for solving elliptic control problems with control and state constraints. II: Distributed control. Comput. Optim. Appl. **18**(2), 141–160 (2001)

11. Utke, J., Hascoët, L., Heimbach, P., Hill, C., Hovland, P., Naumann, U.: Toward adjoinable MPI. In: Proceedings of the 10th IEEE International Workshop on Parallel and Distributed Scientific and Engineering, PDSEC-09 (2009). DOI http://doi.ieeecomputersociety.org/10.1109/IPDPS.2009.5161165
12. Wächter, A., Biegler, L.: On the implementation of a Primal-Dual Interior Point Filter Line Search algorithm for large-scale nonlinear programming. Math. Program. **106**(1), 25–57 (2006)
13. Walther, A., Griewank, A.: Getting started with ADOL-C. In: U. Naumann, O. Schenk (eds.) Combinatorial Scientific Computing. Chapman-Hall (2012). see also http://www.coin-or.org/projects/ADOL-C.xml

Automatic Differentiation Through the Use of Hyper-Dual Numbers for Second Derivatives

Jeffrey A. Fike and Juan J. Alonso

Abstract Automatic Differentiation techniques are typically derived based on the chain rule of differentiation. Other methods can be derived based on the inherent mathematical properties of generalized complex numbers that enable first-derivative information to be carried in the non-real part of the number. These methods are capable of producing effectively exact derivative values. However, when second-derivative information is desired, generalized complex numbers are not sufficient. Higher-dimensional extensions of generalized complex numbers, with multiple non-real parts, can produce accurate second-derivative information provided that multiplication is commutative. One particular number system is developed, termed *hyper-dual numbers*, which produces exact first- and second-derivative information. The accuracy of these calculations is demonstrated on an unstructured, parallel, unsteady Reynolds-Averaged Navier-Stokes solver.

Keywords Hyper-dual numbers • Generalized complex numbers • Operator overloading • Second derivatives • Hessian • Forward mode • C++ • CUDA • MPI

1 Introduction

Techniques for Automatic Differentiation are usually derived based on repeated application of the chain rule of differentiation [8, 19]. These techniques fall into two categories, those that employ source transformation and those that employ operator overloading [1]. In many of them, both forward and reverse modes can be used to maximize efficiency when either computing derivatives for many functions of a few variables or computing derivatives for a few functions of many variables. In

J.A. Fike (✉) · J.J. Alonso
Department of Aeronautics and Astronautics, Stanford University, Stanford, CA 94305, USA
e-mail: jfike@stanford.edu; jjalonso@stanford.edu

S. Forth et al. (eds.), *Recent Advances in Algorithmic Differentiation*, Lecture Notes in Computational Science and Engineering 87, DOI 10.1007/978-3-642-30023-3_15,

any case, the derivatives that are computed are numerically exact and free from truncation error.

An alternative approach is to use alternative number systems whose mathematics inherently produce the desired derivative information. Of particular interest is the family of generalized complex numbers [10], which consist of one real part and one non-real part, $a + bE$. There are three types of generalized complex numbers based on the definition of the non-real part E: ordinary complex numbers $E^2 = i^2 = -1$, double numbers $E^2 = e^2 = 1$, and dual numbers $E^2 = \epsilon^2 = 0$ [3–5, 9]. In these systems, it is important to realize that the E terms are not real valued so it is possible to have $\epsilon^2 = 0$ even though $\epsilon \neq 0$ and $e^2 = 1$ even though $e \neq 1$.

When applied to a real-valued function, the mathematics of these numbers are such that first-derivative information is computed and contained in the non-real part. While some of these methods have truncation errors associated with them, they can be made effectively exact by choosing a step size that is small enough to make the truncation error much less than machine precision. The property of these methods that allows the step size to be made arbitrarily small is that they are free from subtractive cancellation error. Other methods, such as the use of dual numbers, are free from truncation error and thus are exact regardless of step size. Several authors have made the connection between using certain generalized complex numbers and automatic differentiation, in particular for the complex-step approximation [13] and the use of dual numbers [12, 17].

Numbers with one non-real part are sufficient if only first derivative information is required. Second(or higher)-derivative information can be computed using higher-dimensional forms of generalized complex numbers with multiple non-real parts. However, not all higher-dimensional extensions of generalized complex numbers suffice. Only those number systems that possess multiplicative commutivity are free from subtractive cancellation error, and can be made effectively exact through the choice of a small enough step size. In particular, we develop and implement *hyper-dual numbers*, a higher dimensional extension of dual numbers, which is free from both subtractive cancellation error and truncation error enabling exact first- and second-derivative calculations.

2 First Derivative Calculations

As discussed above, first-derivative calculation methods can be created using generalized complex numbers of the form $a + bE$, with one real part and one non-real part. These methods work by taking a real-valued function evaluation procedure and evaluating it subject to a non-real step. First-derivative information is then found by taking the non-real part of the function evaluation and dividing by the step size.

Consider the Taylor Series for a real-valued function with a generalized complex step,

$$f(x + hE) = f(x) + hf'(x)E + \frac{1}{2!}h^2 f''(x)E^2 + \frac{h^3 f'''(x)}{3!}E^3 + \dots . \quad (1)$$

As stated above, there are three types of generalized complex numbers based on the definition of the non-real part E: ordinary complex numbers $E^2 = i^2 = -1$, double numbers $E^2 = e^2 = 1$, and dual numbers $E^2 = \epsilon^2 = 0$.

When using ordinary complex numbers, $E^2 = i^2 = -1$, (1) becomes

$$f(x + hi) = f(x) + hf'(x)i - \frac{1}{2!}h^2 f''(x) - \frac{h^3 f'''(x)}{3!}i + \dots . \quad (2)$$

Like any complex number, this can be separated into its real and non-real parts

$$f(x+hi) = \left(f(x) - \frac{1}{2!}h^2 f''(x) + \dots\right) + h\left(f'(x) - \frac{1}{3!}h^2 f'''(x) + \dots\right)i. \quad (3)$$

The leading term of the imaginary part of (3) is the first derivative. An approximation for the first derivative can be formed simply by taking the imaginary part of $f(x + hi)$ and dividing by the step size,

$$f'(x) = \frac{Imag[f(x + ih)]}{h} + O(h^2). \quad (4)$$

This is the complex-step derivative approximation of [14].

For double numbers, $E^2 = e^2 = 1$, so that (1) becomes

$$f(x + he) = \left(f(x) + \frac{1}{2!}h^2 f''(x) + \dots\right) + h\left(f'(x) + \frac{1}{3!}h^2 f'''(x) + \dots\right)e. \quad (5)$$

This again allows an approximation for the first derivative to be formed simply by taking the non-real part and dividing by the step size.

For dual numbers, $E^2 = \epsilon^2 = 0$, so that (1) simplifies to

$$f(x + h\epsilon) = f(x) + hf'(x)\epsilon. \quad (6)$$

The non-real parts of the expressions for the ordinary complex step and double number approaches, (3) and (5) respectively, contain the first derivative as well as higher order terms. As a result, these approaches do not compute the first derivative exactly due to the truncation error associated with neglecting the higher order terms. While these approaches are subject to truncation error, they are not plagued by the subtractive cancellation error that affects finite-difference formulas. This allows the step size to be chosen to be arbitrarily small in order to make the truncation error much less than machine precision, so that these approaches are effectively exact.

The dual number approach, (6), does not contain any higher order terms so it is free from both truncation error and subtractive cancellation error, yielding a method

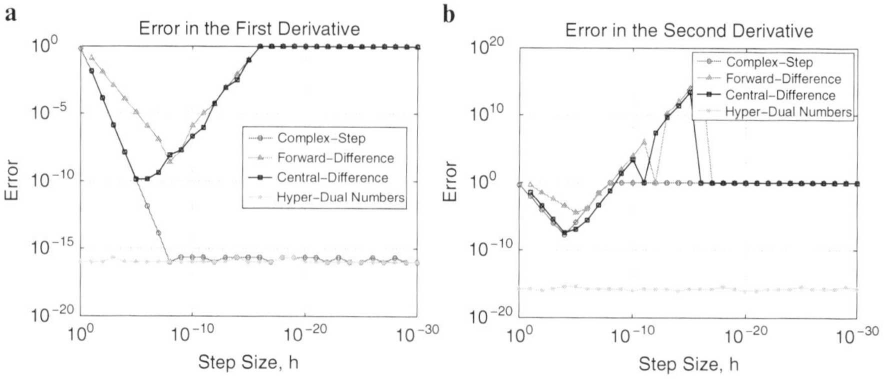

Fig. 1 The accuracies of several derivative calculation methods are presented as a function of step size for the function $f(x) = \frac{e^x}{\sqrt{sin(x)^3 + cos(x)^3}}$. (**a**) First-derivative accuracy. (**b**) Second-derivative accuracy

that produces exact first-derivative information regardless of the step size. These dual numbers function the same as the doublet class defined in [8] or the tapeless calculations in ADOL-C [20]. The only real difference is how they are derived.

Figure 1a shows the error of several first-derivative calculation methods as a function of step size, h. As the step size is initially decreased, the error decreases according to the order of the truncation error of the method. However, after a certain point, the error for the finite-difference approximations begins to grow, while the error for the complex-step approximation continues to decrease until it reaches (and remains at) machine zero. This illustrates the effect of subtractive cancellation error, which affects the finite-difference approximations but not the first-derivative complex-step approximation. The error of the hyper-dual number calculations, which are not subject to truncation error or subtractive cancellation error, is machine zero regardless of step size.

3 Second Derivative Calculations

Complications arise when attempting to use generalized complex numbers to compute second derivatives. The dual number Taylor series (6) does not contain a second-derivative term. One possibility is to define a recursive formulation such as a dual number with dual number components [17] that will produce second-derivative information that is free from truncation error and subtractive cancellation error. This approach is similar to the tangent-on-tangent approach of other automatic differentiation techniques and is identical in function to the hyper-dual number approach that will be developed later in this paper.

The ordinary complex number and double number Taylor series, (3) and (5), contain second-derivative terms but they are in the real part of the expression. Second-derivative information can be obtained using a formula such as

$$f''(x) = \frac{2\,(f(x) - Real[f(x+ih)])}{h^2} + O(h^2). \tag{7}$$

However, this formula involves a difference operation and is therefore subject to subtractive cancellation error, as shown in Fig. 1b. It is possible to create alternative approximations that use multiple, different complex-steps [11], but while these alternative formulations may offer improvements over (7), they still suffer from subtractive cancellation error.

In order to avoid subtractive cancellation error, the second-derivative term should be the leading term of a non-real part. Since the first-derivative term is already the leading term of a non-real part, this suggests that a number with multiple non-real parts is required.

One idea it to use higher dimensional extensions of generalized complex numbers. The best known such numbers are quaternions, which consist of one real part and three non-real parts with the properties $i^2 = j^2 = k^2 = -1$ and $ijk = -1$. The suitability of using quaternions to compute second derivatives can be determined by looking at the Taylor series with a generic step d,

$$f(x+d) = f(x) + df'(x) + \frac{1}{2!}d^2 f''(x) + \frac{1}{3!}d^3 f'''(x) + \dots . \tag{8}$$

For a quaternion step $d = h_1 i + h_2 j + 0k$, and the powers of d in (8) become

$$d^2 = -\left(h_1^2 + h_2^2\right), \tag{9}$$

$$d^3 = -\left(h_1^2 + h_2^2\right)(h_1 i + h_2 j + 0k), \tag{10}$$

$$d^4 = \left(h_1^2 + h_2^2\right)^2, \dots . \tag{11}$$

Ideally, the second-derivative term would be the leading term of the k part. Instead, the k part is always zero and the second-derivative term is only part of the real component of $f(x + h_1 i + h_2 j + 0k)$. An approximation formula for the second derivative can be formed,

$$f''(x) = \frac{2\,(f(x) - Real[f(x + h_1 i + h_2 j + 0k)])}{h_1^2 + h_2^2} + O(h_1^2 + h_2^2), \tag{12}$$

but this approximation is also subject to subtractive cancellation error. The problem with using quaternions is that multiplication is not commutative, $ij = k$ but $ji = -k$.

Instead, consider a number with three non-real components E_1, E_2, and $(E_1 E_2)$ where multiplication is commutative, i.e. $E_1 E_2 = E_2 E_1$. The values of d and its powers from the Taylor series in (8) become:

$$d = h_1 E_1 + h_2 E_2 + 0 E_1 E_2, \tag{13}$$

$$d^2 = h_1^2 E_1^2 + h_2^2 E_2^2 + 2 h_1 h_2 E_1 E_2, \tag{14}$$

$$d^3 = h_1^3 E_1^3 + 3 h_1 h_2^2 E_1 E_2^2 + 3 h_1^2 h_2 E_1^2 E_2 + h_2^3 E_2^3, \tag{15}$$

$$d^4 = h_1^4 E_1^4 + 6 h_1^2 h_2^2 E_1^2 E_2^2 + 4 h_1^3 h_2 E_1^3 E_2 + 4 h_1 h_2^3 E_1 E_2^3 + h_2^4 E_2^4. \tag{16}$$

The first term with a non-zero $(E_1 E_2)$ component is d^2, which means that the second derivative is the leading term of the $(E_1 E_2)$ component. This means that second-derivative approximations can be formed that are not subject to subtractive cancellation error. This is true as long as $E_1 E_2 \neq 0$, regardless of the particular values of E_1^2, E_2^2, and $(E_1 E_2)^2$. The only restriction is that multiplication must be commutative, i.e. $E_1 E_2 = E_2 E_1$.

It must be noted that the values of E_1^2, E_2^2, and $(E_1 E_2)^2$ are not completely independent. The requirement that $E_1 E_2 = E_2 E_1$ produces the constraint

$$(E_1 E_2)^2 = E_1 E_2 E_1 E_2 = E_1 E_1 E_2 E_2 = E_1^2 E_2^2. \tag{17}$$

Satisfying this constraint still leaves many possibilities regarding the definition of E_1 and E_2. One possibility is to use $E_1^2 = E_2^2 = -1$ which results in $(E_1 E_2)^2 = 1$. These are known as circular-fourcomplex [15] or multicomplex [18] numbers. Another approach is to constrain $E_1^2 = E_2^2 = (E_1 E_2)^2$. This leads to two possibilities, $E_1^2 = E_2^2 = (E_1 E_2)^2 = 0$ and $E_1^2 = E_2^2 = (E_1 E_2)^2 = 1$. All of these possibilities are free from subtractive cancellation error, and thus can be made effectively exact by choosing a small enough step size to drive the truncation error below machine precision.

By examining (13)–(16), the best derivative approximation is formed by taking $E_1^2 = E_2^2 = (E_1 E_2)^2 = 0$. To distinguish this situation from other definitions of E_1^2, E_2^2, and$(E_1 E_2)^2$, and to emphasize the connection to dual numbers which use $\epsilon^2 = 0$, in the notation for these hyper-dual numbers we will use ϵ instead of E, and we will require that $\epsilon_1^2 = \epsilon_2^2 = (\epsilon_1 \epsilon_2)^2 = 0$. Using this definition, d^3 and all higher powers of d are identically zero. The Taylor series truncates *exactly* at the second-derivative

$$f(x + h_1 \epsilon_1 + h_2 \epsilon_2 + 0 \epsilon_1 \epsilon_2) = f(x) + h_1 f'(x) \epsilon_1 + h_2 f'(x) \epsilon_2 + h_1 h_2 f''(x) \epsilon_1 \epsilon_2. \tag{18}$$

There is no truncation error since the higher order terms are zero by the definition of ϵ^2. The first and second derivatives are the leading terms of the non-real parts, meaning that these values can simply be found by examining the non-real parts of the number, and therefore the derivative calculations are not subject to subtractive cancellation errors. Therefore, the use of hyper-dual numbers results in first- and second-derivative calculations that are exact, regardless of the step size. The real part is also exactly the same as the function evaluated for a real number, x.

For functions of multiple variables, $f(\mathbf{x})$ where $\mathbf{x} \in \mathbb{R}^n$, first derivatives are found using

$$\frac{\partial f(\mathbf{x})}{\partial x_i} = \frac{\epsilon_1 \ part\left[f(\mathbf{x} + h_1\epsilon_1\mathbf{e}_i + h_2\epsilon_2\mathbf{e}_j + \mathbf{0}\epsilon_1\epsilon_2)\right]}{h_1} \quad (19)$$

and

$$\frac{\partial f(\mathbf{x})}{\partial x_j} = \frac{\epsilon_2 \ part\left[f(\mathbf{x} + h_1\epsilon_1\mathbf{e}_i + h_2\epsilon_2\mathbf{e}_j + \mathbf{0}\epsilon_1\epsilon_2)\right]}{h_2}. \quad (20)$$

Second derivatives are found using

$$\frac{\partial^2 f(\mathbf{x})}{\partial x_i \partial x_j} = \frac{\epsilon_1\epsilon_2 \ part\left[f(\mathbf{x} + h_1\epsilon_1\mathbf{e}_i + h_2\epsilon_2\mathbf{e}_j + \mathbf{0}\epsilon_1\epsilon_2)\right]}{h_1 h_2}, \quad (21)$$

where $\mathbf{e}_i$ and $\mathbf{e}_j$ are unit vectors composed of all zeros except the ith and jth components, respectively.

Figure 1b shows the error of the second-derivative calculation methods as a function of step size, $h = h_1 = h_2$. Again, as the step size is initially decreased, the error of the finite-difference and complex-step approximations behaves according to the order of the truncation error. However, for second derivatives both the finite-difference formulas and the complex-step approximation are subject to subtractive cancellation error, which begins to dominate the overall error as the step size is reduced below 10^{-4} or 10^{-5}. The error of the hyper-dual number calculations is machine zero for any step size.

4 Hyper-Dual Number Implementation and Results

The mathematics of hyper-dual numbers have been implemented as a class using operator overloading. A summary of the mathematical properties of hyper-dual numbers is given in [6]. Implementations are available for C++, CUDA, and MATLAB. A hyper-dual datatype and reduction operations are also available for use in MPI based codes.

4.1 Application to a Computational Fluid Dynamics Code

The hyper-dual number implementation described above has been used to produce first and second derivatives of quantities computed using a Computational Fluid Dynamics(CFD) code. The CFD code used is Joe [16], a parallel, unstructured, 3-D,

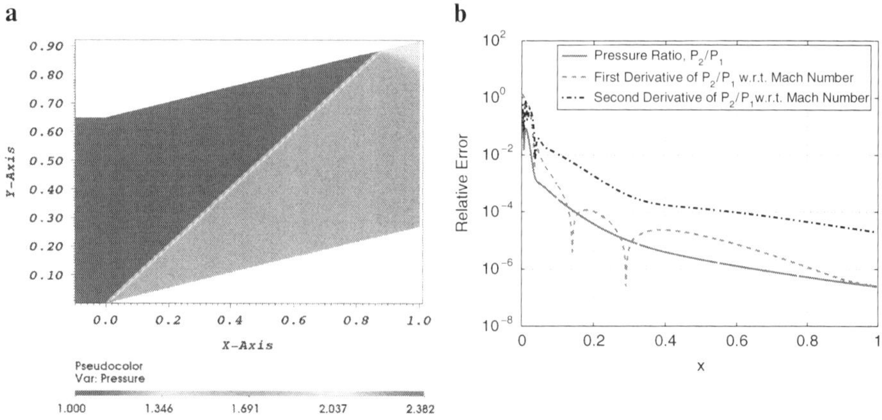

Fig. 2 Flow solution and error plots for inviscid Mach 2.0 flow over a 15° wedge. (**a**) Normalized pressure in the flow field. Flow is from *left* to *right*. (**b**) Relative error of the flow solution and derivative calculations along the wedge

unsteady Reynolds-averaged Navier-Stokes code developed under the sponsorship of the Department of Energy under the Predictive Science Academic Alliance Program. This code is written in C++, which enables straightforward conversion to hyper-dual numbers [6, 7].

To demonstrate the accuracy of the hyper-dual number calculations, the derivative calculations need to be compared to exact values. The example chosen for this demonstration is inviscid, supersonic flow over a wedge. Specifically, we look at derivatives of the pressure ratio, $\frac{P_2}{P_1}$, across the resulting oblique shock with respect to the incoming Mach number. Although no explicit equation exists relating the Mach number and pressure after an oblique shock to the incoming Mach number, the oblique shock relation does provide this relationship implicitly and analytic derivatives can be derived using an adjoint approach [6].

Figure 2a shows the CFD calculations for a Mach 2.0 flow over a 15° wedge. Table 1 compares the derivatives computed using hyper-dual numbers and several other approaches for both the exact oblique shock solution and the CFD solution. The values from the CFD solutions are averaged over the center 60% of the wedge, from $x = 0.2$ to $x = 0.8$. A detailed comparison between the hyper-dual CFD results and the exact values is given in Fig. 2b. This figure shows the error of the pressure ratio, and its first and second derivative, at every point along the wedge. The hyper-dual results are in good agreement with the exact solution over most of the wedge, and the error in the hyper-dual number derivative calculations follows the same trends as the error in the pressure ratio calculation. The accuracy of the derivative values is of roughly the same order as the underlying function evaluation. More precise derivative calculations will require a more accurate flow solution. In particular, derivative values that are numerically exact

Table 1 A comparison of first and second derivatives computed using analytic formulas, hyper-dual numbers, ADOL-C, and finite-difference approximations. The values given are for an inviscid Mach 2.0 flow over a 15° wedge

	$\frac{P_2}{P_1}$	$\frac{d\left(\frac{P_2}{P_1}\right)}{dM_1}$	$\frac{d^2\left(\frac{P_2}{P_1}\right)}{dM_1^2}$
Oblique shock relation, analytic adjoint	2.194653133607664	0.407667273032935	0.863026223964081
Oblique shock relation, hyper-dual	2.194653133607664	0.407667273033135	0.863026223952015
Oblique shock relation, ADOL-C	2.194653133607664	0.407667273033135	0.863026223952014
Joe, hyper-dual	2.194664703661337	0.407666379350755	0.862810467824695
Joe, ADOL-C tapeless	2.194664703661311	0.407666379350701	N/A
Joe, finite-difference	2.194664703661338	0.407665948554126	0.864641691578072

to machine precision are only possible if the flow solution itself is exact to machine precision.

4.2 *Computational Cost*

The use of generalized complex numbers for first-derivative calculations and hyper-dual numbers for second-derivative calculations only works in the forward mode. This means that one derivative calculation needs to be performed for every input variable, which can get very expensive if there are many input variables. In addition, a generalized complex or hyper-dual function evaluation is inherently more expensive than a real-valued function evaluation. Adding two hyper-dual numbers is equivalent to 4 additions of real numbers. Multiplying two hyper-dual numbers is equivalent to 9 real multiplications and 5 real additions. A hyper-dual function evaluation should therefore take between 4 and 14 times the runtime of a real function evaluation. In practice, a hyper-dual CFD run takes roughly ten times that of a real-valued CFD run. The computational cost can be reduced in some situations using techniques that are often used by other AD methods, such as not applying AD directly to an entire iterative procedure [2, 7, 8].

Each second-derivative calculation using hyper-dual numbers is independent. This means that when computing the Hessian there are redundant computations of the function value and first derivatives that could be eliminated by employing a vectorized approach in which the entire gradient and Hessian are propagated at once. However, by keeping the calculations independent, the memory required for the hyper-dual version of the code is limited to only four times that of the real number version. This can make differentiation more tractable for problems where memory is often a limiting factor, such as large CFD calculations. Communication in a parallel function evaluation using MPI also only increases by a factor of 4.

5 Conclusion

Although techniques for Automatic Differentiation are typically derived based on repeated application of the chain rule, other methods can be derived based on the use of generalized complex numbers. The mathematics of generalized complex numbers are such that first derivative information is computed and stored in the non-real part of the number. Methods based on these numbers, such as the complex-step derivative approximation, can be made effectively exact by choosing a small enough step size that the truncation error is much less than machine precision. Other methods, such as the use of dual numbers, are inherently free from truncation error.

When second-derivative information is desired, higher-dimensional extensions of generalized complex numbers can be used to create methods that can be made effectively exact as long as the numbers possess multiplicative commutivity. Hyper-dual numbers are one such number system, which have the additional benefit of being free from truncation error allowing for exact first- and second-derivative computations. Hyper-dual numbers have been implemented as a class, using operator overloading, in C++, CUDA, and MATLAB. This allows a code of arbitrary complexity to be converted to use hyper-dual numbers with relatively minor changes.

Acknowledgements This work was funded, in part, by the United States Department of Energy's Predictive Science Academic Alliance Program (PSAAP) at Stanford University.

References

1. http://www.autodiff.org
2. Bartholomew-Biggs, M.C.: Using forward accumulation for automatic differentiation of implicitly-defined functions. Computational Optimization and Applications **9**, 65–84 (1998)
3. Clifford, W.K.: Preliminary Sketch of Biquaternions. Proc. London Math. Soc. **s1-4**(1), 381–395 (1871). DOI 10.1112/plms/s1-4.1.381. URL http://plms.oxfordjournals.org/cgi/reprint/s1-4/1/381.pdf
4. Deakin, M.A.B.: Functions of a dual or duo variable. Mathematics Magazine **39**(4), 215–219 (1966). URL http://www.jstor.org/stable/2688085
5. Eastham, M.S.P.: 2968. on the definition of dual numbers. The Mathematical Gazette **45**(353), 232–233 (1961). URL http://www.jstor.org/stable/3612794
6. Fike, J.A., Alonso, J.J.: The development of hyper-dual numbers for exact second-derivative calculations. In: AIAA paper 2011-886, 49th AIAA Aerospace Sciences Meeting (2011)
7. Fike, J.A., Jongsma, S., Alonso, J.J., van der Weide, E.: Optimization with gradient and hessian information calculated using hyper-dual numbers. In: AIAA paper 2011-3807, 29th AIAA Applied Aerodynamics Conference (2011)
8. Griewank, A.: Evaluating Derivatives: Principles and Techniques of Algorithmic Differentiation. No. 19 in Frontiers in Appl. Math. SIAM, Philadelphia, PA (2000)
9. Hudson, R.W.H.T.: Review: Geometrie der Dynamen. Von E. Study. The Mathematical Gazette **3**(44), 15–16 (1904). URL http://www.jstor.org/stable/3602894
10. Kantor, I., Solodovnikov, A.: Hypercomplex Numbers: An Elementary Introduction to Algebras. Springer-Verlag, New York (1989)

11. Lai, K.L., Crassidis, J.L.: Extensions of the first and second complex-step derivative approximations. J. Comput. Appl. Math. **219**(1), 276–293 (2008). DOI http://dx.doi.org/10.1016/j.cam.2007.07.026
12. Leuck, H., Nagel, H.H.: Automatic differentiation facilitates OF-integration into steering-angle-based road vehicle tracking. IEEE Computer Society Conference on Computer Vision and Pattern Recognition **2**, 2360 (1999). DOI http://doi.ieeecomputersociety.org/10.1109/CVPR.1999.784659
13. Martins, J.R.R.A., Sturdza, P., Alonso, J.J.: The connection between the complex-step derivative approximation and algorithmic differentiation. In: AIAA paper 2001-0921, 39th Aerospace Sciences Meeting (2001)
14. Martins, J.R.R.A., Sturdza, P., Alonso, J.J.: The complex-step derivative approximation. ACM Transactions on Mathematical Software **29**(3), 245–262 (2003). DOI http://doi.acm.org/10.1145/838250.838251
15. Olariu, S.: Complex Numbers in N Dimensions, *North-Holland Mathematics Studies*, vol. 190. North-Holland, Amsterdam (2002)
16. Pecnik, R., Terrapon, V.E., Ham, F., Iaccarino, G.: Full system scramjet simulation. Annual Research Briefs, Center for Turbulence Research, Stanford University (2009)
17. Piponi, D.: Automatic differentiation, C++ templates, and photogrammetry. Journal of graphics, GPU, and game tools **9**(4), 41–55 (2004)
18. Price, G.B.: An Introduction to Multicomplex Spaces and Functions. Monographs and Textbooks in Pure and Applied Mathematics, New York (1991)
19. Rall, L.B.: Automatic Differentiation: Techniques and Applications, *Lecture Notes in Computer Science*, vol. 120. Springer, Berlin (1981). DOI 10.1007/3-540-10861-0
20. Walther, A., Griewank, A.: ADOL-C: A package for the automatic differentiation of algorithms written in C/C++. User's Manual Version 2.1.12-stable (2010)

Connections Between Power Series Methods and Automatic Differentiation

David C. Carothers, Stephen K. Lucas, G. Edgar Parker, Joseph D. Rudmin, James S. Sochacki, Roger J. Thelwell, Anthony Tongen, and Paul G. Warne

Abstract There is a large overlap in the work of the Automatic Differentiation community and those whose use Power Series Methods. Automatic Differentiation is predominately applied to problems involving differentiation, and Power series began as a tool in the ODE setting. Three examples are presented to highlight this overlap, and several interesting results are presented.

Keywords Higher-order taylor methods • Recursive power series • Projectively polynomial functions

1 Introduction

In 1964, Erwin Fehlberg (best known for the Runge-Kutta-Fehlberg method) wrote:

> Like interpolation methods and unlike Runge-Kutta methods, the power series method permits computation of the truncation error along with the actual integration. This is fundamental to an automatic step size control [and leads to a method that is] far more accurate than the Runge-Kutta-Nystrom method.
>
> ⋮
>
> [Though] differential equations of the [appropriate form] ... are generally not encountered in practice ... a given system can in many cases be transformed into a system of [appropriate form] through the introduction of suitable auxiliary functions, thus allowing solution by power series expansions [3].

Fehlberg, it appears, did not continue work on the approach that he believed to be superior to the methods of the day. In this manuscript (prepared as a NASA technical

D.C. Carothers · S.K. Lucas · G.E. Parker · J.D. Rudmin · J.S. Sochacki · R.J. Thelwell (✉) · A. Tongen · P.G. Warne
James Madison University, Harrisonburg, VA 22807, USA
e-mail: carothdc@jmu.edu; lucassk@jmu.edu; parkerge@jmu.edu; rudminjd@jmu.edu; sochacjs@jmu.edu; thelwerj@jmu.edu; tongenal@jmu.edu; warnepg@jmu.edu

S. Forth et al. (eds.), *Recent Advances in Algorithmic Differentiation*, Lecture Notes in Computational Science and Engineering 87, DOI 10.1007/978-3-642-30023-3_16,

report) Fehlberg was able to efficiently and accurately compute approximations for two important problems: the restricted three-body problem, and the motion of an electron in a field of a magnetic dipole. Introducing auxiliary functions, he recast these problems as a system of first order equations expressed solely as polynomials in existing variables – what we now call *polynomial form*. Although this work was noticed by some in the NASA community [10], Fehlberg's observations remain largely unexploited.

The computation of derivatives lies at the heart of many Automatic Differentiation (AD) routines. AD techniques allow one to generate information about the intrinsic of interest solely in context of other intrinsics. When applied to functions, AD permits efficient evaluation of the derivatives of a given function up to arbitrarily high order, making it ideally suited for higher order Taylor based methods of solution to ODEs. The recursive computation that Fehlberg used is a natural outcome of recursively differentiating polynomial expressions of power series. The trick, then, is to reduce a given problem to one of polynomial form. How does one do so? One answer comes from the so called translator programs from AD.

A higher order Taylor method code is problem specific, requiring the problem to be reduced to one of known recursive relationships. First accomplished by hand, this difficulty was overcome with the advent of automatic translator programs. These programs can parse a given system of ODEs into a form that allows libraries of general recursions to be applied. Nice examples of AD flavored ODE tools are `ATOMFT`, written by Chang and Corliss in 1994. and `AD01`, written by Pryce and Reid in 1990. They automatically parse the original ODE expression using functional dependencies, and then efficiently compute a numeric solution via a recursive recovery of Taylor coefficients to arbitrarily high order. The method has also been applied to differential-algebraic equations (DAEs), with great success. The machinery of generating the polynomial form is distinct from the recursive coefficient recovery, and both `ATOMFT` and `AD01` are wonderful blends of the techniques of AD applied to the ring of power series.

In 1992 Parker and Sochacki discovered that the nth Picard iterate, when applied to non-autonomous polynomial ODEs with initial value at $t = 0$, generates a polynomial whose first n terms match the Maclaurin polynomial. They then looked at how one could use Picard Iteration to classify IVODEs and what was special about the polynomials generated by Picard Iteration. This led Parker and Sochacki to determine which functions can be posed as a solution (to one component of a system) of non-autonomous polynomial ODEs with initial value at $t = 0$, and called this class of functions *projectively polynomial* [8]. Although the computation of successive terms of the Maclaurin polynomial was expensive, the framework allowed theoretic machinery to be applied to ordinary, partial and integral differential equations [5, 6, 9]. In Carothers et al. [2] realized that the Picard iteration in the projectively polynomial system was equivalent to a power series method, allowing an efficient recursive computation of the coefficients of the series. For the remainder of this paper, we will refer to this method as the Power Series Method (PSM). PSM includes power series, Picard Iteration and polynomial projection.

Carothers et al. realized that projectively polynomial functions had many special symbolic and numerical computational properties that were amenable to obtaining qualitative and quantitative results for the polynomial IVODEs and the solution space. Gofen [4] and others also discovered some of these phenomena by looking at Cauchy products and polynomial properties instead of Picard Iteration. Many researchers were able to show large classes of functions were projectively polynomial, that one could uncouple polynomial ODEs and one could do interval analysis with these methods. Carothers et al. generated an a-priori error bound for non-autonomous polynomial ODEs with initial value at $t = 0$. The PSM collaboration at JMU has also shown: the equivalence between power series and non-autonomous polynomial ODEs with initial value at $t = 0$ and Picard Iteration, many of the topological properties of the solutions to these problems, and the structure of the space of polynomial functions. These are summarized in Sochacki [11].

The AD and PSM methods produce equivalent results in the numerical solution of IVODEs. Polynomial form is essential for the simple recursive recovery of series coefficients used by the two groups. Their different history colors the types of problems explored, however. Differentiation forms the backbone of AD, and so problems which involve repeated differentiation are obvious candidates for AD research, with ODE, sensitivity, and root-finding as obvious examples. ODEs lie at the core of the PSM, and so the focus is to re-interpret problems as IVODEs. We present three examples below highlighting this concept.

2 Applying PSM

We believe that many problems can be converted into IVODEs of polynomial form, as demonstrated in the following examples.

2.1 *Example 1: PSM and AD Applied to an IVODE*

Consider the IVODE

$$y' = Ky^{\alpha}, \quad y(x_0 = 0) = y_0, \tag{1}$$

for some complex valued α. This problem highlights the properties of PSM and AD because it has the closed form solution

$$y(x) = \left(\left(Kx - K\alpha x + {y_0}^{1-\alpha}\right)^{(\alpha-1)^{-1}}\right)^{-1}, \tag{2}$$

and because it can be posed in several equivalent polynomial forms. This is not a problem that most would consider immediately amenable to standard power series methods. A simple recursive relationship generates the Taylor coefficients for y^α given by

$$a_n = \frac{1}{n y_0} \sum_{j=0}^{n-1} (n\alpha - j\,(\alpha + 1))\, y_{n-j} a_j, \tag{3}$$

where $a(x) = \sum_{j=0}^{\infty} a_j (x - x_0)^j$, a_n represents the *nth* degree Taylor coefficient of y^α. Similarly, y_n represents the *nth* degree Taylor coefficient of y. Then a series solution of (1) is computed.

Consider the following change of variables: $x_1 = y$, $x_2 = y^\alpha$, and $x_3 = y^{-1}$. Fehlberg called these auxiliary variables. Differentiation of these auxiliary variables provides the system,

$$\begin{aligned} x_1' &= -x_2 & x_1(0) &= y_0, \\ x_2' &= -\alpha x_2^2\, x_3 & x_2(0) &= y_0^\alpha, \\ x_3' &= x_2\, x_3^2 & x_3(0) &= y_0^{-1}. \end{aligned} \tag{4}$$

The solution x_1 to this system is equal to y, the solution to the original system. Note that this augmented system (4) is polynomial, and as such can be easily solved by the PSM with its guaranteed error bound. Computation of the PSM solution requires only additions and multiplications.

However, a better change of variables is obtained by letting $w = y^{\alpha \;\, 1}$. Then (1) can be written as the following system of differential equations

$$\begin{aligned} y' &= Kyw, & y(0) &= y_0 \\ w' &= (\alpha - 1)Kw^2, & w(0) &= y_0^{\alpha-1}, \end{aligned} \tag{5}$$

because the right hand side is quadratic in the variables as opposed to cubic in (4), and subsequently requires fewer series multiplications.

Figure 1 contrasts, on a $\log_{10}$ scale, the absolute error when approximate solutions to

$$y' = y^{\frac{e}{2} + \frac{i}{\pi}} \quad y(0) = 1$$

are computed by the standard Runge-Kutta order 4 method (RK4) for (1), (4), and (5), and automatic differentiation using (3) and PSM using (5) to 4th order. In this example, note that the PSM system (5) recovers the AD recursion (3).

Figure 1 demonstrates that the fixed step solution with automatic differentiation and the power series solution of (5) give the same solution. Of course, we have fixed these methods at fourth order in order to fairly compare with RK4; however, it is straightforward to keep more terms and solve this problem to machine accuracy, as Fehlberg points out. It also demonstrates that by rewriting the equations in polynomial form and solving with a fixed step RK4, the solution to the system of

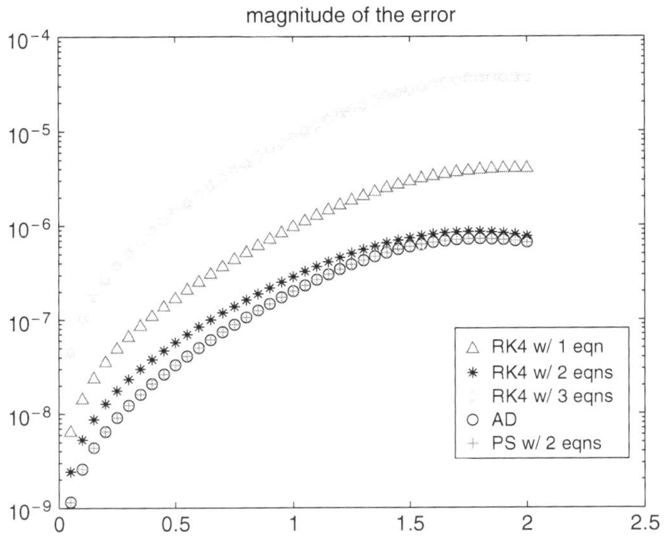

Fig. 1 Solving differential equations (1), (4), and (5) using a fixed step Runge-Kutta on [0,2] with $h = 0.05$ and $y_0 = 1$, $K = 1$, $\alpha = e/2 + i/\pi$

two equations (5) is more accurate than the straightforward implementation (1). Interestingly, not all systems are equal – the system of two equations (5) is more accurate than the system of three equations (4), because the right hand side of (5) is quadratic in the variables on the right hand side.

2.2 *Example 2: Root-Finding*

Newton's Method is a prime example of the efficacy of AD. Consider

$$f(x) = e^{-\sqrt{x}} \sin(x \ln(1 + x^2)), \tag{6}$$

and computing the iteration $x_{i+1} = x_i - f(x_i)/f'(x_i)$, as in the example presented by Neidinger [7] to show the power of AD. The machinery of AD makes the calculation of $f(x_i)$ and $f'(x_i)$ simple, and Neidinger used object oriented programming and overloaded function calls to evaluate both the function and their derivative at a given value.

We take a different approach. We pose the determination of roots as a non-autonomous polynomial IVODE at 0. If one wants to determine the roots of a sufficiently nice function $f : \mathbb{R}^n \to \mathbb{R}^n$ one can define $g : \mathbb{R}^n \to \mathbb{R}$ by

$$g(x) = \frac{1}{2}\langle f(x), f(x)\rangle$$

where $\langle \cdot, \cdot \rangle$ is the standard inner product. Since $g(x)$ is non-negative and $g(x) = 0$ if and only if $f(x) = 0$, we will determine the conditions that make

$$\frac{d}{dt} g(x) < 0.$$

This condition is necessary if one wants to determine $x(t)$ so that $x \to z$, a zero of f (or g). We have

$$\frac{d}{dt} g(x) = \langle \frac{d}{dt} f(x), f(x) \rangle \tag{7}$$

$$= \langle Df(x)x'(t), f(x) \rangle \tag{8}$$

$$= \langle x'(t), Df(x)^T f(x) \rangle, \tag{9}$$

where $Df(x)$ is the Jacobian of f and $Df(x)^T$ is the transpose of $Df(x)$.

If, guided by (8), we let

$$x'(t) = -(Df(x))^{-1} f(x), \tag{10}$$

then certainly $\frac{d}{dt} g(x) < 0$. If we now approximate the solution to this ODE using forward Euler with $h = 1$ we have

$$x_{t+1} = x_t - (Df(x_t))^{-1} f(x_t), \tag{11}$$

which is Newton's method. In (10), we let $x_2 = (Df(x))^{-1}$, and obtain

$$x'(t) = -\, x_2 f(x) \tag{12}$$

$$x_2'(t) = x_2^3 f(x) f''(x). \tag{13}$$

Adding initial conditions $x(0)$ and $x_2(0)$ gives us a non-autonomous IVODE. If f is a polynomial, we can apply PSM directly. If f is not polynomial, we make further variable substitutions to make the ODE polynomial.

Now consider (9). If we choose

$$x'(t) = -(Df(x))^T f(x) \tag{14}$$

then certainly $\frac{d}{dt} g(x) < 0$. Once again, approximating $x'(t)$ with forward Euler we have

$$x_{t+\Delta t} = x_t - \Delta t (Df(x_t))^T f(x_t), \tag{15}$$

which is the method of Steepest Descent.

We note that both (10) and (14) can be approximated using PSM or AD methods to arbitrary order (h^k). These ODEs could also be initially regularized in order for

PSM or AD to converge faster to the root of f. In the case of the Newton form, we would then solve

$$x'(t) = -\alpha(t)(f'(x))^{-1} f(x),$$

where $\alpha(t)$ could be adaptive. Of course, this approach applies easily to higher dimensions and the method of Steepest Descent in a straightforward manner. We now have many options for developing numerical methods to approximate the zeroes of a function f.

In Neidinger's paper [7] he chose the initial condition 5.0 and produced the approximation 4.8871. We used the IVODE (10) and performed a polynomial projection on f to obtain a non-autonomous polynomial IVODE. Using the same initial condition with a step size of 0.0625 and 32nd degree Maclaurin polynomial, our results agree with Neidinger's results (personal communication) to machine epsilon.

This shows that both AD and PSM can be used to efficiently calculate the roots of functions, determine ways (i.e. regularizations) to correct the pitfalls of Newton's Methods, improve the convergence properties of Newton type methods and develop error bounds for these methods.

2.3 Example 3: The Maclaurin Polynomial for an Inverse Function

In 2000, Apostol [1] developed a method for obtaining the power series of the inverse of a polynomial by exploiting the Inverse Function Theorem. To turn this problem into a non-autonomous polynomial ODE with initial value at $t = 0$ we differentiate $f(f^{-1}(t)) = t$ to obtain $f'(x_1)x_1' = 1$, where we let $x_1 = f^{-1}(t)$. We now let $x_2 = [f'(x_1)]^{-1}$ and $x_3 = x_2^2$ to obtain

$$x_1' = \frac{1}{f'(x_1)} = [f'(x_1)]^{-1} = x_2 \tag{16}$$

$$x_2' = -x_2^2 f''(x_1)x_1' = -x_3 f''(x_1)x_1'. \tag{17}$$

$$x_3' = 2x_2x_2' \tag{18}$$

Suppose f is a polynomial. We now outline how to get the power series for its inverse. Let $f(t) = \sum_{i=0}^{n+2} a_i t^i = a_0 + a_1 t + \ldots + a_{n+2}t^{n+2}$. Using the above polynomial ODE we now have

$$\begin{aligned} x_2' &= -x_2^2 f''(x_1)x_1 = -x_3 f''(x_1)x_1' = -x_3 p_n x_1' = -x_3 x_2 p_n \\ x_3' &= 2x_2x_2' \\ p_n' &= f'''(x_1)x_1' = p_{n-1}x_2 \end{aligned} \tag{19}$$

$$p'_{n-1} = f^{(iv)}(x_1)x'_1 = p_{n-2}x_2$$

$$\vdots$$

$$p'_1 = f^{(n+2)}(x_1)x'_1 = (n+2)!a_{n+2}x_2,$$

where $p_n = f''(x_1)$. We have ignored the x'_1 equation since $x'_1 = x_2$. Now we use Cauchy products and find

$$x_2 = \sum_{i=0}^{K} x_{2i}t^i, \qquad x_3 = \sum_{i=0}^{K} x_{3i}t^i \tag{20}$$

$$p_{n-k} = \sum_{i=0}^{K} p_{(n-k),i}t^i, \qquad \text{for } k = 0, \ldots, n-1. \tag{21}$$

Substituting these power series into (19) gives us a simple algorithm for generating the power series for the derivative of the inverse of a polynomial. One integration gives the power series for the inverse.

Of course, the auxiliary variables can be chosen in many ways to make an IVODE polynomial. It is usually straightforward to 'parse' from the inside of composition of functions outward. However, it is an open question as to what is the most efficient algorithm for making an IVODE polynomial.

These three examples are meant to show the similarities and differences of PSM and AD and how PSM can be applied to many problems of applied and computational mathematics by posing them as non-autonomous polynomial IVODEs. These examples have also raised questions of interest. For example; (1) Is it more efficient to pose the problem as a non-autonomous polynomial IVODE or solve it in the existing form using AD? (2) Does the structure and topology of non-autonomous polynomial IVODEs lead to answers in applied and computational mathematics? (3) What are the symbolic and numerical differences and similarities between PSM and AD? (4) How can the PSM, AD and polynomial communities come together to answer these questions?

3 PSM Theory and AD

Picard Iteration and polynomial projection for IVODEs have led to an interesting space of functions and some interesting results for polynomial ODEs. We present the basic definitions and important theorems arising from Picard Iteration and polynomial projections. Gofen and others have obtained some of these results through the properties of polynomials and power series.

We begin with the question of which ODEs may be transformed into an autonomous polynomial system as in Example 1; that is, a system of the form:

$$\mathbf{x}'(t) = \mathbf{h}(\mathbf{x}(t)), \quad \mathbf{x}(a) = \mathbf{b}, \tag{22}$$

noting that a non-autonomous system $\mathbf{y}'(t) = \mathbf{h}(\mathbf{y}(t), t))$ may be recast by augmenting the system with an additional variable whose derivative is 1.

To this end the class of *projectively polynomial* functions consists of all real analytic functions which may be expressed as a component of the solution to (22) with $\mathbf{h}$ a polynomial. The following properties of this class of functions are summarized in [2] and elsewhere. It may be shown that any polynomial system, through the introduction of additional variables, may be recast as a polynomial system of degree at most two.

The projectively polynomial functions include the so-called elementary functions. The class of projectively polynomial functions is closed under addition, multiplication, and function composition. A local inverse of a projectively polynomial function f is also projectively polynomial (when $f'(a) \neq 0$), as is $\frac{1}{f}$. The following theorem illustrates the wide range of ODEs that may be recast as polynomial systems.

Theorem 1. *(Carothers et al. [2]) Suppose that f is projectively polynomial. If y is a solution to*

$$y'(t) = f(y(t)); \quad y(a) = b \tag{23}$$

then y is also projectively polynomial.

As an interesting consequence, it is possible for a very wide range of systems of ODEs to provide an algorithm by which the system may be "de-coupled" by applying standard Gröbner basis techniques.

Theorem 2. *(Carothers et al. [2]) A function u is the solution to an arbitrary component of a polynomial system of differential equations if and only if for some n there is a polynomial Q in $n + 1$ variables so that $Q(u, u', \cdots, u^{(n)}) = 0$.*

That is, for any component x_i of the polynomial system $\mathbf{x}' = \mathbf{h}(\mathbf{x})$ the component x_i may be isolated in a single equation involving x_i and its derivatives. This implies, for example, that the motion of one of the two masses in a double pendulum may be described completely **without** reference to the second mass.

Of very special practical and theoretical interest is the existence of explicit a-priori error bounds for PSM solutions to ODEs of this type which depends only on immediate observable quantities of the polynomial system.

We consider again a polynomial system (at $a = 0$) of the form $\mathbf{x}'(t) = \mathbf{h}(\mathbf{x}(t))$, $\mathbf{x}(0) = \mathbf{b}$. In the following $K = (m-1)c^{m-1}$, where m is the degree of $\mathbf{h}$ (the largest degree of any single term on the right hand side of the system), c the larger of unity and the magnitude of $\mathbf{b}$ (the largest of the absolute value of the elements of the initial condition), and $\sum_{k=0}^{n} \mathbf{x}_k t^k$ is the nth degree Taylor approximation of $\mathbf{x}(t)$. As an example we have the following error bound with $m \geq 2$:

Theorem 3. *(Warne et al. [12])*

$$\left\| \mathbf{x}(t) - \sum_{k=0}^{n} \mathbf{x}_k t^k \right\|_\infty \leq \frac{\|\mathbf{b}\|_\infty |Kt|^{n+1}}{1 - |Mt|} \quad \textit{for} \quad m \geq 2 \tag{24}$$

for any $n \in \mathbb{N}$, with $|t| < \frac{1}{K}$, where M is the larger of unity and the maximum row sum of the absolute values of the constant coefficients of the system.

It can be shown that no universally finer error bound exists for all polynomial systems than one that is stated in a tighter but slightly more involved version of this theorem.

4 Conclusion

Clearly, there is a large overlap in the work of the AD community and the PSM group. However, while AD is predominately applied to problems involving differentiation, PSM began as a tool in the ODE setting. There are numerous benefits to sharing the tool-sets of recursive computation of Taylor coefficients between these two communities. Some of these are: (1) There are methods that easily compute *arbitrarily high order Taylor coefficients*, (2) The tools can *solve highly nonlinear IVODEs*, and *automatically solve stiff problems*, (3) There are numerical and symbolic computational tools that lead to semi-analytic methods and (4) Evaluation of functions can be *interpolation free* to machine capability (error and calculation).

Acknowledgements The authors would like to thank the reviewers and the editor for improving this paper with their comments and suggestions.

References

1. Apostol, T.: Calculating higher derivatives of inverses. Amer. Math. Monthly **107**(8), 738–741 (2000)
2. Carothers, D., Parker, G.E., Sochacki, J., Warne, P.G.: Some properties of solutions to polynomial systems of differential equations. Electronic Journal of Differential Equations **2005**, 1–18 (2005)
3. Fehlberg, E.: Numerical integration of differential equations by power series expansions, illustrated by physical examples. Tech. Rep. NASA-TN-D-2356, NASA (1964)
4. Gofen, A.: The ordinary differential equations and automatic differentiation unified. Complex Variables and Elliptic Equations **54**, 825–854 (2009)
5. Liu, J., Parker, G.E., Sochacki, J., Knutsen, A.: Approximation methods for integrodifferential equations. Proceedings of the International Conference on Dynamical Systems and Applications, III pp. 383–390 (2001)

6. Liu, J., Sochacki, J., Dostert, P.: Chapter 16: Singular perturbations and approximations for integrodifferential equations. In: S. Aizicovici, N.H. Pavel (eds.) Differential Equations and Control Theory. CRC press (2001). ISBN: 978-0-8247-0681-4
7. Neidinger, R.: Introduction to automatic differentiation and matlab object-oriented programming. SIAM Review **52**(3), 545–563 (2010)
8. Parker, G.E., Sochacki, J.: Implementing the Picard iteration. Neural, Parallel Sci. Comput. **4**(1), 97–112 (1996)
9. Parker, G.E., Sochacki, J.: A Picard-Maclaurin theorem for initial value PDE's. Abstract Analysis and its Applications **5**, 47–63 (2000)
10. Perlin, I., Reed, C.: The application of a numerical integration procedure developed by Erwin Fehlberg to the restricted problem of three bodies. Tech. Rep. NAS8-11129, NASA (1964)
11. Sochacki, J.: Polynomial ordinary differential equations – examples, solutions, properties. Neural Parallel & Scientific Computations **18**(3-4), 441–450 (2010)
12. Warne, P.G., Warne, D.P., Sochacki, J.S., Parker, G.E., Carothers, D.C.: Explicit a-priori error bounds and adaptive error control for approximation of nonlinear initial value differential systems. Comput. Math. Appl. **52**(12), 1695–1710 (2006). DOI http://dx.doi.org/10.1016/j.camwa.2005.12.004

Hierarchical Algorithmic Differentiation A Case Study

Johannes Lotz, Uwe Naumann, and Jörn Ungermann

Abstract This case study in Algorithmic Differentiation (AD) discusses the semi-automatic generation of an adjoint simulation code in the context of an inverse atmospheric remote sensing problem. In-depth structural and performance analyses allow for the run time factor between the adjoint generated by overloading in C++ and the original forward simulation to be reduced to 3.5. The dense Jacobian matrix of the underlying problem is computed at the same cost. This is achieved by a hierarchical AD using adjoint mode locally for preaccumulation and by exploiting interface contraction. For the given application this approach yields a speed-up over black-box tangent-linear and adjoint mode of more than 170. Furthermore, the memory consumption is reduced by a factor of 1,000 compared to applying black-box adjoint mode.

Keywords Atmospheric remote sensing • Inverse problems • Algorithmic differentiation

1 Introduction

Algorithmic differentiation (AD) [4, 9] is the preferred method for computing derivatives of mathematical functions $\mathbf{y} = F(\mathbf{x})$, where $\mathbf{x} \in \mathbb{R}^n$ (inputs) and $\mathbf{y} \in \mathbb{R}^m$ (outputs), that are implemented as computer programs. AD offers two fundamental

J. Lotz (✉) · U. Naumann
Lehr- und Forschungsgebiet Informatik 12: Software and Tools for Computational Engineering, RWTH Aachen University, 52056 Aachen, Germany
e-mail: lotz@stce.rwth-aachen.de; naumann@stce.rwth-aachen.de

J. Ungermann
Institute of Energy and Climate Research – Stratosphere (IEK-7), Research Center Jülich GmbH, 52425 Jülich, Germany
e-mail: j.ungermann@fz-juelich.de

S. Forth et al. (eds.), *Recent Advances in Algorithmic Differentiation*, Lecture Notes in Computational Science and Engineering 87, DOI 10.1007/978-3-642-30023-3_17,

modes, the tangent-linear or forward mode and the adjoint or reverse mode. This paper discusses an adjoint version of an atmospheric simulation code. An automatically (black-box) generated adjoint code turns out to be infeasible due to violated memory constraints. A semi-automatic (targeted; hierarchical) application of AD is crucial to achieve the desired level of efficiency and robustness. It is based on the structural analysis of the underlying numerical simulation code (also referred to as the *forward simulation model*) that requires a close collaboration between domain (Computational Physics in our case) and AD experts. The resulting adjoint code contributes to a better understanding of the physical phenomenon through computer-based experiments. At the same time it allows the AD tool to be further developed driven by a highly relevant real-world application.

The Institute of Energy and Climate Research – Stratosphere at Research Center Jülich has been developing the *Juelich Rapid Spectral Simulation Code Version 2* (JURASSIC2) [16] for deriving atmospheric constituents from measurements taken remotely from aircrafts or satellites. This ill-posed inverse problem is highly sensitive with respect to errors in the measurements. It is commonly solved by gradient based methods [11] applied to an appropriately regularized problem formulation. An efficient adjoint version of the forward simulation model is required. The exploitation of problem structure turns out to be essential. This paper is based on prior work reported in [15]. Its focus is on the structural and performance analyses of JURASSIC2 in the context of adjoint mode AD.

This paper proceeds with a brief description of the physical problem and the corresponding mathematical model in Sect. 2. It follow structural (in Sect. 3) and formal run time performance (in Sect. 4) analyses of JURASSIC2. The numerical results reported in Sect. 5 demonstrate impressive speed-up rates in good correlation with the theoretical results.

2 Problem Description

JURASSIC2 is a retrieval processor used in the field of atmospheric remote sensing. It aims to derive atmospheric variables such as temperature or trace gas volume mixing ratios from emitted or scattered radiation. These can then be used to improve long-term climate models and short-term weather-forecast. JURASSIC2 has been optimized to evaluate measurements in the infrared part of the spectrum made by airborne or satellite-borne limb-sounders, which receive radiation tangent to the surface of the earth.

The derivation of such quantities from infrared measurements is an inverse problem. Initially, only a forward simulation model is available that maps a state of atmospheric quantities onto simulated measurements. The *retrieval process* numerically inverts this process and – depending on the method – requires first and/or second derivatives of the forward simulation model. The time required for the retrieval is often dominated by the evaluation of these derivatives.

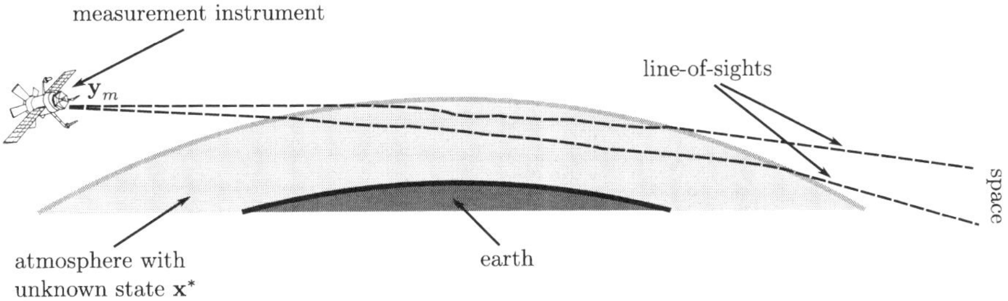

Fig. 1 Limb sounding configuration. $\mathbf{y}_m$ are the measurements, $\mathbf{x}^*$ the unknown atmospheric state and the *line-of-sights* illustrate the ray tracing approach

Given a forward simulation model $\mathbf{y} = F(\mathbf{x})$, $F : \mathbb{R}^n \to \mathbb{R}^m$ with $\mathbf{x} \in \mathbb{R}^n$ denoting the atmospheric state and $\mathbf{y} \in \mathbb{R}^m$ denoting the simulated measurements, the inverse problem consists of identifying a particular atmospheric state $\mathbf{x}^*$ that generates a set of measurements $\mathbf{y}_m$ using the forward simulation model. An exemplary measurement configuration is shown in Fig. 1. The model first performs a ray tracing step to determine the line-of-sight of the measurement and then numerically integrates radiation emitted by the excited molecules along this line-of-sight. As such an idealized calculation does not accommodate for the field-of-view of the instrument, multiple such line-of-sights are cast for each measurement and combined in a weighted manner. Hence, the forward simulation model has a structure that yields $F(\mathbf{x}) = H(G(\mathbf{x}))$, where $G : \mathbb{R}^n \to \mathbb{R}^{m_G}$ integrates along the line-of-sights and $H : \mathbb{R}^{m_G} \to \mathbb{R}^m$ maps those linearly onto the simulated measurements. As this is an ill-posed inverse problem, a straightforward inversion would be too sensitive to measurement errors. The problem is therefore approximated by a well-posed one by including a regularizing component. A solution to this problem is given by the minimum of the cost function

$$J(\mathbf{x}) = \overbrace{(F(\mathbf{x}) - \mathbf{y}_m)^T S_\epsilon^{-1} (F(\mathbf{x}) - \mathbf{y}_m)}^{(I)} + \overbrace{(\mathbf{x} - \mathbf{x}_\alpha)^T S_a^{-1} (\mathbf{x} - \mathbf{x}_\alpha)}^{(II)} .$$

This quadratic form aims to fit the simulated measurements to the actual measurements in (I) under the side condition (II) that allows for the inclusion of a priori information, such as typical distributions of trace gases taken from given climatological data. The covariance matrix S_ϵ models available knowledge of the measurement error and S_a^{-1} can, e.g., either be the inverse of a covariance matrix coming from climatological data or a Tikhonov regularization matrix [14].

For this class of problems, typically Quasi-Newton methods such as Gauss-Newton are employed as minimizers [10]. JURASSIC2 implements a truncated Quasi-Newton method that yields the following iterative algorithm [16]:

$$\mathbf{x}_{i+1} = \mathbf{x}_i - \underbrace{\left(S_a^{-1} + \nabla F(\mathbf{x}_i)^T S_\epsilon^{-1} \nabla F(\mathbf{x}_i)\right)^{-1}}_{\approx \frac{1}{2} \nabla^2 J(\mathbf{x}_i)^{-1}} \cdot \underbrace{\left(S_a^{-1}(\mathbf{x}_i - \mathbf{x}_\mathbf{a}) + \nabla F(\mathbf{x}_i)^T S_\epsilon^{-1} (F(\mathbf{x}_i) - \mathbf{y})\right)}_{= \frac{1}{2} \nabla J(\mathbf{x}_i)} .$$

Matrix-matrix multiplications are avoided by a dedicated implementation of a conjugate gradient (CG) algorithm to approximately solve the involved linear equation system. For these problems CG typically requires hundreds to thousands of iterations. Therefore it is important to explicitly calculate and maintain the Jacobian matrix $\nabla F \in \mathbb{R}^{m \times n}$ of F. The focus of this paper is on the computation of this Jacobian, where $m \approx n$. As the Jacobian is possibly dense, black-box AD suggests tangent-linear mode in this case. However, an in-depth structural analysis of JURASSIC2 yields significant run time performance gains if adjoint mode is used as shown in the following.

3 Structure of JURASSIC2

Figure 2 summarizes the structure of JURASSIC2. It can be written as $F(\mathbf{x}) = H(G(\mathbf{x}))$. A similar structure can, for example, also be found in Monte Carlo methods. $G(\mathbf{x})$ is decomposed into $q-1$ multivariate vector functions $G_i : \mathbb{R}^n \to \mathbb{R}^{m_i}$ taking the global $\mathbf{x} \in \mathbb{R}^n$ as input. The G_i do therefore not have a limited support. The output of $G : \mathbb{R}^n \to \mathbb{R}^{m_G}$ is the input of the linear function $H : \mathbb{R}^{m_G} \to \mathbb{R}^m$, with $m_G = \sum_{i=1}^{q-1} m_i$, $\nabla_G H$ dense. The linearity will be exploited later but is in general not necessary for the gain in efficiency. This structure yields a Jacobian $\nabla_{\mathbf{x}} F$, which is also dense and the underlying function is neither partially value separable nor partially argument separable. In this specific case, G_i is the same function for all i with different passive arguments and different quantitative dependency on the input $\mathbf{x}$. Nevertheless, for the gain in performance, this is irrelevant. Similar discussions of the exploitation of structure (different from the one found in JURASSIC2) within simulation code can be found, for example, in [6, 13] for a scalar output, and in [12] for the Jacobian assembly with a compact stencil yielding a limited support of each output as well as partial separability.

The $G_j(\mathbf{x})$ are mutually independent, i. e., do not share any intermediate program variables. For the output sets $\mathbf{Y}^j = \{i : v_i \text{ output of } G^j\}$ with v_i the i-th intermediate variable we get $\mathbf{Y}^j \cap \mathbf{Y}^k = \emptyset$ for $j \neq k$. Additionally $\cup_{j=1}^{q-1} \mathbf{Y}^j$ are the inputs for the function H. The mutual independence of the G_i is the key prerequisite for the chosen preaccumulation method that computes the local Jacobians $\nabla G_i(\mathbf{x})$ in plain adjoint mode. The Jacobian $\nabla_G H$ of the linear function H is readily available within JURASSIC2. In case H were nonlinear, its Jacobian could also be computed via AD.

Preaccumulation yields the local Jacobian matrices $\nabla_{\mathbf{x}} G_i(\mathbf{x})$. This step is followed by the evaluation of the matrix product $\nabla_{\mathbf{x}} F = \nabla_G H \cdot \nabla_{\mathbf{x}} G$. The preaccumulation of the $\nabla_{\mathbf{x}} G_i(\mathbf{x})$ is done in plain adjoint mode, as the number of local outputs m_i is typically very small compared to the number of global inputs n for all G_i, for example $m_i = 13 \ll 895 = n$ in the numerical example presented in Sect. 5. Further approaches include compression techniques for exploiting sparsity

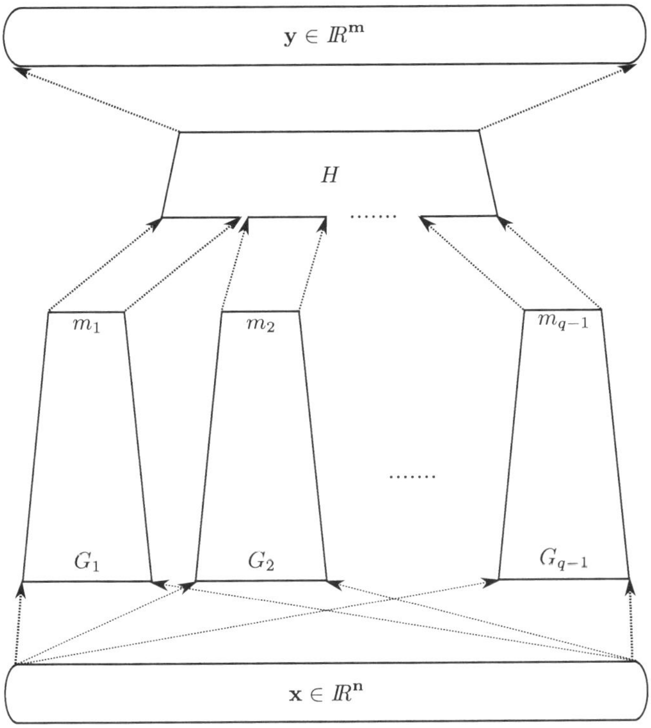

Fig. 2 Structure of the JURASSIC2 forward simulation model: $m_i \ll n$

of the local Jacobians [2] as well as the application of elimination techniques to the respective linearized DAGs [8]. Preaccumulation in plain adjoint mode turns out to be a reliable and the most efficient method in the given case.

4 Performance Analysis

In this section we estimate the gains of preaccumulation in comparison with plain black-box mode for a code structure as described in Sect. 3. We develop theoretical run time performance ratios for the computation of the Jacobian matrix $\nabla F(\mathbf{x})$ using the preaccumulation approach with respect to black-box tangent-linear and adjoint modes. A similar approach was taken in [1] where interface contraction was exploited in the context of tangent-linear mode AD. The following run time analysis can be seen as a generalization to a more complex structure using adjoint mode AD. Additionally we analyze the memory requirements.

The following estimates are valid for black-box tangent-linear as well as adjoint modes. Only the pre-factor varies. We therefore introduce

$$t = \begin{cases} n & \text{for tangent-linear mode} \\ m & \text{for adjoint mode} \end{cases} .$$

No preaccumulation (np) yields

$$\begin{aligned}
\text{Cost}^{\text{np}}(\nabla F(\mathbf{x})) &= \mathcal{O}(t) \cdot \text{Cost}(F) \\
&= \mathcal{O}(t) \cdot \left(\text{Cost}(H) + \sum_{j=1}^{q-1} \left(\text{Cost}(G_j)\right) \right) \\
&\geq \mathcal{O}(t) \cdot \left(\text{Cost}(H) + (q-1) \cdot \text{Cost}(\check{G}) \right) ,
\end{aligned}$$

with

$$\check{G} = \arg \min_{j=1}^{q-1} (\text{Cost}(G_j)) .$$

With preaccumulation (p) and for Cost(M) denoting the cost for computing the product of ∇H with the local Jacobian matrices ∇G_j for $j = 1, \ldots, q-1$ we get

$$\begin{aligned}
\text{Cost}^{\text{p}}(\nabla F(\mathbf{x})) &= \overbrace{\text{Cost}(\nabla H)}^{=0} + \sum_{j=1}^{q-1} \mathcal{O}(m_j) \cdot \text{Cost}(G_j) + \text{Cost}(M) \\
&\leq \mathcal{O}(\hat{m}) \cdot (q-1) \cdot \text{Cost}(\hat{G}) + \text{Cost}(M)
\end{aligned}$$

$$\text{with} \quad \hat{m} = \max_{j=1}^{q-1} \left(m_j\right) \quad \text{and} \quad \hat{G} = \arg \max_{j=1}^{q-1} (\text{Cost}(G_j)) .$$

Cost(∇H) is zero due to the linearity of the mapping.

For the ratios

$$\hat{Q} = \frac{\text{Cost}(\hat{G})}{\text{Cost}(H)} , \quad \check{Q} = \frac{\text{Cost}(\check{G})}{\text{Cost}(H)} , \quad Q^M = \frac{\text{Cost}(M)}{\text{Cost}(H)}$$

this observation yields

$$\begin{aligned}
\frac{\text{Cost}^{\text{p}}(\nabla F(\mathbf{x}))}{\text{Cost}^{\text{np}}(\nabla F(\mathbf{x}))} &\leq \frac{\mathcal{O}(\hat{m}) \cdot (q-1) \cdot \hat{Q} + Q^M}{\mathcal{O}(t) \left(1 + (q-1) \cdot \check{Q}\right)} \\
&= \mathcal{O}(1) \cdot \left(\underbrace{\frac{t^{-1} \cdot \hat{m} \cdot (q-1) \cdot \hat{Q}}{1 + (q-1) \cdot \check{Q}}}_{I} + \underbrace{\frac{t^{-1} \cdot Q^M}{1 + (q-1) \cdot \check{Q}}}_{II} \right) .
\end{aligned}$$

Term (I) reflects the speed-up generated by the exploitation of the small number of outputs of the G_j. Its value decreases with a growing number of global inputs or outputs t as well as with a falling number of local outputs or inputs $\hat{m}$. Term *(II)* on the other hand is the slow-down induced by the additional work due to the multiplication of the local Jacobian matrices. In case that the ratios $\hat{Q} = \check{Q} = Q^M \approx 1$ we get

$$\frac{\mathrm{Cost}^{\mathrm{p}}(\nabla F(\mathbf{x}))}{\mathrm{Cost}^{\mathrm{np}}(\nabla F(\mathbf{x}))} = \mathcal{O}(1) \cdot \left(\frac{\hat{m} \cdot (q-1) + 1}{t \cdot q}\right) \approx \mathcal{O}\left(\frac{\hat{m}}{t}\right)$$

for $q \gg 1$, which directly shows the gain due to interface contraction.

It is well known that the memory requirement of adjoint code is orders of magnitude lower when preaccumulating the ∇G_j [5]. The following derivation is only valid for black-box adjoint mode as tangent-linear mode is typically implemented tapeless. A *tape*[1] of a generalized elemental function G_j occupies memory of size $\mathrm{M}(G_j) = \sum_i \mathrm{M}(n_i^j)$, where n_i^j are the constituents of G_j, for example, the individual vertices in the computational graph. A single vertex n occupies memory of constant size $\mathrm{M}(n) = c$, whereas the memory consumption of a subgraph is defined recursively.

Without preaccumulation the tape includes all generalized elemental functions yielding

$$\begin{aligned}\mathrm{M}^{\mathrm{np}}(F(\mathbf{x})) = \sum_{j=1}^{q-1} \mathrm{M}(G_j) + \mathrm{M}(H) &\geq (q-1) \cdot \min_{j=1}^{q-1}(\mathrm{M}(G_j)) + \mathrm{M}(H) \\ &\geq (q-1) \cdot \mathrm{M}(\check{G}) + \mathrm{M}(H)\,,\end{aligned}$$

where $\check{G} = \arg\min_{j=1}^{q-1}\left(\mathrm{M}(G_j)\right)$.

With preaccumulating the memory consumption consists of the preaccumulated local Jacobians M_J and the tape of a single generalized elemental function yielding

$$\begin{aligned}\mathrm{M}^{\mathrm{p}}(F(\mathbf{x})) &\leq \sum_{j=1}^{q-1} \mathrm{M}_J(G_j) + \mathrm{M}_J(H) + \max_{j=1}^{q}\left(\mathrm{M}(G_j)\right) \\ &\leq (q-1) \cdot \mathrm{M}_J(\hat{G}) + \mathrm{M}_J(H) + \max_{j=1}^{q}(\mathrm{M}(G_j))\,,\end{aligned}$$

with $\hat{G} = \arg\max_{j=1}^{q-1}\left(\mathrm{M}_J(G_j)\right)$. Moreover, the product of ∇H with the local Jacobian matrices ∇G_j can be performed implicitly line by line. Therefore only a single gradient of a local output of G_j, $j = 1, \ldots, q-1$ needs to be stored at any given time.

---

[1]We use AD overloading tool `dco/c++` [7] that uses a *tape* to store a representation of the computational graph. Similar approaches are taken by other AD overloading tools, for example, by ADOL-C [3].

This observation yields the memory ratio

$$\frac{\mathrm{M}^{\mathrm{p}}(F(\mathbf{x}))}{\mathrm{M}^{\mathrm{np}}(F(\mathbf{x}))} \leq \frac{\mathrm{M}_J(H) + \frac{\mathrm{M}_J(\hat{G})}{\check{m}} + \max_{j=1}^{q}(\mathrm{M}(G_j))}{\mathrm{M}^{\mathrm{np}}(F(\mathbf{x}))},$$

with $\check{m} = \min_{j=1}^{q-1}(m_j)$.

5 Numerical Results

The numerical example processes measurements taken by the airborne infrared limb-sounder CRISTA-NF [17] aboard the high-flying Russian research aircraft M55-Geophysica during the Arctic RECONCILE campaign from January to March 2011 [15]. The instrument was configured to repeatedly take 60 infrared spectra in altitude steps roughly 250 m apart. Thirteen selected integrated regions from these spectra are used to derive vertical profiles of temperature, aerosol, and trace gas volume mixing ratios of nine trace gasses. A single representative profile retrieval has been selected as an example. The forward simulation model of this retrieval setup was configured to use $m_G = 139$ line-of-sights for the simulation of the 60 separate measurements. Together with the 13 integrated spectral regions, this adds up to 780 independent measurements, of which nine have been flagged as faulty, so that $m = 771$ valid measurements remain. These are used to derive $n = 895$ target quantities. This retrieval requires four iterations of the Gauss-Newton algorithm and consumes about 150 s of computation time, split roughly in half between actual minimization and the production of diagnostic information.

In Table 1 the results for the case study are shown. The Cost()-operator used in Sect. 4 is interpreted as the run time of one evaluation of a generalized elemental function. We show the run time of a single evaluation of $F(\mathbf{x})$ as well as the different possibilities for the computation of the Jacobian $\nabla F(\mathbf{x})$: using black-box tangent-linear mode, using black-box adjoint mode and using hierarchical adjoint mode implying the previously described preaccumulation scheme. The real gain in performance exceeds the theoretical conservative estimate by a factor of 6–12. The actual speed-up is roughly equal to 300 yielding a ratio of 3.5 between the run time of the adjoint code producing a full Jacobian and the run time of a single evaluation of the forward simulation model. Additionally one can see that the black-box adjoint mode is twice as fast as the black-box tangent-linear mode, even though $m \approx n$. This fact points out that one tape interpretation is more efficient compared to one tapeless forward run. We have also observed that for bigger test cases the speed-up even increases.

In Table 2 the memory requirement of running the black-box adjoint is compared with the memory requirement of the hierarchical approach. The estimated gain overestimates the real gain by a factor of approximately 6.

Table 1 Run time measurements for the test case, $n = 895, m = 771$. The run times are presented as multiples of the run time of $F(\mathbf{x})$

$F(\mathbf{x})$	∇F TLM black-box	∇F ADM black-box	∇F ADM Preacc.	Estimated speed-up	Real Speed-Up over TLM	Real speed-up over ADM
4.56 s	1,112	599	3.5	27.8	~320	~170

Table 2 Memory requirements for the test case

Tape memory, black-box	Tape memory, Preacc.	Estimated gain	Real gain
6,591 MB	7.1 MB	0.0059	0.00108

6 Summary and Conclusion

The application of adjoint mode AD to complex real-word numerical simulation codes is typically not automatic. Even if the black-box approach results in an executable adjoint,[2] the latter likely violates the given memory constraints and thus is infeasible. Hierarchical AD resolves this problem for certain program structures at the expense of sometimes nontrivial software analysis and restructuring effort. Nevertheless, this initial investment is likely to pay off in form of a robust and efficient adjoint simulation code as illustrated by the case study presented in this paper as well as by other applications discussed in the literature.

We were able to generate an efficient and scalable adjoint version of the spectral simulation code JURASSIC2 used for the solution of a highly relevant inverse atmospheric remote sensing problem at Forschungszentrum Juelich. Concurrency and interface contraction in adjoint mode AD yield run times that are close to the theoretical minimum for the required dense Jacobian. Insight into the given problem is combined with an optimized implementation of adjoint mode AD by overloading in `dco/c++`.

Acknowledgements Johannes Lotz is supported by the German Science Foundation (DFG grant No. 487/4-1).

References

1. Bücker, H.M., Rasch, A.: Modeling the performance of interface contraction. ACM Transactions on Mathematical Software **29**(4), 440–457 (2003). DOI http://doi.acm.org/10.1145/962437.962442
2. Gebremedhin, A., Manne, F., Pothen, A.: What color is your Jacobian? Graph coloring for computing derivatives. SIAM Review **47**(4), 629–705 (2005)

[2] In most cases significant manual preprocessing is required to make today's AD tools "digest" real-word application codes.

3. Griewank, A., Juedes, D., Utke, J.: Algorithm 755: ADOL-C: A package for the automatic differentiation of algorithms written in C/C++. ACM Transactions on Mathematical Software **22**(2), 131–167 (1996). URL http://doi.acm.org/10.1145/229473.229474
4. Griewank, A., Walther, A.: Evaluating Derivatives: Principles and Techniques of Algorithmic Differentiation, 2nd edn. No. 105 in Other Titles in Applied Mathematics. SIAM, Philadelphia, PA (2008). URL http://www.ec-securehost.com/SIAM/OT105.html
5. Hascoët, L., Fidanova, S., Held, C.: Adjoining independent computations. In: G. Corliss, C. Faure, A. Griewank, L. Hascoët, U. Naumann (eds.) Automatic Differentiation of Algorithms: From Simulation to Optimization, Computer and Information Science, chap. 35, pp. 299–304. Springer, New York, NY (2002)
6. Hovland, P.D., Bischof, C.H., Spiegelman, D., Casella, M.: Efficient derivative codes through automatic differentiation and interface contraction: An application in biostatistics. SIAM Journal on Scientific Computing **18**(4), 1056–1066 (1997). DOI 10.1137/S1064827595281800. URL http://link.aip.org/link/?SCE/18/1056/1
7. Leppkes, K., Lotz, J., Naumann, U.: dco/c++ – derivative code by overloading in C++. Tech. Rep. AIB-2011-05, RWTH Aachen (2011)
8. Naumann, U.: Optimal accumulation of Jacobian matrices by elimination methods on the dual computational graph. Math. Prog. **99**(3), 399–421 (2004). DOI 10.1007/s10107-003-0456-9
9. Naumann, U.: The Art of Differentiating Computer Programs. An Introduction to Algorithmic Differentiation. SIAM (2011)
10. Nocedal, J., Wright, S.: Numerical optimization, series in operations research and financial engineering (2006)
11. Rodgers, C.: Inverse Methods for Atmospheric Sounding. World Scientific (2000)
12. Tadjouddine, M., Forth, S., Qin, N.: Elimination ad applied to jacobian assembly for an implicit compressible cfd solver. International journal for numerical methods in fluids **47**(10–11), 1315–1321 (2005)
13. Tadjouddine, M., Forth, S.A., Keane, A.J.: Adjoint differentiation of a structural dynamics solver. In: H.M. Bücker, G. Corliss, P. Hovland, U. Naumann, B. Norris (eds.) Automatic Differentiation: Applications, Theory, and Implementations, *Lecture Notes in Computational Science and Engineering*, vol. 50, pp. 309–319. Springer, New York, NY (2005). DOI 10.1007/3-540-28438-9_27
14. Tikhonov, A.N., Arsenin, V.Y.: Solutions of ill-posed problems. Winston, Washington D.C., USA (1977)
15. Ungermann, J., Blank, J., Lotz, J., Leppkes, K., Hoffmann, L., Guggenmoser, T., Kaufmann, M., Preusse, P., Naumann, U., Riese, M.: A 3-D tomographic retrieval approach with advection compensation for the air-borne limb-imager GLORIA. Atmos. Meas. Tech. **4**(11), 2509–2529 (2011). DOI 10.5194/amt-4-2509-2011
16. Ungermann, J., Hoffmann, L., Preusse, P., Kaufmann, M., Riese, M.: Tomographic retrieval approach for mesoscale gravity wave observations by the premier infrared limb-sounder. Atmos. Meas. Tech. **3**(2), 339–354 (2010). DOI 10.5194/amt-3-339-2010
17. Ungermann, J., Kalicinsky, C., Olschewski, F., Knieling, P., Hoffmann, L., Blank, J., Woiwode, W., Oelhaf, H., Hösen, E., Volk, C.M., Ulanovsky, A., Ravegnani, F., Weigel, K., Stroh, F., Riese, M.: CRISTA-NF measurements with unprecedented vertical resolution during the RECONCILE aircraft campaign. Atmos. Meas. Tech. **4**(6), 6915–6967 (2011). DOI 10.5194/amtd-4-6915-2011

Storing Versus Recomputation on Multiple DAGs

Heather Cole-Mullen, Andrew Lyons, and Jean Utke

Abstract Recomputation and storing are typically seen as tradeoffs for checkpointing schemes in the context of adjoint computations. At finer granularity during the adjoint sweep, in practice, only the store-all or recompute-all approaches are fully automated. This paper considers a heuristic approach for exploiting finer granularity recomputations to reduce the storage requirements and thereby improve the overall adjoint efficiency without the need for manual intervention.

Keywords Source transformation • Reverse mode • Storage recomputation trade-off • Heuristics

1 Introduction

Computing derivatives of a numerical model $\boldsymbol{f}: \mathbb{R}^n \mapsto \mathbb{R}^m$, $\boldsymbol{x} \mapsto \boldsymbol{y}$, given as a computer program P, is an important but also computation-intensive task. Automatic differentiation (AD) [6] in *adjoint* (or *reverse*) mode provides the means to obtain gradients and is used in many science and engineering contexts (refer to the recent conference proceedings [1, 2]). Two major groups of AD tool implementations are operator overloading tools and source transformation tools. The latter are the focus of this paper. As a simplified rule, for each intrinsic floating-point operation ϕ (e.g., addition, multiplication, sine, cosine) that is executed during runtime in P as the sequence

$$[\ldots, j : (\, u = \phi(v_1, \ldots, v_k)), \ldots], \quad j = 1, \ldots, p, \tag{1}$$

H. Cole-Mullen (✉) · J. Utke
Argonne National Laboratory, The University of Chicago, Chicago, IL, USA
e-mail: hlcm@mcs.anl.gov; utke@mcs.anl.gov

A. Lyons
Dartmouth College, Hanover, NH, USA
e-mail: lyonsam@gmail.com

S. Forth et al. (eds.), *Recent Advances in Algorithmic Differentiation*, Lecture Notes in Computational Science and Engineering 87, DOI 10.1007/978-3-642-30023-3_18,

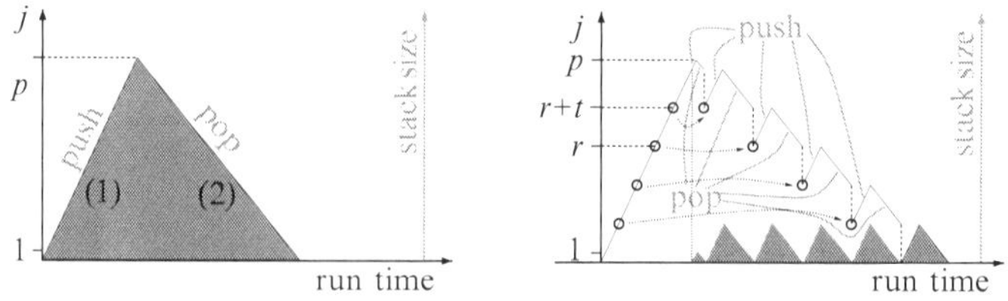

Fig. 1 Tape space for phases (1) and (2) without (*left*) and with (*right*) checkpointing

of p such operations, the generated adjoint code has to implement the following sequence that reverses the original sequence in j:

$$[\ldots, j : (\bar{v}_1 += \frac{\partial \phi}{\partial v_1}\bar{u}, \ldots, \bar{v}_k += \frac{\partial \phi}{\partial v_k}\bar{u}), \ldots], \quad j = p, \ldots, 1, \tag{2}$$

with incremental assignments of adjoint variables $\bar{v}$ for each argument v of the original operation ϕ. If $m = 1$ and we set $\bar{\boldsymbol{y}} = 1$, then the adjoint sequence yields $\bar{\boldsymbol{x}} = \nabla \boldsymbol{f}$. The two phases are illustrated in Fig. 1; note that to compute $\frac{\partial \phi}{\partial v_i}$ in phase (2), one needs the values of the variables v_i from phase (1).

The need to store and restore variable values for the adjoint sweep requires memory, commonly referred to as *tape*, for the derivative computation. This tape storage can be traded for recomputations in a checkpointing scheme. In theory, the storage for the tape and the checkpoints may be acquired from one common pool, as was considered in [7]. However, practical differences arise from the typical in-memory stack implementation of the tape, in contrast to the possible bulk writes and reads to and from disk for checkpoints. Furthermore, one may nest checkpoints or do hierarchical checkpointing [5], while the tape access is generally stack-like. The size of the checkpointed segment of the program execution, which is limited by the available memory, impacts the checkpointing scheme and therefore the overall adjoint efficiency. *Reducing the storage requirements for taping permits a larger checkpointed segment, which implies fewer checkpoints written and read, which implies fewer recomputations in the hierarchical checkpointing scheme.*

The goal of source code analysis has been the reduction of taping storage [8] for the "store-all" approach and the reduction of recomputation [3] for the "recompute-all" approach. The recompute-all approach replaces the tape altogether, at least initially, whereas the adjoint sweep requires the values to be available in the reverse order of the original computation. Recomputing the values in reverse order can carry a substantial cost. Consider a loop with k iterations and loop-carried dependencies for the values to be recomputed. The cost of computing the loop itself is k times the cost of a single iteration. Recomputing the values in reverse order has a complexity of $\mathcal{O}(k^2)$ times the cost of a single iteration. In tool implementations [4], this problem is mitigated by allowing the user to manually force certain, expensive-to-recompute values to be stored on a tape. This manual intervention can achieve

```
o=sin(a);
p=cos(b);
q=o*p;
```

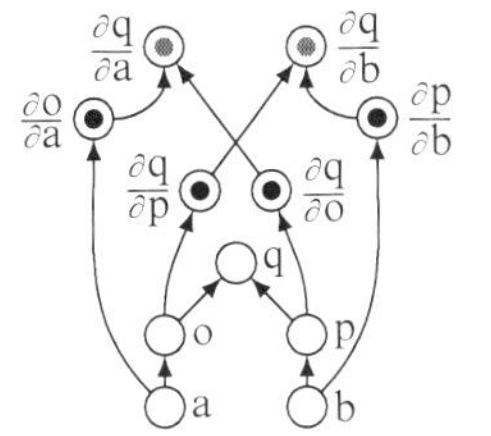

$$\left.\begin{aligned} \frac{\partial o}{\partial a} &= \cos(a); \\ \frac{\partial p}{\partial b} &= -\sin(b); \\ \frac{\partial q}{\partial p} &= o; \\ \frac{\partial q}{\partial o} &= p; \end{aligned}\right\} \text{linearization}$$

$$\left.\begin{aligned} \frac{\partial q}{\partial a} &= \frac{\partial q}{\partial o} * \frac{\partial o}{\partial a}; \\ \frac{\partial q}{\partial b} &= \frac{\partial q}{\partial p} * \frac{\partial p}{\partial b}; \end{aligned}\right\} \text{preaccumulation}$$

Fig. 2 An example for G^{*i}, with the nodes for the partials marked by • and the nodes for the preaccumulation marked by ●

an excellent tradeoff between taping and recomputation, but it requires deep insight into the code and is fragile in models that are subject to frequent changes.

Static source code analysis often cannot reliably estimate the complexity of recomputing values when this computation includes control flow or subroutine calls. On the other hand, one can safely assume that re-executing a fixed, moderate-length sequence of built-in operations and intrinsic calls to recompute a given value will be preferable to storing and restoring said value. Such fixed, moderate-length sequences are given naturally by the computational graphs already used for the elimination heuristics in OpenAD [10].

Following the approach in [10], we denote the computational graph representing a section of straight-line code (i.e., sequence of assignments) with $G^i = (V^i, E^i)$. The G^i are directed acyclic graphs with vertices $v_j \in V^i = V^i_{min} \cup V^i_{inter} \cup V^i_{max}$, where V^i_{min} are the minimal nodes (no in-edges), V^i_{max} are the maximal nodes (no out-edges) and V^i_{inter} are the intermediate nodes of G^i. The direct predecessors, $\{\ldots, v_i, \ldots\}$, of each intermediate or maximal node v_j represent the arguments to a built-in operation or intrinsic $\phi(\ldots, v_i, \ldots) = v_j$. In the usual fashion, we consider the partials $\frac{\partial \phi}{\partial v_i}$ as labels c_{ji} to edge (v_i, v_j).

Generally, these partials have a closed-form expression in terms of the predecessors v_i, and we can easily add them to the G^i. A more flexible approach than the rigid order suggested by (2) is the use of elimination techniques (vertex, edge or face elimination) on the computational graph to preaccumulate the partial derivatives. The elimination steps performed on the G^i reference the edge labels as arguments to fused multiply-add operations, which can be represented as simple expressions whose minimal nodes are edge labels or maximal nodes of preceding multiply-add operations. They too can be easily added to the G^i, and we denote the computational graph with the partial expressions and the preaccumulation operations as the extended computational graph G^{*i}. For the propagation of the adjoint variables, the edge labels of the remainder graph are required. In the example in Fig. 2, the required set is the maximal nodes $\{\frac{\partial q}{\partial a}, \frac{\partial q}{\partial b}\}$, but not node q, even though it too is maximal. The question now is how the values for the required nodes are provided: by storing, by recomputation from the minimal nodes, or by a mix of the two.

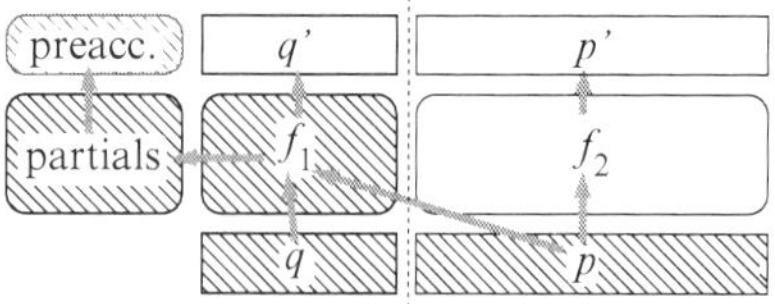

Fig. 3 Use case illustration, the shaded areas make up G^*

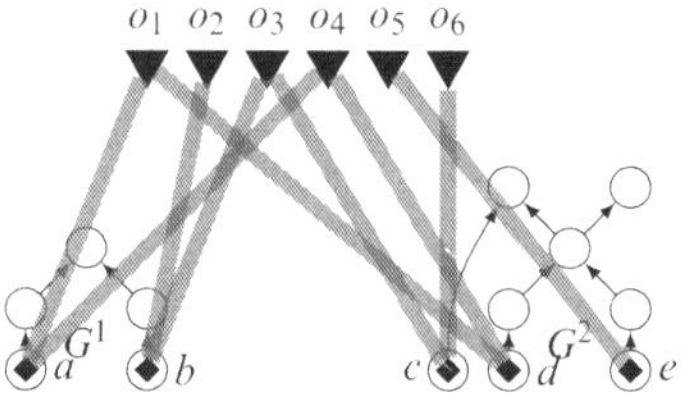

Fig. 4 An example for the graph G_b with respect to two computational graphs G^1 and G^2. The two node sets for G_b are shown as ♦ and ▼ symbols, respectively

2 A Use Case for Storing Edge Labels

We limit the scope of the recomputation to the respective G^{*i}, and if we decide to always recompute from the minimal nodes, we replicate the To-Be-Recorded (TBR) [8] behavior. However, this may not be optimal, and in some cases one may prefer to store preaccumulated partials. Consider the computation of a coupled model $(q', p') = f(q, p)$, where we consider the q part of the model for differentiation, and the coupling is done as a one-way forcing, such that $q' = f_1(q, p)$ and $p' = f_2(p)$, leaving the p portion of the model passive. The scenario is illustrated in Fig. 3. Recomputing f would require the whole model state (q, p), while propagating the adjoint values requires only the scarcity-preserving remainder graph edges. The original TBR analysis would store at least the portions of p that impact f_1 nonlinearly. Here, we have not even considered the cost of (re)evaluating f_1 and f_2. If they are particularly expensive, then one may prefer to store edge labels or certain intermediate values as a tradeoff for the evaluation cost.

3 Computational Graphs Share Value Restoration

For storing the required values, we can follow the TBR approach [8] by storing the values before they are overwritten. This information can be expressed as a bipartite graph $G_b = ((\bigcup V^i_{min}) \cup O, E_b)$, where O is the set of overwrite locations. An example for G_b associated with two computational graphs G^1 and G^2 is given in Fig. 4. In the example, one can see that recovering the value for node a requires restores in overwrite locations o_1 and o_4. This implies that the value for node d is restored; hence the value restoration benefits both graphs. Multiple overwrite locations for a given use are caused by aliasing, which can result from the use of array indices, pointers, or branches in the control flow. The overwrite locations $o_k \in O$ can be vertices in $\bigcup(V^i_{inter} \cup V^i_{max})$ or "placeholder" vertices for variables

```
t=5*d+4*e;
p=sin(c)+t;
r=cos(t);
```

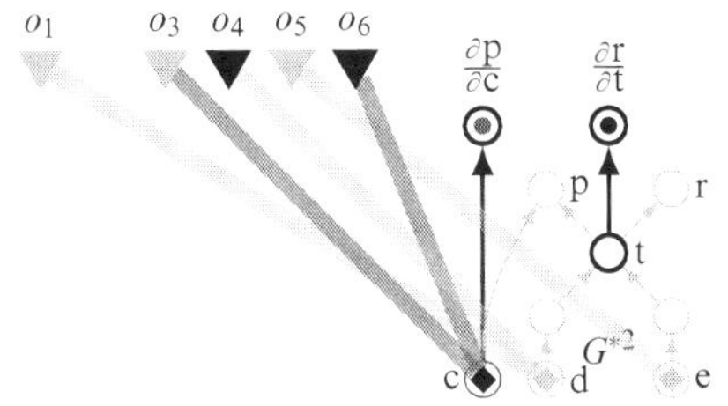

$$\frac{\partial t}{\partial d} = 4*1;$$
$$\frac{\partial t}{\partial d} = 5*1;$$
$$\frac{\partial p}{\partial t} = 1;$$
$$\frac{\partial r}{\partial t} = -\sin(t);$$
$$\frac{\partial p}{\partial c} = \cos(c);$$

Fig. 5 An example code (*left*) for the graph G^{*2} (*center*) highlighted for the subgraph with required nodes $\frac{\partial r}{\partial t}$ and $\frac{\partial p}{\partial c}$ that are computed from nodes $\{c, t\}$ according to the partials listing (*right*)

(unique for each program variable) that go out of scope without the value in question being overwritten by an actual statement. That association is essential only for the final code generation and not for the formulation of the combinatorial problem.

4 Problem Formulation

We assume a set $\mathscr{G} = \{G^{*i}\}$ of extended computational graphs G^{*i}, as introduced in Sect. 1, along with their required sets $\mathscr{R}^i$ and one common bipartite use-overwrite graph G_b as introduced in Sect. 3. For each G^i, there is a bipartite use-overwrite subgraph $G_b[V^i_{min}] = G^i_b = ((\bigcup V^i_{min}) \cup O^i, E^i_b)$ containing only the edges and vertices adjacent to V^i_{min}. The goal is to determine sets $S \subseteq O$ and $U \subseteq \bigcup V^i$ such that we minimize a static estimate for the number of values to be stored on tape.

Given these sets S and U of values to be restored, we need to be able to recompute values in the remaining subgraph of G^{*i} such that all required nodes in $\mathscr{R}^i$ can be recomputed. To impose this as a formal condition on S and U, we denote with $G^*_{\mathscr{R}}$ the subgraph of G^* induced by all the nodes preceding $\mathscr{R}$.

Condition for Recomputation (CR): The sets S and U are sufficient to allow recomputation of nodes in all $\mathscr{R}_i$ if $\forall G^{*i}_{\mathscr{R}_i}$: $\exists$ vertex cut C^i with respect to $\mathscr{R}_i$ such that $\forall c_j \in C^i : (c_j \in U) \veebar \big((c_j \in V^i_{min}) \wedge ((c_j, o) \in O \Rightarrow o \in S)\big)$.

In other words, if we know the values of all the vertices in the vertex cut C^i we are guaranteed to be able to recompute the values of the nodes in $\mathscr{R}_i$ by re-executing the computational graph between C^i and $\mathscr{R}_i$. A vertex in any of the cuts is either in U or it is not in U, in which case it must be a minimal node; if there is an overwrite of that value, then that overwrite location must be in S.

Consider the example shown in Fig. 5, where we reuse G^2 from Fig. 4 but add some example code for it and accordingly extend G^2 to G^{*2}. For scarcity preservation, during elimination we stop after an incomplete elimination sequence

[9], such that five edges from G^{*2} remain, of which only two are non-constant. These two are easily computable from nodes $\{\mathrm{c},\mathrm{t}\}$ representing our vertex cut for the subgraph and for which values are to be restored. Therefore, we can choose $S = \{o_3, o_6\}$ and $U = \{\mathrm{t}\}$. This choice emulates the TBR behavior [8], whereas choosing $U = \{\frac{\partial \mathrm{r}}{\partial \mathrm{t}}\}$, for example, would be outside of the options TBR considers. A simple example of the benefits of a non-TBR choice is a sequence of assignments $v_i = \phi_i(v_{i-1}), i = 1, \ldots, n$ with non-linear ϕ_i for which one would prefer storing the single preaccumulated scalar $\frac{\partial v_n}{\partial v_0} = \prod_i \frac{\partial v_i}{\partial v_{i-1}}$ over the TBR choice of storing all arguments $v_i, i = 0, \ldots, n-1$. On the other hand, for the example in Fig. 5, adding $\frac{\partial \mathrm{p}}{\partial \mathrm{c}}$ to U in exchange for $S = \emptyset$ would prevent any shared benefits that restoring node c has on restoring node b in G^1 as shown in Fig. 4.

4.1 A Cost Function

Because the decision about S and U has to be made during the transformation (aka compile time), any estimate regarding the runtime tape size cannot be more than a coarse indicator. Most significantly, the problem formulation disregards control flow, which may lead to an overestimate if the overwrite location is in mutually exclusive branches, or an underestimate if it is in a loop. On the other hand, this problem formulation allows for a very simple formulation of the cost function as $|S| + |U|$.

4.2 A Search Strategy

Because the choice of S impacts multiple graphs in $\mathscr{G}$, and thereby their contributions to U, there is no obvious best choice for a single $G^{*i}_{\mathscr{R}_i}$ that necessarily implies an optimal choice for all the other reduced combined graphs in general. For all but the simplest cases, the size of the search space implies that an exhaustive search is not a practical option. Therefore, we need a heuristic search strategy, and this strategy is crucial to obtaining useful, practical results from the proposed problem formulation.

One difficulty in devising a heuristic search strategy stems from the fact that while changing U or S is the elementary step, we have to adjust S or U respectively to satisfy (CR) so that we may compute a valid cost function value on a consistent pair S, U. Adapting the sets to satisfy (CR) involves the determination of vertex cuts and therefore is rather costly. In addition to determining vertex cuts, one also has to satisfy that all the overwrite locations of the minimal nodes in the respective cuts are in S.

Therefore, it appears plausible to choose a search strategy that adds or removes elements in S in small but consistent groups. The two important special cases establishing upper bounds for the cost function are: (i) TBR, i.e. $U = \emptyset$ and

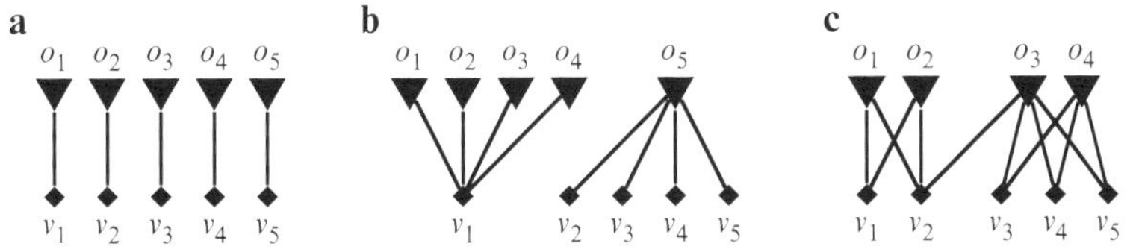

Fig. 6 Example scenarios for G_b

S determined according to (CR), and (ii) saving preaccumulated partials, i.e. $U = \bigcup \mathscr{R}_i$, $S = \emptyset$. We pick the one with the lower cost, which gives us an initial pair (S, U). Note that case (i) is the original TBR case as presented in [8] only if the graphs G^i each represent one individual assignment statement. As soon as multiple assignments are flattened into a single graph G^i, the computed cost for case (i) will be less than or equal to that of the original TBR.

While removing or adding the elements of S, we aim at making a change that will plausibly have the desired effect on the cost function. To get an indication of the effect on the cost function, we may limit our consideration to G_b. The most obvious observations for different scenarios in G_b which inform changes to S are shown in Fig. 6. It is clear that for (a) no particular preference of $v_i \in U$ vs. $o_i \in S$ can be deduced, while for (b) $U = \{v_1\}$ and $S = \{o_5\}$ are preferred.

To make consistent changes to S, we consider maximal bicliques covering G_b. For the moment, let us assume the bicliques are not overlapping, i.e., do not share nodes in V or O, as in Fig. 6a, b, for example. For each biclique $B = (V_B, O_B)$, we can evaluate the node ratios for removing from and adding to S.

$$r_B^- = \frac{|O_B^-|}{|v_B|}; \quad r_B^+ = \frac{|v_B|}{|O_B^+|} \tag{3}$$

where $O_B^- = O_B \cap S$ and $O_B^+ = O_B \setminus S$. Obviously, the r^+ is only meaningful for $O_B^+ \neq \emptyset$ and otherwise we set r^+ to 0. A biclique $B^* = (v_B^*, o_B^*)$ with the maximal ratio has the potential for the largest impact on the cost function when applied as follows

$$S := \begin{cases} S \cup O_B^+ \text{ if maximal ratio is } r_B^+ \\ S \setminus O_B^- \text{ if maximal ratio is } r_B^- \end{cases} \tag{4}$$

If $S = \emptyset$, then all bicliques in Fig. 6a have ratio 1. In Fig. 6b the biclique for v_1 has $r^- = 0$ and $r^+ = 1/4$ while the one for o_5 has $r^- = 1/4$ and $r^+ = 4$. So, we start by adding o_5 to S as our first step. In this setup, only ratios greater than one are hinting at an improvement of the cost function. Ratios equal to one are expected to be neutral and those less than one are expected to be counterproductive.

After updating S, we apply (CR) to determine U and evaluate the cost function, compare this to the minimum found so far, and accept or reject the step. If the step is rejected, we mark the biclique as rejected, remove it from further consideration for changes to S, restore the previous S, take the next biclique from the list order

Algorithm 1 Apply (CR) to a single combined, reduced DAG to update U

Given $G^*_{\mathscr{R}} = (V_{min} \cup V_{inter} \cup V_{max}, E), \mathscr{R}, S, V_S, U$

$U := U \setminus V; \quad C := V_S \cap V_{min}$

form the subgraph $G^{*'}$ induced by all paths from $V_{min} \setminus C$ to $\mathscr{R}$

determine a minimal vertex cut C' in $G^{*'}$ using as tie breaker the minimal distance from C' to $\mathscr{R}$.

set $C := C \cup C'$ as the vertex cut for G^* and set $U := U \cup C'$.

by ratio, and so on. If we accept the step, we mark the biclique as accepted and removed it from further consideration for changes to S.

Before we formalize the search algorithm, we have to address the case of overlapping bicliques as illustrated in Fig. 6c. There, biclique $(\{v_2\}, \{o_3\})$ overlaps with $(\{v_1, v_2\}.\{o_1, o_2\})$ and $(\{v_3, v_4, v_5\}, \{o_3, o_4\})$. If we consider a biclique (V_B, O_B) with an overlap to another biclique in the V_B, then we need to add to O_B all nodes connected to the nodes in the overlap, in order to obtain a consistent change to S. In our example, this means that $(\{v_2\}, \{o_3\})$ is augmented to $(\{v_2\}, \{o_1, o_2, o_3\})$. After the augmentation, we no longer have a biclique cover, so one may question whether starting with a biclique cover is appropriate to begin with. However, the purpose of starting with the minimal biclique cover (maximal bicliques) is to identify large groups of program variables whose overwrite locations are shared and for whom the store on overwrite yields a benefit to multiple uses. At the same time, using the minimal biclique cover implies a plausible reduction of the search space, compared to any other collection of bicliques which do not form a cover. While it is certainly possible to consider the case where one starts with a biclique collection that is not a cover, we currently have no rationale that prefers any such collection over the one where the V_B are the singletons $\{v_i\}$.

4.3 Algorithm

We formalize the method in Algorithms 1 and 2. Assume from here on that $G_b = (V_b, O)$ is the subgraph induced by the vertices occurring in the $G^{*i}_{\mathscr{R}^i}$. For a given S, the subset of restored vertices $V_S \subseteq V_b$ contains the vertices whose successors are all in S. A choice of $\delta < 1$ permits a cut off in the search which disregards bicliques not expected to improve the cost function.

5 Observations and Summary

As pointed out in Sect. 4.1, the principal caveat to estimating a runtime memory cost by counting instructions (i.e., value overwrite locations or uses) as done here is the lack of control flow information. Conversely, for straight-line code, one will have

Algorithm 2 Search algorithm for pair (S, U)

Given $\delta \in [0,1]$, $\mathscr{R} = \bigcup \mathscr{R}^i$, $G^{*i}_{\mathscr{R}^i}$ for all $G^i \in \mathscr{G}$ and $G_b = (V_b, O, E_b)$;
01 if $|O| < |\mathscr{R}|$ then $(S, U) := (O, \emptyset)$; $\quad c := |O|$
02 else $(S, U) := (\emptyset, \mathscr{R})$; $\quad c := |\mathscr{R}|$
03 compute minimal biclique cover $\mathscr{C}$ for G_b
04 $\forall B = (V_B, O_B) \in \mathscr{C}$ set $O_B := O_B \cup \{o : ((v,o) \in E_b \wedge v \in V_B)\}$
05 while $\mathscr{C} \neq \emptyset$
06 $\quad\forall B \in \mathscr{C}$ compute ratios r^-_B and r^+_B according to (3) and sort
07 $\quad$if maximal ratio is less than $1-\delta$ exit with current (S, U)
08 $\quad$update S according to (4)
09 $\quad\forall G^{*i}_{\mathscr{R}^i}$ update U using Algorithm 1
10 $\quad$if $c \geq |S| + |U|$ then set $c := |S| + |U|$
11 $\quad$else reset S to the value it had before line 07
12 $\quad$set $\mathscr{C} := \mathscr{C} \setminus \{B\}$

either a single DAG when there is no aliasing or multiple DAGs with aliasing. In these cases, the algorithm presented here will produce a result better than or on par with the cases (i) and (ii) of Sect. 4.2, which are used for initialization in lines 01 and 02 of Algorithm 2. The instruction count accurately reflects the runtime memory cost for a single execution of the straight-line code segment in question.

In the presence of control flow, the elements in U are correctly accounted for in the cost function by $|U|$ for a single execution of the DAG in which the respective vertices occur. In contrast, the runtime memory requirements for the elements in S are generally not related to the execution count of the DAGs for which the values are stored. It has been observed for the store-on-overwrite approach that the $|S|$ undercounts if it contains instructions in a loop and overcounts if its instructions are spread over mutually exclusive branches. Research related to the incorporation of the control flow is ongoing, but given the complexity of our flow-insensitive problem formulation, a detailed discussion is beyond the scope of this paper. Results that yield $U = \emptyset$ are on par or better than TBR (see Sect. 4.2). The problem formulation does not limit the number of DAGs in $\mathscr{G}$ to a single procedure, as long as the reaching definitions analysis that forms G_b is interprocedural. However, going beyond the scope of a single procedure increases the possibility of loop nesting and thus increases the error in the runtime cost estimate when Algorithm 2 yields both S and U as non-empty.

While this formulation is not the final answer to the general problem of storing versus recomputation, we view it as a stepping stone that widens the reach of automatic decisions by combining the information for multiple DAGs and permitting more recomputation through instructions flattened into DAGs. For a practical implementation, two things are needed. Firstly, one has to add logic to exclude subgraphs of the combined graphs that evaluate to constant values. Secondly, to correctly match the memory references represented as vertices in the heuristically determined bipartite subgraph, all computed address and index values become required as well (i.e., the proposed algorithm is to be additionally applied to them). This is already necessary for the original TBR algorithm. It is quite plausible

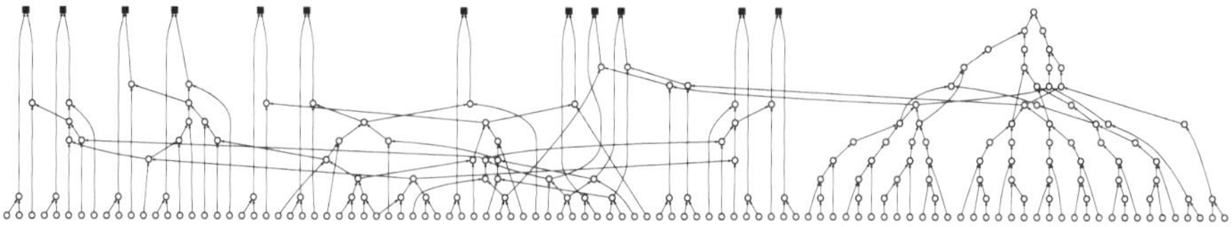

Fig. 7 Combined graph in OpenAD

to add expressions computing addresses or control flow conditions to the combined computational graphs and to add appropriate vertices to the set $\mathscr{R}$ of required values so that they become part of the automatic restore-recompute decisions. Then, the ordering for the adjoint code generation has to abide by certain dependencies that memory references in the vertices v have upon addresses or indices that occur as required values in the same graph. These are technical refinements that do not change the approach of the paper and are therefore left out. An implementation of the algorithms is forthcoming in OpenAD [11]. An example for G^* from a practical code using the experimental OpenAD implementation is shown in Fig. 7.

Acknowledgements This work was supported by the U.S. Department of Energy, under contract DE-AC02-06CH11357.

References

1. Bischof, C.H., Bücker, H.M., Hovland, P.D., Naumann, U., Utke, J. (eds.): Advances in Automatic Differentiation, *Lecture Notes in Computational Science and Engineering*, vol. 64. Springer, Berlin (2008). DOI 10.1007/978-3-540-68942-3
2. Bücker, H.M., Corliss, G.F., Hovland, P.D., Naumann, U., Norris, B. (eds.): Automatic Differentiation: Applications, Theory, and Implementations, *Lecture Notes in Computational Science and Engineering*, vol. 50. Springer, New York, NY (2005). DOI 10.1007/3-540-28438-9
3. Giering, R., Kaminski, T.: Recomputations in reverse mode AD. In: G. Corliss, C. Faure, A. Griewank, L. Hascoët, U. Naumann (eds.) Automatic Differentiation: From Simulation to Optimization, Computer and Information Science, chap. 33, pp. 283–291. Springer, New York (2002). URL http://www.springer.de/cgi-bin/search_book.pl?isbn=0-387-95305-1
4. Giering, R., Kaminski, T.: Applying TAF to generate efficient derivative code of Fortran 77-95 programs. Proceedings in Applied Mathematics and Mechanics **2**(1), 54–57 (2003). URL http://www3.interscience.wiley.com/cgi-bin/issuetoc?ID=104084257
5. Griewank, A., Walther, A.: Algorithm 799: Revolve: An implementation of checkpoint for the reverse or adjoint mode of computational differentiation. ACM Transactions on Mathematical Software **26**(1), 19–45 (2000). URL http://doi.acm.org/10.1145/347837.347846. Also appeared as Technical University of Dresden, Technical Report IOKOMO-04-1997.
6. Griewank, A., Walther, A.: Evaluating Derivatives: Principles and Techniques of Algorithmic Differentiation, 2nd edn. No. 105 in Other Titles in Applied Mathematics. SIAM, Philadelphia, PA (2008). URL http://www.ec-securehost.com/SIAM/OT105.html
7. Hascoët, L., Araya-Polo, M.: The adjoint data-flow analyses: Formalization, properties, and applications. In: Bücker et al. [2], pp. 135–146. DOI 10.1007/3-540-28438-9_12

8. Hascoët, L., Naumann, U., Pascual, V.: "To be recorded" analysis in reverse-mode automatic differentiation. Future Generation Computer Systems **21**(8), 1401–1417 (2005). DOI 10.1016/j.future.2004.11.009
9. Lyons, A., Utke, J.: On the practical exploitation of scarsity. In: Bischof et al. [1], pp. 103–114. DOI 10.1007/978-3-540-68942-3_10
10. Utke, J.: Flattening basic blocks. In: Bücker et al. [2], pp. 121–133. DOI 10.1007/3-540-28438-9_11
11. Utke, J., Naumann, U., Fagan, M., Tallent, N., Strout, M., Heimbach, P., Hill, C., Wunsch, C.: OpenAD/F: A modular, open-source tool for automatic differentiation of Fortran codes. ACM Transactions on Mathematical Software **34**(4), 18:1–18:36 (2008). DOI 10.1145/1377596.1377598

Using Directed Edge Separators to Increase Efficiency in the Determination of Jacobian Matrices via Automatic Differentiation*

Thomas F. Coleman, Xin Xiong, and Wei Xu

Abstract Every numerical function evaluation can be represented as a directed acyclic graph (DAG), beginning at the initial input variable settings, and terminating at the output or corresponding function value(s). The "reverse mode" of automatic differentiation (AD) generates a "tape" which is a representation of this underlying DAG. In this work we illustrate that a directed edge separator in this underlying DAG can yield space and time efficiency gains in the application of AD. Use of directed edge separators to increase AD efficiency in different ways than proposed here has been suggested by other authors (Bischof and Haghighat, Hierarchical approaches to automatic differentiation. In: Berz M, Bischof C, Corliss G, Griewank A (eds) Computational differentiation: techniques, applications, and tools, SIAM, Philadelphia, PA, pp 83–94, 1996; Bücker and Rasch, ACM Trans Math Softw 29(4):440–457, 2003). In contrast to these previous works, our focus here is primarily on space. Furthermore, we explore two simple algorithms to find good directed edge separators, and show how these ideas can be applied recursively to great advantage. Initial numerical experiments are presented.

*This work was supported in part by Ophelia Lazaridis University Research Chair (held by Thomas F. Coleman), the National Sciences and Engineering Research Concil of Canada and the Natural Science Foundation of China (Project No: 11101310).

T.F. Coleman (✉) · X. Xiong
Department of Combinatorics and Optimization, University of Waterloo, Waterloo, ON, N2L 3G1, Canada
e-mail: tfcoleman@uwaterloo.ca; x6xiong@uwaterloo.ca

W. Xu
Department of Mathematics, Tongji University, Shanghai, 200092, China
e-mail: wdxu@tongji.edu.cn

S. Forth et al. (eds.), *Recent Advances in Algorithmic Differentiation*, Lecture Notes in Computational Science and Engineering 87, DOI 10.1007/978-3-642-30023-3_19,

Keywords Automatic differentiation • Reverse mode • Adjoint method • Directed acyclic graph • Computational graph • Edge separator • Jacobian matrix • Ford-Fulkerson algorithm • Minimum cutset • Newton step

1 Introduction

Many scientific and engineering computations require the repeated calculation of matrices of derivatives. The repeated calculation of these derivative matrices often represents a significant portion of the overall computational cost of the computation.

Automatic differentiation (AD) can deliver matrices of derivatives given a source code to evaluate the function F (or in the case of minimization, the objective function f). Good methods that exploit sparsity, constant values, or duplicate values, have also been developed, e.g. [3, 17]. In addition, if the objective function exhibits certain kinds of structures, and this structure is conveniently noted in the expression of the objective function, then the efficiency of the automatic differentiation process can be greatly enhanced [1, 6, 7, 9, 12, 15].

This paper is concerned with the case where the problem structure is *not noted* á priori and AD may subsequently be regarded as too costly either in time or space.

1.1 Automatic Differentiation and the Edge Separator

Let us consider a nonlinear mapping

$$F : \mathbb{R}^n \mapsto \mathbb{R}^m$$

where $F(x) = [f_1(x), \cdots, f_m(x)]^T$, and each component function $f_i : \mathbb{R}^n \mapsto \mathbb{R}^1$ is differentiable. The Jacobian matrix $J(x)$ is the $m \times n$ matrix of first derivatives: $J_{ij} = \frac{\partial f_i}{\partial x_j}(i = 1, \cdots, m; j = 1, \cdots, n)$. Given the source code to evaluate $F(x)$, automatic differentiation can be used to determine $J(x)$. Generally, the work required to evaluate $J(x)$ via a *combination* of the forward and reverse modes of AD, and in the presence of sparsity in $J(x)$, is propotional to $\chi_B(G^D(J)) \cdot \omega(F)$ where χ_B is the bi-chromatic number of the double intersection graph $G^D(J)$, and $\omega(\cdot)$ is the work required, (i.e., number of basic computational steps) to evaluate $F(x)$ – see [9]. We note that when reverse mode AD is invoked the space required to compute the Jacobian is proportional to $\omega(F)$, and this can be prohibitively large. If AD is restricted to forward mode then the space required is much less, i.e., it is proportional to $\sigma(F)$, the space required to evaluate $F(x)$, and typically $\omega(F) \gg \sigma(F)$; however, forward mode alone can be much more costly than a combination of forward and reverse modes [9, 12].

Consider now the (directed) computational graph that represents the structure of the program to evaluate $F(x)$:

$$\mathbf{G}(F) = (V, \mathbf{E}) \tag{1}$$

where V consists of three sets of vertices. Specifically, $V = \{V_x, V_y, V_z\}$ where vertices in V_x represent the input variables; a vertex in V_y represent ***both*** a basic or elementary *operation* receiving one or two inputs, producing a single output variable ***and*** the output *intermediate variable*; vertices in V_z represent the output variables. So input variable x_i corresponds to vertex $v_{x_i} \in V_x$, intermediate variable y_k corresponds to vertex $v_{y_k} \in V_y$, and output $z_j = [F(x)]_j$ corresponds to vertex $v_{z_j} \in V_z$. Note that the number of vertices in V_y, i.e., $|V_y|$, is the number of basic operations required to evaluate $F(x)$. Hence $\omega(F) = |V_y|$.

The edge set $\mathbf{E}$ represents the traffic pattern of the variables. For example, there is a directed edge $e_k = (v_{y_i}, v_{y_j}) \in \mathbf{E}$ if intermediate variable y_i is required by computational node v_{y_j} to produce intermediate variable y_j. If $e_k = (v_{y_i}, v_{y_j}) \in \mathbf{E}$ is a directed edge from vertex v_{y_i} to vertex v_{y_j} then we refer to vertex v_{y_i} as the *tail node* of edge e_k and vertex v_{y_j} as the *head node* of edge e_k. It is clear that if F is well-defined then $\mathbf{G}(F)$ is an acyclic graph.

Definition 1. $E_d \subset \mathbf{E}$ is a *directed edge separator* in directed graph $\mathbf{G}$ if $\mathbf{G} - \{E_d\}$ consists of disjoint components $\mathbf{G}_1$ and $\mathbf{G}_2$ where all edges in E_d have the same orientation relative to $\mathbf{G}_1, \mathbf{G}_2$.

Suppose $E_d \subset \mathbf{E}_y$ is an edge separator of the computational graph $\mathbf{G}(F)$ with orientation forward in time. Then the nonlinear function $F(x)$ can be broken into two parts:

$$\left.\begin{array}{r} \text{solve for } y\text{: } F_1(x, y) = 0 \\ \text{solve for } z\text{: } F_2(x, y) - z = 0 \end{array}\right\} \tag{2}$$

where y is the vector of intermediate variables defined by the *tail vertices* of the edge separator E_d, and z is the output vector, i.e., $z = F(x)$. Let p be the number of tail vertices of edge set E_d, i.e., $y \in \mathbb{R}^p$. Note: $|E_d| \geq p$. The nonlinear function F_1 is defined by the computational graph above E_d, i.e., G_1, and nonlinear function F_2 is defined by the computational graph below E_d, i.e., G_2. See Fig. 1b. We note that the system (1) can be differentiated wrt (x, y) to yield an 'extended' Jacobian matrix [8, 10, 14]:

$$J_E \triangleq \begin{bmatrix} (F_1)_x & (F_1)_y \\ (F_2)_x & (F_2)_y \end{bmatrix} \tag{3}$$

Since y is a well-defined unique output of function $F_1 : \mathbb{R}^{n+p} \mapsto \mathbb{R}^p$, $(F_1)_y$ is a $p \times p$ nonsingular matrix. The Jacobian of F is the Schur-complement of (2), i.e.,

$$J(x) = (F_2)_x - (F_2)_y (F_1)_y^{-1} (F_1)_x \tag{4}$$

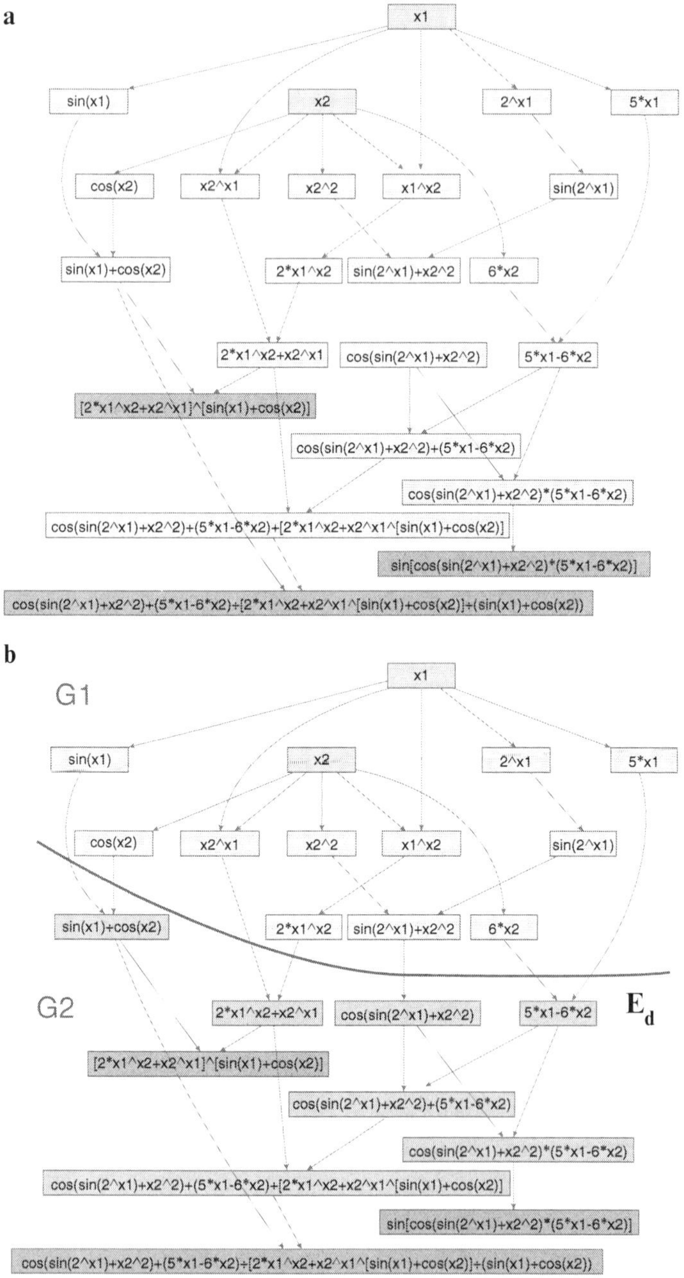

Fig. 1 An example of computational graphs and a sample directed edge separator. (**a**) A computational graph G. (**b**) An example of graph G's directed edge separator E_d

There are two important computational issues to note. The first is that the work to evaluate J_E is often less than that required to evaluate $J(x)$ directly. The second is that less space is often required to calculate and save J_E relative to calculating and saving J directly by AD (when the AD technique involves the use of "reverse mode" as in the bi-coloring technique).

It is usually less expensive, in time and space, to compute $J_E(x)$ rather than $J(x)$, using a combination of forward and reverse modes of automatic differentiation [11]. However, what is the utility of $J_E(x)$? The answer is that $J_E(x)$ can often be used directly to simulate the action of J and this computation can often be less expensive (due to sparsity in J_E that is not present in J) than explicitly forming and using J. For example, the Newton system 'solve $Js = -F$' can be replaced with

$$\text{solve } J_E \begin{bmatrix} s \\ t \end{bmatrix} = \begin{bmatrix} 0 \\ -F \end{bmatrix}. \tag{5}$$

The main points are that calculating matrix J_E can be less costly than calculating matrix J, and solving (5) can also be relatively inexpensive given sparsity that can occur in J_E that may not be present in J.

The ideas discussed above can be generalized to the case with multiple mutually independent directed edge separators, $E_{d_1}, \cdots, E_{d_k} \in \mathbf{E}$, where we assume $G - \{E_{d_1}, \cdots, E_{d_k}\} = \{G_1, \cdots, G_{k+1}\}$. The connected graphs $G_1, \cdots, G_{k+1}$ are pairwise disjoint and are ordered such that when evaluating F, G_i can be fully evaluated before $G_{i+1}, i = 1, \cdots, k$.

Suppose $E_{d_1}, \cdots, E_{d_k} \in \mathbf{E}$ are pairwise disjoint separators of the computational graph $\mathbf{G}(F)$ with orientation forward in time (as indicated above). Then the evaluation of nonlinear function $F(x)$ can be broken into $k + 1$ steps:

$$\left.\begin{array}{rr} \text{solve for } y_1 : & F_1(x, y_1) = 0 \\ \vdots & \vdots \\ \text{solve for } y_k : & F_k(x, y_1, \cdots, y_k) = 0 \\ \text{solve for } z : & F_{k+1}(x, y_1, \cdots, y_k) - z = 0 \end{array}\right\} \tag{6}$$

where y_i is the vector of intermediate variables defined by the *tail vertices* of the edge separator E_{d_i}, for $i = 1, \cdots, k + 1$ and z is the output vector, i.e., $z = F(x)$.

2 On Finding Separators to Increase Efficiency in the Application of Automatic Differentiation

In Sect. 1.1 we observed that if a small directed edge separator divides the computational graph G into roughly two equal components G_1 and G_2, then the space requirements are minimized (roughly halved). Moreover, the required work will not increase, and due to increasing sparsity, will likely decrease.

Therefore, our approach is to seek a small directed edge separator that will (roughly) bisect the fundamental computational graph. In this section, we present two algorithms to find good separators.

2.1 Minimum Weighted Separator

This minimum weighted separator approach is based on the Ford Fulkerson (FF) algorithm [13], a well known max-flow/min-cut algorithm. The Ford Fulkerson algorithm finds the minimum $s - t$ cut, a set of edges whose removal separates specified node s and node t, two arbitary nodes in the graph. A minimum cut does not always correspond to a directed separator; we "post process" the min-cut solution to obtain a directed separator.

We desire that the determined separator (roughly) divide the fundamental computational graph in half. To add this preference into the optimization, we assign capacities to edges to reflect distance from the input or output nodes, whichever is closer. With this kind of weight distribution, a 'small' cut will likely be located towards the middle of the fundamental computational graph.

To determine the weights we first calculate depth of nodes and edges.

Definition 2. We define the *depth* of a node v in a DAG to be the shorter of shortest directed path from an input node (source) to v and the shortest directed path from v to an output node (sink). We define the *depth* of an edge y in a DAG in an analogous fashion.

Our proposed method is as follows:

1. Assign weights to edges to reflect depth of an edge.
2. Solve the weighted mincut problem, e.g. using the Ford-Fulkerson method.
3. If the cut is not a directed separator, modify according to Algorithm 1.

Algorithm 1 *Let $E \subset \mathbf{E}$ such that graph $\mathbf{G} - E$ consists of two components $\mathbf{G}_1$ and $\mathbf{G}_2$, where source nodes are in $\mathbf{G}_1$ and sink nodes are in $\mathbf{G}_2$. If E is not a directed separator, then E contains both edges from $\mathbf{G}_2$ to $\mathbf{G}_1$ and edges from $\mathbf{G}_1$ to $\mathbf{G}_2$. Let $S = V(\mathbf{G_1})$ and $T = V(\mathbf{G_2})$. A directed separator (S, T) can be generated either by moving tail nodes of $T \to S$ edges from T to S recursively, or by moving head nodes of $T \to S$ edges from S to T recursively. The formal description is stated as follows:*

1. $T_1 \leftarrow \{v : v \in T\} \cup \{v :$ *there exists a directed uv-path in* $\mathbf{G}, u \in T\}$.
2. $S_1 \leftarrow V(\mathbf{G}) - T_1, E_1 = E(\mathbf{G}) - E(\mathbf{G}(S_1)) - E(\mathbf{G}(T_1))$.
3. $S_2 \leftarrow \{v : v \in S\} \cup \{v :$ *there exists a directed vu-path in* $\mathbf{G}, u \in S\}$.
4. $T_2 \leftarrow V(\mathbf{G}) - S_2, E_2 = E(\mathbf{G}) - E(\mathbf{G}(S_2)) - E(\mathbf{G}(T_2))$.
5. Pick the smaller between E_1 and E_2 as the desired separator.

2.2 *Natural Order Edge Separator*

A second method to generate directed separators comes from the observation that if the 'tape' generated by reverse-mode AD is snipped at any point then effectively a directed separator is located.

Suppose we are given a computational graph $\mathbf{G}$ and the corresponding computational tape T with length $|V(\mathbf{G})|$. A natural partition ($\mathbf{G}_1$,$\mathbf{G}_2$) of $\mathbf{G}$ is $\mathbf{G}_1$=$\mathbf{G}$ $(T(1 : i))$, $\mathbf{G}_2 = \mathbf{G}(T(i+1 : |V(\mathbf{G})|))$, where i is some integer between 1 and $|V(\mathbf{G})| - 1$. Since cells in the tape are in chronological order, all basic operations represented in $\mathbf{G}_1$ are evaluated before those represented in $\mathbf{G}_2$, therefore all edges between $\mathbf{G}_1$, $\mathbf{G}_2$ are directed from $\mathbf{G}_1$ to $\mathbf{G}_2$. Since these edges form a directed edge separator, we can then choose i to get the preferred edge separator in terms of separator size and partition ratio.

2.3 *Multiple Separators*

Either of the proposed directed separator methods can be applied, recursively, to yield multiple separators. We do exactly this in our code and in our computational experiments below, always working on the largest remaining subgraph (details will be provided in [11]).

3 Experiments

In this section we provide computational results on some preliminary experiments to automatically reveal 'AD-useful' structure using the separator idea. These experiments are based on the minimum weighted separator algorithm and natural order separator algorithm described in previous section, to find directed edge separators that bisect the fundamental computational graph.

We use the AD-tool, ADMAT [5], to generate the computational graphs. However, for efficiency reasons, ADMAT sometimes condenses the fundamental computational graph to produce a condensed computational graph. In a condensed computational graph nodes may represent matrix operations such as matrix-multiplication. Therefore our weighting heuristic is adjusted to account for this.

In our numerical experiments we focus on two types of structures that represent the two shape extreme cases.

3.1 *Thin Computational Graphs*

A function involving recursive iterations usually produces a "thin" computational graph.

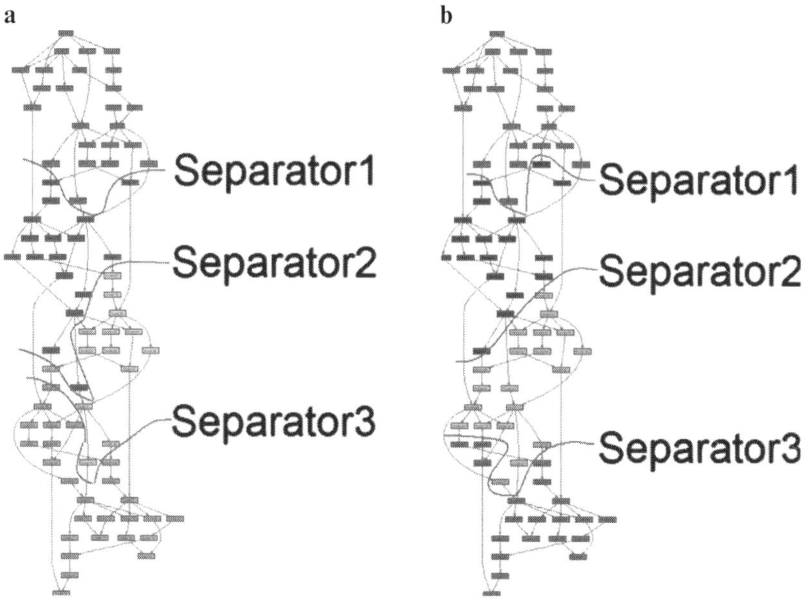

Fig. 2 Obtained separators of F_1's condensed computational graph by the two different algorithms. (**a**) Minimum weighted separator. (**b**) Natural order separator

Example. Define

$$F\left(\begin{bmatrix} x_1 \\ x_2 \\ x_3 \end{bmatrix}\right) = \begin{bmatrix} x_3 \cdot \cos(\sin(2^{x_1} + x_2^2)) \\ 5x_1 - 6x_2 \\ 2x_2^{x_2} + x_2^{x_1} \end{bmatrix} \tag{7}$$

and

$$F_1 = F \circ F \circ F \circ F \circ F \circ F$$

Note that F_1's computational graph is long and Narrow (i.e. 'thin').

After three interactions, three separators in Fig. 2 are found. The graph is divided into four subgraphs. Visually, these edge separator are good in terms of size and evenly dividing the graph.

3.2 *Fat Computational Graphs*

A "fat" computational graph is produced when macro-computations are independent of each other. A typical example is:

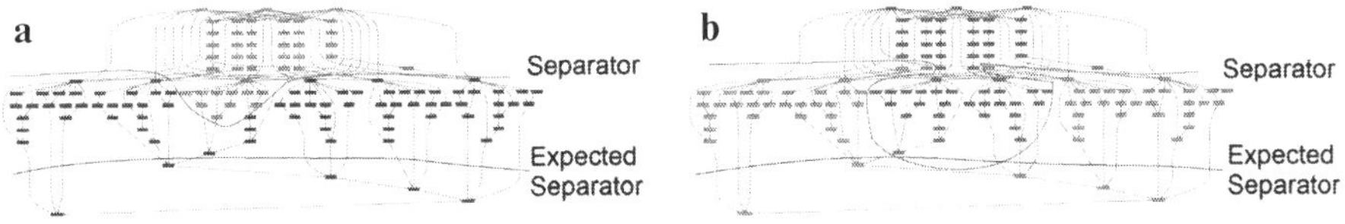

Fig. 3 Obtained separators of F_2's condensed computational graph by the two different algorithms. (**a**) Minimum weighted separator. (**b**) Natural order separator

$$F_2 = \sum_{i=1}^{6} F(x + \text{rand}_i(3,1))$$

where F is defined by (7) in the previous experiment (Fig. 2).

The separators (Fig. 3) found by our two algorithms on this example are useful but are less than ideal in contrast to the separators found in the "long thin" class. Additional experiments using different weighting schemes, are ongoing.

4 Accelerating the Calculation of the Jacobian matrix

To illustrate how separators accelerate computation, we construct the following numeric example:

Let

$$f\left(\begin{bmatrix} x_1 \\ x_2 \\ x_3 \end{bmatrix}\right) = \begin{bmatrix} \frac{x_2+3x_3}{4} \\ \sqrt{x_1 x_3} \\ \frac{x_1+2x_2+x_3}{4} \end{bmatrix}$$

and

$$F_k = f \circ f \circ \cdots \circ f \qquad \text{where there are } k \ f\text{'s} \tag{8}$$

It is obvious that $F_n \equiv F_{k_1} \circ F_{k_2} \circ \cdots \circ F_{k_m}$ provided $n = \sum_{i=1}^{m} k_i$.

We calculate the Jacobian matrix $J \in \mathbb{R}^{3\times 3}$ of $F_{2400}(x_0)$ at $x_0 = [6, 9, 3]^T$. We will use ADMAT reverse mode to obtain J both directly and by constructing directed separators.

The performance plot in Fig. 4 does not count in time used to locate separators. The 'running time' refers to the time used to obtain Jacobian matrix once the separators are found. Work is ongoing to perform the separator determine step efficiently, in space and time. We note that this separator structure can (typically) be re-used over many iterations.

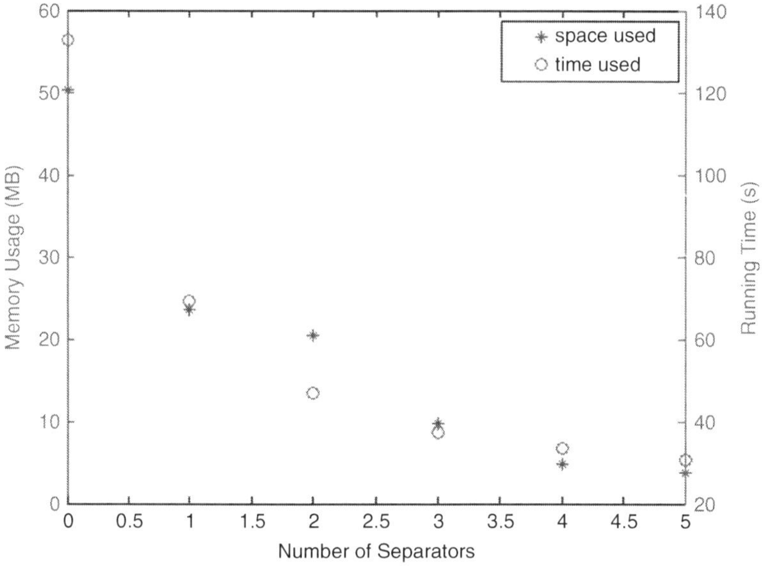

Fig. 4 Acceleration of separator method

5 Concluding Remarks

Our initial experiments and analysis indicate that separation of nonlinear systems with use of directed separators can significantly reduce the space and time requirements. Directed separators have also been proposed to improve the performance of hierarchical preaccumulation strategies [2, 16]. Issues to be investigated include:

- The amortization remarks above assume that the structure of F is invariant with x. This is not always the case.
- To further reduce memory usage, we are investigating use of an "online" algorithm, i.e., generation of separators with only partial information.

References

1. Bischof, C.H., Bouaricha, A., Khademi, P., Moré, J.J.: Computing gradients in large-scale optimization using automatic differentiation. INFORMS J. Computing **9**, 185–194 (1997)
2. Bischof, C.H., Haghighat, M.R.: Hierarchical approaches to automatic differentiation. In: M. Berz, C. Bischof, G. Corliss, A. Griewank (eds.) Computational Differentiation: Techniques, Applications, and Tools, pp. 83–94. SIAM, Philadelphia, PA (1996)
3. Bischof, C.H., Khademi, P.M., Bouaricha, A., Carle, A.: Efficient computation of gradients and Jacobians by dynamic exploitation of sparsity in automatic differentiation. Optimization Methods and Software **7**, 1–39 (1997)

4. Bücker, H.M., Rasch, A.: Modeling the performance of interface contraction. ACM Transactions on Mathematical Software **29**(4), 440–457 (2003). DOI http://doi.acm.org/10.1145/962437.962442
5. Cayuga Research Associates, L.: ADMAT-2.0 Users Guide (2009). URL http://www.cayugaresearch.com/
6. Coleman, T.F., Jonsson, G.F.: The efficient computation of structured gradients using automatic differentiation. SIAM Journal on Scientific Computing **20**(4), 1430–1437 (1999). DOI 10.1137/S1064827597320794
7. Coleman, T.F., Santosa, F., Verma, A.: Efficient calculation of Jacobian and adjoint vector products in wave propagational inverse problem using automatic differentiation. J. Comp. Phys. **157**, 234–255 (2000)
8. Coleman, T.F., Verma, A.: Structure and efficient Jacobian calculation. In: M. Berz, C. Bischof, G. Corliss, A. Griewank (eds.) Computational Differentiation: Techniques, Applications, and Tools, pp. 149–159. SIAM, Philadelphia, PA (1996)
9. Coleman, T.F., Verma, A.: The efficient computation of sparse Jacobian matrices using automatic differentiation. SIAM J. Sci. Comput. **19**(4), 1210–1233 (1998). DOI 10.1137/S1064827595295349. URL http://link.aip.org/link/?SCE/19/1210/1
10. Coleman, T.F., Verma, A.: Structure and efficient Hessian calculation. In: Y. Yuan (ed.) Proceedings of the 1996 International Conference on Nonlinear Programming, pp. 57–72. Kluwer Academic Publishers (1998)
11. Coleman, T.F., Xiong, X.: New graph approaches to the determination of Jacobian and Hessian matrices, and Newton steps, via automatic differentiation (in preparation)
12. Coleman, T.F., Xu, W.: Fast (structured) Newton computations. SIAM Journal on Scientific Computing **31**(2), 1175–1191 (2008). DOI 10.1137/070701005. URL http://link.aip.org/link/?SCE/31/1175/1
13. Ford, L., Fulkerson, D.: Maximal flow through a network. Canadian Journal of Mathematics **8**, 399–404 (1956)
14. Griewank, A.: Evaluating Derivatives: Principles and Techniques of Algorithmic Differentiation. No. 19 in Frontiers in Appl. Math. SIAM, Philadelphia, PA (2000)
15. Rall, L.B.: Automatic Differentiation: Techniques and Applications, *Lecture Notes in Computer Science*, vol. 120. Springer, Berlin (1981). DOI 10.1007/3-540-10861-0
16. Tadjouddine, E.M.: Vertex-ordering algorithms for automatic differentiation of computer codes. The Computer Journal **51**(6), 688–699 (2008). DOI 10.1093/comjnl/bxm115. URL http://comjnl.oxfordjournals.org/cgi/content/abstract/51/6/688
17. Xu, W., Coleman, T.F.: Efficient (partial) determination of derivative matrices via automatic differentiation (to appear in SIAM journal on Scientific Computing, 2012)

An Integer Programming Approach to Optimal Derivative Accumulation

Jieqiu Chen, Paul Hovland, Todd Munson, and Jean Utke

Abstract In automatic differentiation, vertex elimination is one of the many methods for Jacobian accumulation and in general it can be much more efficient than the forward mode or reverse mode (Forth et al. ACM Trans Math Softw 30(3):266–299, 2004; Griewank and Walther, Evaluating derivatives: principles and techniques of algorithmic differentiation, SIAM, Philadelphia, 2008). However, finding the optimal vertex elimination sequence of a computational graph is a hard combinatorial optimization problem. In this paper, we propose to tackle this problem with an integer programming (IP) technique, and we develop an IP formulation for it. This enables us to use a standard integer optimization solver to find an optimal vertex elimination strategy. In addition, we have developed several bound-tightening and symmetry-breaking constraints to strengthen the basic IP formulation. We demonstrate the effectiveness of these enhancements through computational experiments.

Keywords Vertex elimination • Combinatorial optimization • Integer programming

1 Introduction

Automatic differentiation (AD) is a family of methods for obtaining the derivatives of functions computed by a program [4]. AD couples rule-based differentiation of language intrinsics with derivative accumulation according to the chain rule. The associativity of the chain rule leads to many possible "modes" of combining partial

J. Chen (✉) · P. Hovland · T. Munson · J. Utke
Mathematics and Computer Science Division, Argonne National Laboratory,
Argonne, IL 60439, USA
e-mail: jieqchen@mcs.anl.gov; hovland@mcs.anl.gov; tmunson@mcs.anl.gov;
utke@mcs.anl.gov

S. Forth et al. (eds.), *Recent Advances in Algorithmic Differentiation*, Lecture Notes in Computational Science and Engineering 87, DOI 10.1007/978-3-642-30023-3_20,

derivatives, such as the forward mode and reverse mode. Exponentially many cross-country modes are possible, and finding the optimal Jacobian accumulation strategy is NP-hard [8]. Therefore, all AD tools employ some heuristic strategies. The most popular heuristics are pure forward mode, pure reverse mode, and a hierarchical strategy using the forward mode overall but "preaccumulating" the derivatives of small program units (often statements or basic blocks). A simplified version of the optimal Jacobian accumulation problem is to find an optimal vertex elimination strategy, where a vertex is eliminated by combining all in-edges with all out-edges, requiring $|\text{in}| \times |\text{out}|$ multiplications; see Sect. 2 for more details. It is well known that vertex elimination can be much more efficient than the incremental forward and the incremental reverse modes [3, 4]. Yet vertex elimination is a hard combinatorial optimization problem. Although, to the best knowledge of the authors, the complexity of this problem is still undetermined, it is speculated to be NP-complete.

A variety of heuristic strategies have been studied for the vertex elimination problem. Albrecht et al. [2] proposed several Markowitz-type heuristics for vertex eliminationx. Naumann and Gottschling [9] applied simulated annealing technique to this problem. The optimality preserving eliminations technique developed for the optimal Jacobian accumulation problem [6] also applies to the optimal vertex elimination problem and it features search space reduction in a branch-and-bound (B&B) framework. In this paper, we propose to use an IP technique to tackle the vertex elimination problem. The advantages of using IP technique include that its modeling capability is powerful and there are many high quality solvers for its solution.

IP deals with problems of minimizing a function of many variables subject to (1) linear inequality and equality constraints and (2) restrictions that the variables are integers [12]. IP is usually stated as

$$\min\{c^T x : Ax \leq b, x \in Z_+^n\}, \tag{1}$$

where Z_+^n is the set of nonnegative integer n-dimensional vectors and $x = (x_1, \ldots, x_n)$ are the *variables*. The generality of (1) allows it to model a wide variety of combinatorial optimization problems, for example, the traveling salesman problem [5], the minimum-weight spanning tree problem, and the set partitioning problem [12].

We develop an IP formulation of the vertex elimination problem, which enables us to use existing IP solvers to find the minimum number of multiplications for vertex elimination. Our objective is not to replace the elimination heuristics used in AD tools, since finding the optimal elimination strategy for all basic blocks would be prohibitively expensive. Rather, we aim to use the optimization formulation to evaluate the effectiveness of heuristics and find an optimal strategy for certain key computational kernels. In particular, the optimal computational cost of the vertex elimination can be used to measure whether the heuristic solution is close enough to the optimal one. In addition to the basic IP formulation, we develop bound-tightening and symmetry-breaking constraints to help computationally solve the problem.

The remaining paper is organized as follows. Section 2 discusses the IP formulation for the vertex elimination. Section 3 presents computational results of solving the IP formulation of several small problems. Section 4 summarizes our work and briefly describes future areas for research.

2 Integer Programming Formulations

In this section, we first briefly introduce vertex elimination. Next, we describe how we model vertex elimination as an integer program. In the last subsection, we discuss computational considerations in solving the integer program.

Consider the computational graph $G = (V, E)$ induced by a computer program that implements $F : \Re^n \rightarrow \Re^m$. The vertex set $V = \{1-n, \ldots, p+m\}$ and can be partitioned into $X = \{1-n, \ldots, 0\}$, $Z = \{1, \ldots, p\}$, and $Y = \{p+1, \ldots, p+m\}$, representing the set of n independent variables, p intermediate variables, and m dependent variables of the computer program F, respectively. The edge set E encodes the dependence relationship among the variables. The partial derivative of the jth variable with respect to the ith variable is assigned as an edge weight for (i, j), $\forall (i, j) \in E$. Vertex elimination transforms G into a bipartite graph $G' = (\{X, \emptyset, Y\}, E')$ by eliminating all intermediate vertices and their associate edges according to certain rules and appropriately updating the edge weights. After eliminating all intermediate vertices, the edge weights of the resulting bipartite graph are exactly the entries of F'; the readers can refer to [4] and [10] for more information. For a given intermediate vertex k, let P_k and S_k denote the set of predecessors and successors of k, respectively. Then eliminating vertex k involves removing vertex k and its incident edges from the graph and updating the edge weights, $C_{ij} + = C_{ik} * C_{kj}$, $\forall i \in P_k$, $j \in S_k$, where $C_{ij}, \forall (i, j) \in E$ denote the edge weights. Eliminating vertex k thus requires $|P_k| \times |S_k|$ number of floating-point multiplications; if $C_{ij} = 0$ before the elimination, then the addition need not be performed. Here we assume the number of additions is relatively small as compared to the number of multiplications, and thus the total number of multiplications in vertex elimination is a good approximation of the computational cost of accumulating the Jacobian matrix.

2.1 IP Formulation

Define $T = \{1, \ldots, p\}$ to be a time set. For any $t \in T$, we use $G(t) = (V(t), E(t))$ to represent the computational graph after eliminating t intermediate vertices. Denote $G(0) = G$. Note that $G(t)$ is undetermined unless a vertex elimination sequence is provided. To model this fact, we use x to define the elimination sequence and c_{ijt} to represent the edge (i, j) at time t as follows:

$$x_{it} = \begin{cases} 1, & \text{if vertex } i \text{ is eliminated at time } t \\ 0, & \text{otherwise} \end{cases}, \quad c_{ijt} = \begin{cases} 1, & \text{if } (i, j) \in E(t) \\ 0, & \text{otherwise} \end{cases}.$$

Similarly, let variable d denote the edge deleted, and variable f denote the edge filled or updated:

$$d_{ijt} = \begin{cases} 1, & \text{if } (i, j) \text{ is deleted on eliminating the } t\text{-th vertex} \\ 0, & \text{otherwise} \end{cases}$$

$$f_{ijt} = \begin{cases} 1, & \text{if } (i, j) \text{ is filled or updated on eliminating the } t\text{-th vertex} \\ 0, & \text{otherwise} \end{cases}.$$

Since we do not know $E(t), \forall t \in T$ exactly unless an elimination sequence is specified, the first two indices of d, f and c have to be defined over a different set than $E(t)$. Let $\tilde{G} = (V, \tilde{E})$ be the transitive closure of G. Obviously $E(t) \subseteq \tilde{E}$, and we index d, f and c over the set $\tilde{E} \times T$. With this notation, the vertex elimination problem is formulated as (MinFlops).

$$\text{minimize} \quad F = \sum_{t \in T} \sum_{(i,j) \in \tilde{E}} f_{ijt} \qquad \text{(MinFlops)}$$

$$\text{subject to} \quad \sum_{i \in Z} x_{it} = 1 \quad \forall\, t \in T \qquad (2)$$

$$\sum_{t \in T} x_{it} = 1 \quad \forall\, i \in Z \qquad (3)$$

$$x_{it} = 0 \quad \forall\, i \in X \cup Y, \ \forall\, t \in T \qquad (4)$$

$$\left.\begin{array}{l} d_{ijt} \geq x_{it} + c_{ij(t-1)} - 1 \\ d_{ijt} \geq x_{jt} + c_{ij(t-1)} - 1 \\ d_{ijt} \leq x_{it} + x_{jt} \\ d_{ijt} \leq c_{ij(t-1)} \end{array}\right\} \quad \forall\, (i, j) \in \tilde{E}, \ \forall\, t \in T \qquad (5)$$

$$f_{ijt} \geq d_{ikt} + d_{kjt} - 1 \quad \forall\, (i, j, k, t) \in \mathscr{I} \qquad (6)$$

$$\left.\begin{array}{l} c_{ijt} \geq f_{ijt} \\ c_{ijt} \leq 1 - d_{ijt} \\ c_{ijt} \leq c_{ij(t-1)} + f_{ijt} \\ c_{ijt} \geq c_{ij(t-1)} - d_{ijt} \end{array}\right\} \quad \forall\, (i, j) \in \tilde{E}, \ \forall\, t \in T \qquad (7)$$

$$x_{it} \in \{0, 1\} \quad \forall\, i \in V, \ \forall\, t \in T \qquad (8)$$

$$c_{ijt}, \ d_{ijt}, \ f_{ijt} \in \{0, 1\} \quad \forall\, (i, j) \in \tilde{E}, \ \forall\, t \in T, \qquad (9)$$

where

$$\mathscr{I} := \{(i,j,k,t) : (i,j) \in \tilde{E}, \forall k \in Z,\ t \geq l_{ij} - 1,\ \forall t \in T, \},$$

and l_{ij} denotes the length of the shortest path between i and j in $G(0)$, $\forall(i,j) \in \tilde{E}$. The rationale behind indexing (6) over $\mathscr{I}$ is that for any $(i,j) \in \tilde{E}$ and $l_{ij} \geq 2$, one needs to eliminate at least $l_{ij} - 1$ vertices in order to form an direct edge between i and j.

The objective function of (MinFlops) is the sum of the number of edges filled or updated in all time periods, which is equal to the total number of multiplications. Constraints (2) ensure that at any time period we eliminate exactly one vertex, and constraints (3) ensure that every intermediate vertex is eliminated once. Constraint (4) enforces that independent or dependent vertices cannot be eliminated.

Constraints (5) mean that d_{ijt} takes the value 1 if and only if edge (i,j) is present in time $t-1$ and i or j is eliminated in time t. Constraints (6) ensure that if edge (i,k) and (k,j) are deleted, then (i,j) must be filled or updated by combining (i,k) and (k,j). Constraints (7) enforce the proper relationship between $G(t-1)$ and $G(t)$. In particular, if (i,j) is filled or updated at time t, then $(i,j) \in E(t)$, which is enforced through the first inequality of (7). Similarly, if (i,j) is deleted at time t, then $c_{ijt} \leq 1 - d_{ijt}$ forces $(i,j) \notin E(t)$. The last two inequalities of (7) ensure that all the other edges that are not incident to the eliminated vertex at time $t-1$ also exist in time t. Constraints (8) and (9) restrict all variables to be binary. Overall, constraints (2)–(9) model the vertex elimination process.

By construction of the IP model, for any elimination sequence, there exists a solution (x,d,f,c) that encodes the corresponding elimination and satisfies (2)–(9). For any (x,d,f,c) satisfying (2)–(9), it may or may not refer to a valid elimination. However, we next show that any optimal solution to (MinFlops) corresponds to a valid elimination sequence with minimum number of multiplications. To show this, it suffices to prove Proposition 1 and to do so requires Lemma 1.

Lemma 1. *Let $G_1 = (V_1, E_1)$ be a directed acyclic graph with p intermediate vertices and $(i,j) \in E_1$. Define $\bar{G}_1 = (V_1, E_1 \backslash \{(i,j)\})$. The cost of vertex elimination on G_1 is at least at expensive as that of $\bar{G}_1$.*

Proof. First, for any elimination sequence applying to both G_1 and $\bar{G}_1$, we claim that $\bar{G}_1(t) = (V_1(t), \bar{E}_1(t))$ is a subgraph of $G_1(t) = (V_1(t), E_1(t))$, that is, $\bar{E}_1(t) \subseteq E_1(t)$, $\forall t \in \{1, \ldots, p\}$. This can be proved by directly using the result of Corollary 3.13 of [6]. Second, it follows that any vertex of $G_1(t)$ has at least as many predecessors and successors as the same vertex on $\bar{G}_1(t)$, and thus the cost of eliminating a particular vertex on $\bar{G}_1(t)$ cannot be more than that of the same vertex on $G_1(t)$. By induction, the desired result follows.

Proposition 1. *Let (x^*, d^*, f^*, c^*) be an optimal solution of (MinFlops). $f^*_{ijt} = 1$ if and only if there exists $k \in Z$, $d^*_{ikt} = d^*_{kjt} = 1$, $\forall(i,j,t) \in \tilde{E} \times T$.*

Proof. The "if" direction is ensured by (6). Now we show the "only if" direction. Suppose this is not the case, then $d^*_{i'kt'} + d^*_{kj't'} \leq 1, \forall k \in Z$ and $f^*_{i'j't'} = 1$ for some (i', j', t'). The optimal solution implies that $(i', j') \in E(t')$, although it does not correspond to any filled or updated edge. Case I: $c^*_{i'j'(t'-1)} = 1$. In this case, (6) and the third inequality of (7) holds even if $f^*_{i'j't'} = 0$. We construct a new solution from the optimal solution by only changing $f^*_{i'j't'}$ from 1 to 0. The new solution satisfies (2)–(9), but its corresponding objective value is smaller than the optimal value by 1, which leads to contradiction.

Case II: $c^*_{i'j'(t'-1)} = 0$. Let $G(t') = (V(t'), E(t'))$ represents the graph modeled by $c^*_{ijt'}, \forall (i, j) \in \tilde{E}$. Obviously, the graph is not supposed to have edge (i', j') after eliminating the t'-th vertex. In other words, $\bar{G}(t') = (V(t'), E(t') \backslash \{(i', j')\})$ should be the right graph after t'-th elimination. We construct a new elimination sequence as follows: it is the same as the elimination sequence specified by x^* in the first t' time periods. The remaining $p - t'$ vertices are eliminated in the optimal way with $\bar{G}(t')$ being the graph at time t'. Since $G(t')$ has one more edge than $\bar{G}(t')$, it follows from Lemma 1 that the cost of eliminating the remaining $(p - t')$ vertices on $G(t')$ should be at least as large as that of $\bar{G}(t')$. For the new elimination sequence, let $(\bar{x}, \bar{d}, \bar{f}, \bar{c})$ be the corresponding solution. In particular, $f_{i'j't'} = 0$. Therefore, $\sum_{t \in T} \sum_{(i,j) \in \tilde{E}} \bar{f}_{ijt}$ is smaller than the optimal value $F^* = \sum_{t \in T} \sum_{(i,j) \in \tilde{E}} f^*_{ijt}$ by at least 1, which also leads to contradiction.

One can easily recover the elimination sequence from the optimal solution of (MinFlops) by checking the value of x^*.

2.2 *Computational Consideration for Solving (MinFlops)*

A standard integer program (1) is usually solved by a B&B method. In each node of the B&B tree, a *linear programming* (LP) relaxation of the original IP is solved to obtain a lower bound of the objective function, where a LP relaxation is (1) without the integral restriction. See Sect. 7 of [12] and references therein for a detailed description of the B&B method. Next we discuss two methods that help computationally solve (MinFlops) with a standard IP solver: developing valid lower bounds, and developing symmetry-breaking constraints.

Developing valid lower bounds. We state two known results concerning the cost of vertex elimination and transform them into valid bound-tightening constraints. Let $[X \mapsto k]$ denote the set of paths connecting the independent vertices and k, and let $[k \mapsto Y]$ be the set of paths connecting k and the dependent vertices. The first known result is as follows.

Observation 1 ([7]). *For any $k \in Z$, the cost of eliminating vertex k last, regardless of the elimination sequence of all the other vertices, is $|X \mapsto k| \cdot |k \mapsto Y|$.*

From now on, we use γ_k to represent $|X \mapsto k| \cdot |k \mapsto Y|$. Although we do not know which vertex is eliminated last, variable x allows the flexibility of choosing any vertex k as the last one to eliminate, and $\sum_{k \in Z} \gamma_k \, x_{kp}$ represents the cost of eliminating the last vertex. The following valid inequality for vertex elimination can be added to (MinFlops) to strengthen the formulation:

$$F \geq \sum_{t \in T \setminus \{p\}} \sum_{(i,j) \in \tilde{E}} f_{ijt} + \sum_{k \in Z} \gamma_k \, x_{kp}, \tag{10}$$

where F is the variable representing the total cost of vertex elimination, and the first summation on the right-hand side is the cost of removing the first $p-1$ vertices. At first glance, the terms on both sides of the inequality seem to represent the same quantity. However, when computationally solving IP and the associated LP relaxations, all the integral restrictions on the variables are dropped, and so the right-hand side becomes a valid lower bound.

The second known result is established in [10]. Using the same notation as in [10], let X-k be the minimum vertex cut between X and k, and let k-Y be the minimum vertex cut between k and Y.

Observation 2 (Lemma 3.1 and 3.2 [10]). *The number of multiplications required to eliminate vertex k, among all possible vertex elimination sequences, is underestimated by* $|X\text{-}k| \cdot |k\text{-}Y|$*; the minimal number of multiplications required for the transformation* $G \to G'$ *is greater than or equal to* $\sum_{k \in Z} |X\text{-}k| \cdot |k\text{-}Y|$.

From now on, we use λ_k to denote $|X\text{-}k| \cdot |k\text{-}Y|$, $\forall k \in Z$. One immediate implication of Observation 2 is $F \geq \sum_{k \in Z} \lambda_k$. Although this inequality is valid, computationally it is not useful. The reason is that only one variable F is involved in this inequality, and this inequality does not cut off any fractional solutions of the LP relaxations, and thus cannot improve the lower bound. Instead, we express the results in Observation 2 as

$$F \geq \sum_{t<s} \sum_{(i,j) \in \tilde{E}} f_{ijt} + \sum_{s \leq t' \leq p} \sum_{k \in Z} \lambda_k \, x_{kt'}, \quad \forall\, s \in T, \tag{11}$$

where the second term on the right-hand side is a lower bound on the cost of removing the last $p-s+1$ vertices. We comment that (11) defines p constraints that have the same meaning but might have different effects computationally because each constraint involves different components of x and f and so might cut off different fractional solutions of the LP relaxations.

Developing symmetry-breaking constraints. Conceivably some elimination sequences may result in the same number of multiplications. We consider such elimination sequences equivalent or "symmetric" in the sense that they have the same cost. However, we need only one optimal sequence. The standard IP solvers cannot recognize the equivalence of two sequences and will waste considerable time exploring different branches of the B&B tree with the same optimal values. We consider one situation in which two elimination sequences give the same cost. This is when vertex i and j are not adjacent at time $t-1$, $\forall\, i, j \in Z,\ i < j,\ \forall\, t \in T\backslash\{p\}$, and we either (1) eliminate vertex i at time t and vertex j at time $t+1$, or (2) eliminate vertex j at period t and vertex i at period $t+1$, with all the other vertices eliminated in the same order.

Observation 3. *(a) Elimination sequences (1) and (2) result in the same computational graph at time $t+1$; (b) The cost of the two elimination sequences (1) and (2) are the same.*

Proof. The first statement is proved in Lemma 4.3 of [6]. Since i and j are not adjacent at time $t-1$, eliminating one of them does not change the predecessors and successors of the other, and the second statement follows.

We develop symmetry-breaking constraints that permit only one of the two sequences.

Proposition 2. *The following constraints are violated by one of the two symmetric sequences in Observation 3:*

$$x_{jt} + x_{i(t+1)} - c_{ij(t-1)} \leq 1, \quad \forall\, i, j \in Z,\ i < j,\ \forall\, t \in T\backslash\{p\}. \tag{12}$$

Proof. If i and j are not adjacent at time $t-1$, then $c_{ij(t-1)} = 0$. Thus (12) becomes $x_{jt} + x_{i(t+1)} \leq 1$, which is invalid for the sequence that eliminates j at time t and then i at time $t+1$, in the other words, $x_{jt} = x_{i(t+1)} = 1$. Obviously the sequence that eliminates i at timc t and j at time $t+1$, namely, $x_{it} = x_{j(t+1)} = 1$, is permitted.

We do not know beforehand whether two vertices i and j are adjacent at time $t-1$. Constraints (12) should also be valid in the case where i and j are adjacent at time $t-1$. In this case, $c_{ij(t-1)}$ is expected to have value 1. Then (12) becomes

$$x_{jt} + x_{i(t+1)} - 1 \leq 1 \quad \text{or} \quad x_{jt} + x_{i(t+1)} \leq 2,$$

which permits all four possible combinations of x_{jt} and $x_{i(t+1)}$.

3 Computational Experiments

In this section, we present computational results for solving (MinFlops). We collect several small problems from [4], where the optimal value of (MinFlops) can be verified by hand. Here the main goals are twofold: to compare the optimal values found by the IP solver with heuristic solutions, and to demonstrate the effects of the computational methods developed in Sect. 3. The IP model is formulated with GAMS [11] and solved with XPRESS 22.01 [1] on a standard computer under the Linux system.

Table 1 Comparison of CPU times (in seconds) of solving model (M_0)–(M_2) for five small problems. The first three are from [4]. A *t* indicates failure to find a provably optimal solution within 600 s

	Vertices			Number of multiplications				Times			LP bounds
Problem	X	Y	Z	Forward	Backward	Markowitz	F^*	M_0	M_1	M_2	of M_1
Fig. 10.4	4	3	3	22	18	22	18	0.04	0.02	0.11	18
Exercise 10.8	4	3	5	28	24	26	22	0.26	0.37	0.77	16
Fig. 10.1	4	2	5	20	18	16	15	0.71	0.35	0.29	13
Revbound	1	1	10	10	19	10	10	t	42.33	31.42	10
Butterfly	4	4	8	48	48	48	48	t	t	337.97	36

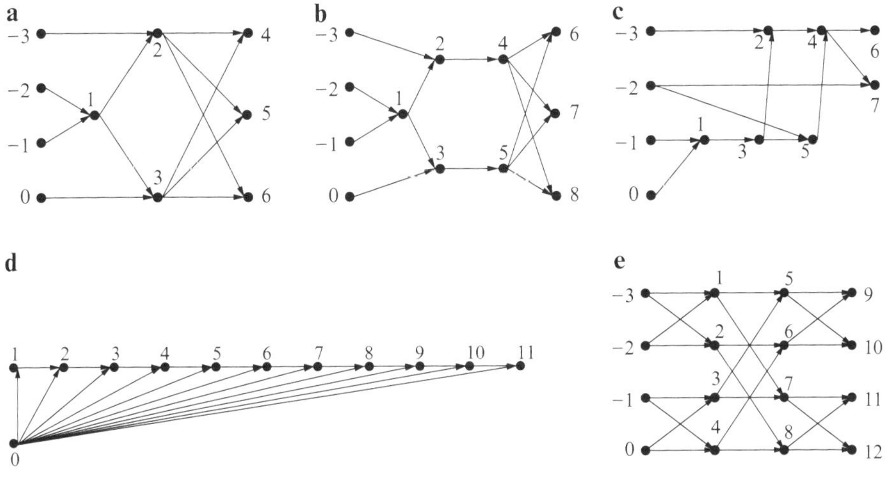

Fig. 1 Computational graphs of test problems in Table 1. (**a**) Fig. 10.4 of [4] (**b**) Exercise 10.8 of [4] (**c**) Fig. 10.1 of [4] (**d**) revbound (**e**) butterfly

We compare three IP models for the vertex elimination problem,

$$\min\{F : (2)-(9)\} \tag{M_0}$$

$$\min\{F : (2)-(9), (10), (11)\} \tag{M_1}$$

$$\min\{F : (2)-(9), (10), (11), (12)\}, \tag{M_2}$$

where (M_0) is the basic model (MinFlops), and (M_1)–(M_2) gradually incorporate the two methods proposed in Sect. 3 into the basic model. To solve (M_0)–(M_2) via IP solvers, we calculate $\tilde{E}$, $l_{ij}, \forall (i, j) \in \tilde{E}$, and the values of γ_k and λ_k, $\forall k \in Z$ in the pre-processing step. The pre-processing times are negligible compared to the expected run time of an IP solver, and thus are not included in our computational times. Table 1 presents the computational times of these three models on five small problems, and Fig. 1 shows their computational graphs. We also include in the table the number of multiplications required by the forward mode, the backward mode, and minimum Markowitz degree heuristic for comparison.

By comparing the CPU times for (M_0) and (M_1), we see that "revbound" could not be solved within the time limit (600 s) if modeled by (M_0) but can if modeled by (M_1); the lower bound constraints tighten the LP relaxations significantly. We also list LP bounds of (M_1) (at the root node) in the last column of the table. We observed that the bounds of (M_0) are all zeros, and the LP bounds of (M_2) are the same as those of (M_1), which indicate that the LP relaxation of (M_0) is not quite useful, and the bounds are tightened only by (10) and (11). Observe that "butterfly" is a computational graph with many equivalent elimination sequences. Although (M_2) does not provide tighter LP bounds over (M_1), its symmetry-breaking constraints prevents the IP solver from exploring many nodes, which is evidenced by the fact that the solution time of "butterfly" under (M_2) is much shorter than that of (M_1). We point out that for the three instanses where time limit was exceeded, the IP solvers ran out of time when exploring nodes in the B&B tree, and were not timed out at the root node.

4 Conclusion

We have developed an IP model for the vertex elimination problem for the optimal Jacobian accumulation. The model allows us to use IP technology to solve the problem, so that one can evaluate the effectiveness of heuristics used in AD tools and find an optimal strategy for certain computational kernels. In addition, we have developed several methods to help computationally solve the IPs. The effectiveness of the proposed methods are demonstrated by preliminary computational results.

Several directions remain for research. First, we hope to create a benchmark of many small to medium-sized instances with known optimal (MinFlops) values, to be used to evaluate different heuristics. Second, conceptually it is easy to see that extending this approach to other combinatorial optimization problems in AD is possible, for example, the edge elimination problem and the scarcity problem. How to computationally solve the corresponding IP models is nevertheless challenging.

Acknowledgements We thank Robert Luce for help on deriving an earlier IP formulation of the vertex elimination problem. This work was supported by the Office of Advanced Scientific Computing Research, Office of Science, U.S. Dept. of Energy, under Contract DE-AC02-06CH11357.

References

1. Xpress-Optimizer Reference Manual (2009). URLhttp://fico.com/xpress
2. Albrecht, A., Gottschling, P., Naumann, U.: Markowitz-type heuristics for computing Jacobian matrices efficiently. In: Computational Science – ICCS 2003, *LNCS*, vol. 2658, pp. 575–584. Springer (2003). DOI 10.1007/3-540-44862-4_61

3. Forth, S.A., Tadjouddine, M., Pryce, J.D., Reid, J.K.: Jacobian code generated by source transformation and vertex elimination can be as efficient as hand-coding. ACM Transactions on Mathematical Software **30**(3), 266–299 (2004). URLhttp://doi.acm.org/10.1145/1024074.1024076
4. Griewank, A., Walther, A.: Evaluating Derivatives: Principles and Techniques of Algorithmic Differentiation, 2nd edn. No. 105 in Other Titles in Applied Mathematics. SIAM, Philadelphia, PA (2008). URLhttp://www.ec-securehost.com/SIAM/OT105.html
5. Miller, C.E., Tucker, A.W., Zemlin, R.A.: Integer programming formulation of traveling salesman problems. J. ACM **7**, 326–329 (1960)
6. Mosenkis, V., Naumann, U.: On optimality preserving eliminations for the minimum edge count and optimal jacobian accumulation problems in linearized dags. Optimization Methods and Software **27**, 337–358 (2012)
7. Naumann, U.: An enhanced Markowitz rule for accumulating Jacobians efficiently. In: K. Mikula (ed.) ALGORITHMY'2000 Conference on Scientific Computing, pp. 320–329 (2000)
8. Naumann, U.: Optimal Jacobian accumulation is NP-complete. Math. Prog. **112**, 427–441 (2006). DOI 10.1007/s10107-006-0042-z
9. Naumann, U., Gottschling, P.: Simulated annealing for optimal pivot selection in Jacobian accumulation. In: A. Albrecht, K. Steinhöfel (eds.) Stochastic Algorithms: Foundations and Applications, no. 2827 in Lecture Notes in Computer Science, pp. 83–97. Springer (2003). DOI 10.1007/b13596
10. Naumann, U., Hu, Y.: Optimal vertex elimination in single-expression-use graphs. ACM Transactions on Mathematical Software **35**(1), 1–20 (2008). DOI 10.1145/1377603.1377605
11. Rosenthal, R.E.: GAMS – A User's Guide (2011)
12. Wolsey, L.A., Nemhauser, G.L.: Integer and Combinatorial Optimization. Wiley-Interscience (1999)

The Relative Cost of Function and Derivative Evaluations in the CUTEr Test Set

Torsten Bosse and Andreas Griewank

Abstract The CUTEr test set represents a testing environment for nonlinear optimization solvers containing more than 1,000 academic and applied nonlinear problems. It is often used to verify the robustness and performance of nonlinear optimization solvers. In this paper we perform a quantitative analysis of the CUTEr test set. As a result we see that some paradigms of nonlinear optimization and Automatic Differentiation can be verified whereas others need to be questioned. Furthermore, we will show that the CUTEr test set is probably biased, i.e., solvers that use exact derivatives and sparse linear algebra are likely to perform advantageously compared to solvers employing directional derivatives and low-rank updating.

Keywords CUTEr test set • Automatic Differentiation • Run-time • Linear algebra • Numerical analysis

1 Introduction

The purpose of the following analysis is to get a better overview of the structural and computational behavior of general nonlinear problems (cf. [12]) in terms of size, sparsity and evaluation time. The observations might then be used to increase the performance of NLOP solvers. Therefore, we considered 1,086 out of 1,094

T. Bosse (✉)
Humboldt-Universität zu Berlin, Unter den Linden 6, 10099 Berlin, Germany
e-mail: bosse@math.hu-berlin.de

A. Griewank
Humboldt-Universität zu Berlin, Unter den Linden 6, 10099 Berlin, Germany
e-mail: griewank@math.hu-berlin.de

S. Forth et al. (eds.), *Recent Advances in Algorithmic Differentiation*, Lecture Notes in Computational Science and Engineering 87, DOI 10.1007/978-3-642-30023-3_21,

nonlinear problems from the CUTEr test set (see large and small mastsif archive [6]), which represents the largest standardized collection of nonlinear problems.

At first let us fix some notation before presenting the numerical results. The nonlinear problems under investigation are encoded in the general form

$$\begin{aligned} \min_{x\in\mathbb{R}^n} f(x) \text{ s.t.}& \\ c_i(x) = 0, &\quad i \in \mathscr{E} \\ c_i(x) \leq 0, &\quad i \in \mathscr{I} \\ l_i \leq x_i \leq u_i, &\quad i \in \{1,\dots,n\} \end{aligned}$$

with an *objective function* $f : \mathbb{R}^n \to \mathbb{R}$ depending on the variable vector $x = (x_1,\dots,x_n) \in \mathbb{R}^n$ and *general equality* and *inequality constraint* functions $c_i : \mathbb{R}^n \to \mathbb{R}$ which can be uniquely identified by their corresponding index i from one of the two disjoint index sets $\mathscr{E}$ and $\mathscr{I}$. The finite entries of the vectors $l = (l_1,\dots,l_n) \in \bar{\mathbb{R}}^n$ and $u = (u_1,\dots,u_n) \in \bar{\mathbb{R}}^n$ are referred to as lower and upper *box constraints* where $\bar{\mathbb{R}} = \mathbb{R} \cup \{-\infty, \infty\}$. Furthermore, we call all general equality constraints and all general inequality constraints that are not strictly satisfied at a given point x *(currently) active general constraints*. In other words, inequality constraints are active if $c_i(x) \geq 0, i \in \mathscr{I}$ at a given point x, which applies equivalently to box constraints.

2 Numerical Results

CUTEr is based on the so-called SIF format, a representation of functions in group partial separable form. These data are used to evaluate function and first or second derivatives in an interpretive fashion. There are even procedures for evaluating directional derivatives and first and second order adjoint vectors in the reverse mode. The resulting run time ratios are quite favorable because the SIF structure is used and the underlying function evaluations are also performed by interpretation. Applying standard AD tools would require the generation of source code in a standard procedural language and probably yield higher run-time ratios for the derivatives. It certainly would make no sense for a general CUTEr user to go this route.

All tests were performed on a Laptop with a 2*2.0 GHz Intel Pentium DualCore processor (T7250, FSB: 800 MHz, L2-Cache: 2 MB shared) using 3 GB of memory and operating system OpenSUSE Linux 11.4 (64 bit). The following measurements were obtained as averages over ten test runs using the latest version of CUTEr interface (cf. [6, 7]). The run-times needed for the evaluation of the underlying quantities are measured as CPU clock time (seconds) for one processor and are pure, i.e., the initialization of a problem and necessary memory de-/allocation was not included. Also, we neglected the time for evaluating box constraints and

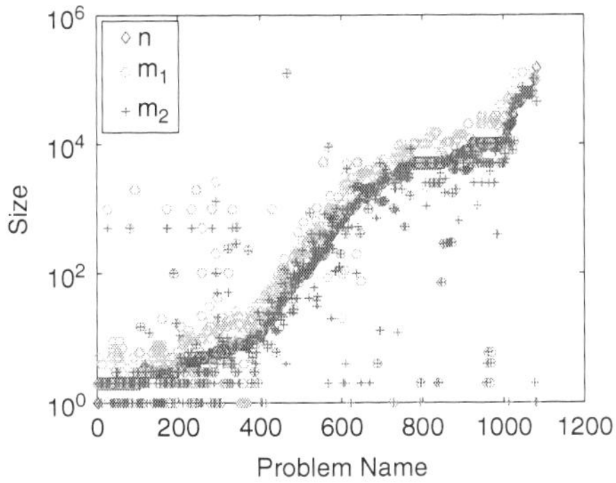

Fig. 1 Dimension – sorted by the number of variables (*left*) or data (*right*)

their corresponding derivatives due to the simplicity of their special structure. Furthermore, all functions were evaluated at the given initial point x within the SIF file (cf. [2]) containing the problem description, and the active constraints were defined accordingly.

The resulting data is visualized by pairs of plots (left, right), which should be inspected simultaneously. All problems were presorted according to their number of independent variables and consecutively named by numbers ranging from 1 to 1,086. In the left plots these integer names represent the horizontal axis and the corresponding quantities of interest is plotted on the vertical axis. In the plots on the right these quantities are sorted (increasing) on the vertical axis while the number of problems are plotted on the horizontal axis irrespective of the name ordering.

2.1 *Problem Dimension and Ratio*

In the first two figures we visualized the number of variables n, the corresponding number of the general and the box constraints m_1, and the number of the active general and box constraints m_2.

As can be seen in Fig. 1 (right) the problems from the CUTEr test set represent a rather evenly spread sample of small (number of variables/general constraints between 0 and 100), medium (number of variables/general constraints between 101 and 10,000) and large sized problems (number of variables/general constraints greater than 10,000). Furthermore, we see from Fig. 1 (left) that the number of (active) general and box constraints strongly correlates with the number of variables. This result can be validated by looking at Fig. 2 (left and right) where we depicted for each problem the ratio $\rho_{m_1} = m_1/n$ of the number of general and box constraints divided by the number of variables as well as the ratio $\rho_{m_2} = m_2/n$ of the number of active general and box constraints divided by the number of variables.

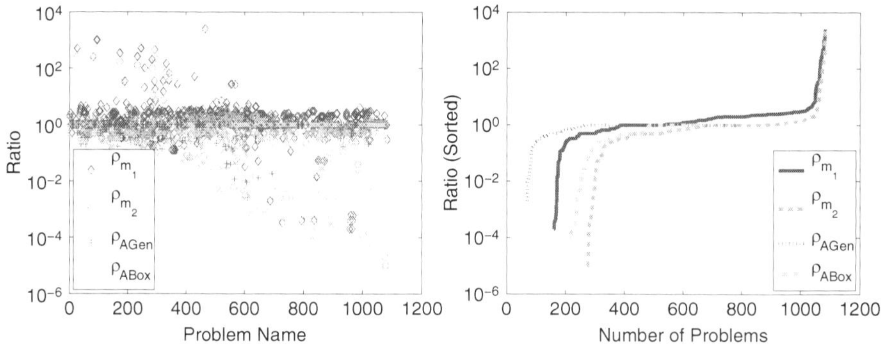

Fig. 2 Dimension ratio – sorted by number of variables (*left*) or data (*right*)

Also, we see by looking at the ratios ρ_{AGen} (= number of active general constraints normalized by the number of general constraints) and ρ_{ABox} (= number of active box constraints normalized by the number of box constraints) that for over 50% out of the 782 general constraint problems nearly all ($\geq$ 99.99%) of the general constraints are active at the first iterate. On the other hand we find that for about 40% out of the 668 box constraint problems nearly all ($\geq$ 99.99%) of the box constraints are inactive at the first iterate.

2.2 Run-Time and Run-Time Ratios

In the following two sections we examine the quantities that mainly dominate the run-times for nonlinear problem solvers. Namely, we analyze the run-times needed to evaluate the problem functions and the corresponding first/second order derivatives. The derivative information is usually used to approximate the nonlinear problem by a sequence of simpler models, as in SQP or IP methods, which then are solved by some linear algebra package. Therefore, we also quantified the sparsity degree of the corresponding first/second order derivative matrices.

The time measurements in Fig. 3 show that the run-time t_{Obj} needed to evaluate the objective function and the corresponding derivatives (t_{Grad} Gradient, t_{SHess} sparse Hessian, t_{DHess} dense Hessian[1]) strongly correlates with the number of variables. Similar observations apply (see Fig. 4) for the run-time t_{Con} needed to evaluate the general constraint function and the corresponding derivatives (t_{For} Jacobian-Vector product, t_{Back} Vector-Jacobian product, t_{SJac} sparse Jacobian, t_{DJac} dense Jacobian).

Furthermore, we see by looking at the run-time ratios $\rho_{Grad} = t_{Grad}/t_{Obj}$, $\rho_{SHess} = t_{SHess}/t_{Obj}$, $\rho_{DHess} = t_{DHess}/t_{Obj}$ for the objective function in Fig. 5 and the run-time

[1]The run-time t_{DHess} needed to evaluate the dense Hessian of the objective function is not depicted for problems with more than 10^4 variables due to memory shortage.

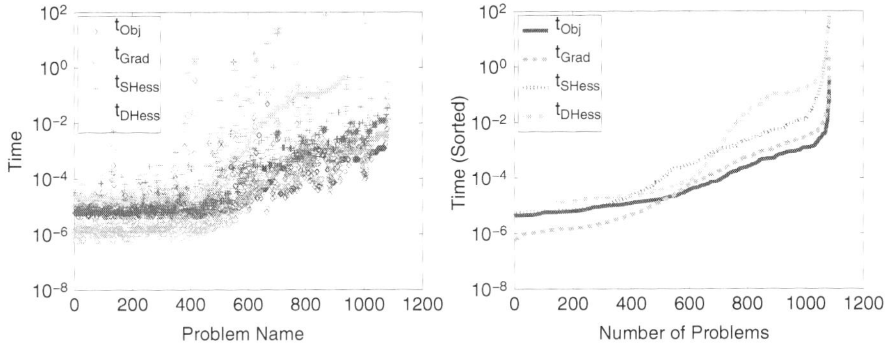

Fig. 3 Evaluation time for the objective function and its derivatives

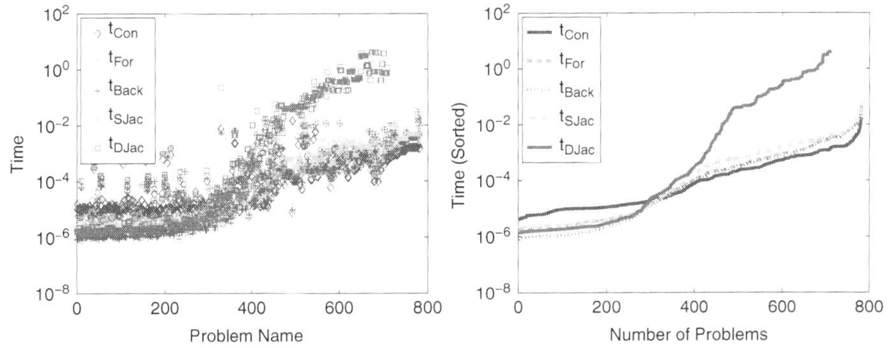

Fig. 4 Evaluation time for the constraint function and its derivatives

ratios $\rho_{For} = t_{For}/t_{Con}$, $\rho_{Back} = t_{Back}/t_{Con}$, $\rho_{SJac} = t_{SJac}/t_{Con}$ and $\rho_{DJac} = t_{DJac}/t_{Con}$ for the constraint function in Fig. 6 that the paradigms of Automatic Differentiation (cf. [8]) are valid:

Let $F : \mathbb{R}^n \rightarrow \mathbb{R}^m \in \mathscr{C}^1$, $\bar{y} \in \mathbb{R}^m$ *and* $x, y \in \mathbb{R}^n$ *then it holds*

$$OPS(eval(\bar{y}F'(x))) \leq c * OPS(eval(F)) and$$

$$OPS(eval(F'(x)y)) \leq c * OPS(eval(F)),$$

where $c \in \mathbb{R}_+$ *is a small constant depending only on the computing platform.*

In other words, we have numerical evidence that for a differentiable function $F : \mathbb{R}^n \rightarrow \mathbb{R}^m$ the vector-Jacobian product $\bar{y}\ F'(x)$ ($\bar{y} \in \mathbb{R}^m$) and the Jacobian-vector product $F'(x)y$ ($y \in \mathbb{R}^n$) can be obtained within the same complexity as a function evaluation itself times a small constant c that is independent of the dimensions. In fact the numerical results imply that c is in average smaller than the theoretical upper bound given in [8] since the mean value of the (nonzero) ratios ρ_{Grad}, ρ_{For}, and ρ_{Back} were found to be $\rho_{Grad}^{avg} = 1.3301$, $\rho_{For}^{avg} = 1.5176$, and $\rho_{Back}^{avg} = 1.4005$. Moreover,

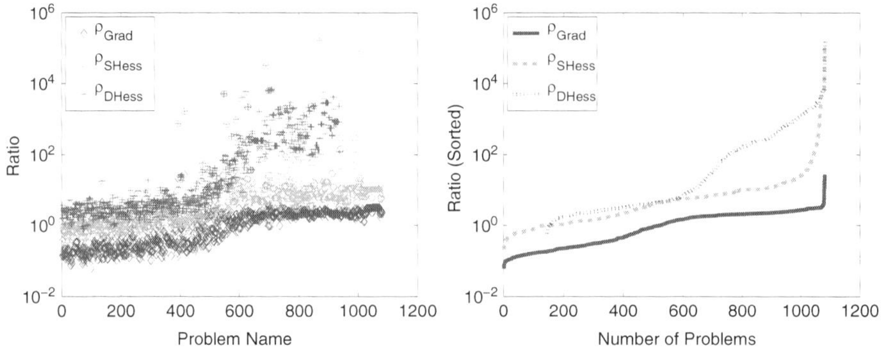

Fig. 5 Evaluation time ratios for the objective function and up to second order derivatives

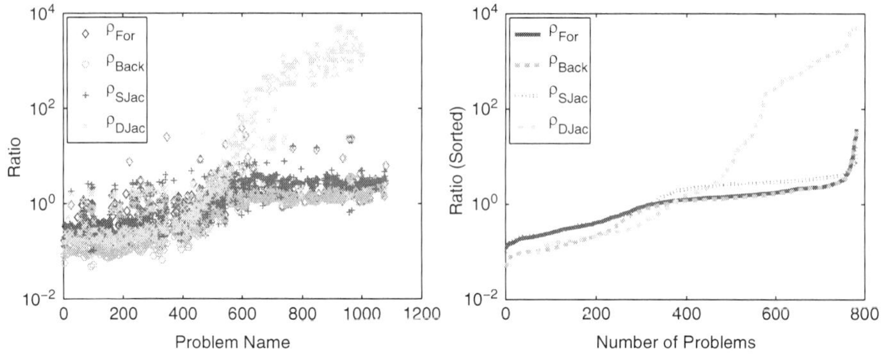

Fig. 6 Evaluation time ratios for the constraint function and first order derivatives information

we observed that the average value of the (nonzero) ratios ρ_{SHess} and ρ_{SJac} were surprisingly low: $\rho_{SHess}^{avg} = 284.1010$ and $\rho_{SJac}^{avg} = 1.9338$. The latter observation indicates that most of the objective and constraint functions have sparse derivative matrices.

2.3 Sparsity

The sparsity of the derivative matrix has a major impact on the performance of linear algebra packages (see for example LAPACK [1]) used within an NLOP solver. Here, sparsity refers to the measure reflecting how many entries of the underlying derivative matrices are nonzero. Therefore, we computed the sparsity measure of a matrix as a dimension independent probability of an entry being nonzero, i.e., we normalized the number of nonzero entries by dividing through the number of

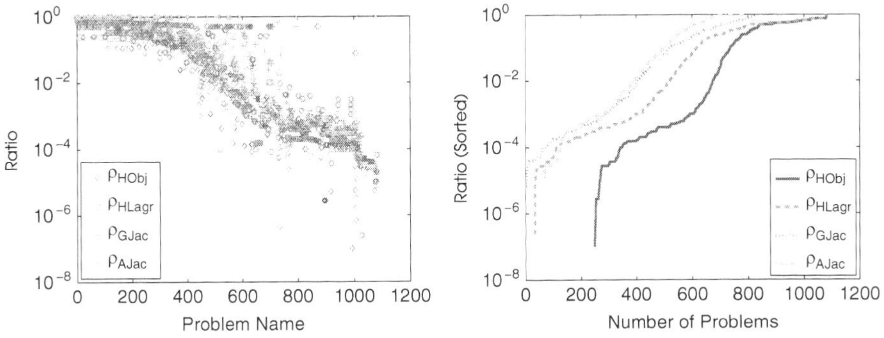

Fig. 7 Sparsity measure for the derivative matrices

all entries within the given matrix. The corresponding probability ρ_{HObj} for the Hessian of the objective function, ρ_{HLagr} for the Hessian of the Lagrangian, ρ_{GJac} for the general and box constraints Jacobian, and ρ_{AJac} for the active general and box constraint Jacobian are depicted in Fig. 7.

The sparsity ratio plotted on the right suggests an evenly spread distribution of sparse problems within the CUTEr test set. However, we find that this is not really true by looking at Fig. 7 (left). Here, the decreasing curve shows that the CUTEr test set is biased in terms of sparsity, the problems become more sparse for growing number of variables. Hence, NLOP solvers that use a sparse linear algebra library are likely to perform advantageously for large scale optimization compared to solvers using dense linear algebra.

3 Conclusion

In this article we provided a numerical overview of the CUTEr test set. We analyzed the structural and computational behavior of general nonlinear problems in terms of size, sparsity and evaluation time. As a result we found that the theoretical results from Automatic Differentiation concerning the evaluation time for tangents and gradients in the forward and backward mode are valid. Furthermore, we presented results indicating that for most of the nonlinear problems one can compute exact sparse Jacobian and sparse Hessian at comparatively low cost. Hence, our previous strategy of designing optimization algorithms based exclusively on evaluating derivative vectors rather than matrices cannot bear fruit on the test set considered (see for example [3, 9, 11]). In contrast the more main stream approach of constructing nonlinear optimization tools using exact sparse derivative matrices and sparse linear algebra works not surprisingly rather well on CUTEr.

However, it is not known to the author whether the sparsity structure of the CUTEr test set problems is representative of real applications or if the problem selection is somewhat biased. A closer investigation by studying run-time ratios and sparsity patterns from other solver interfaces (e.g. from the NEOS server [4, 5, 10]) should be undertaken.

References

1. LAPACK - Linear Algebra PACKage. Website (2011). URL http://www.netlib.org
2. Andrew R. Conn, N.I.M.G., Toint, P.L.: Sif reference document (revised version). Website (2003). URL http://www.numerical.rl.ac.uk/lancelot/sif/sifhtml.html
3. Bosse, T., Lehmann, L., Griewank, A.: On Hessian- and Jacobian-free SQP methods - a total quasi-Newton scheme with compact storage. In: M. Diehl (ed.) Recent Advances in Optimization and its Applications in Engineering (BFG 2009). Springer (2010)
4. Czyzyk, J., Mesnier, M.P., Moré, J.J.: The NEOS server. IEEE Computational Science & Engineering **5**(3), 68–75 (1998). URL http://www.neos-server.org
5. Dolan, E.: The neos server 4.0 administrative guide. Tech. rep., Mathematics and Computer Science Division, Argonne National Laboratory (2001)
6. Gould, N., Toint, P., Orban, D.: CUTEr Test set, small and large test set. Website (2002). URL http://www.cuter.rl.ac.uk. Used Version: After 1.12.2011
7. Gould, N., Toint, P., Orban, D.: General CUTEr documentation. Website (2005). URL http://www.cuter.rl.ac.uk
8. Griewank, A., Walther, A.: Evaluating Derivatives: Principles and Techniques of Algorithmic Differentiation, 2nd edn. No. 105 in Other Titles in Applied Mathematics. SIAM, Philadelphia, PA (2008). URL http://www.ec-securehost.com/SIAM/OT105.html
9. Griewank, A., Walther, A., Korzec, M.: Maintaining factorized KKT systems subject to rank-one updates of Hessians and Jacobians. Optim. Methods Softw. **22**(2), 279–295 (2007)
10. Gropp, W., Moré, J.J.: Optimization environments and the NEOS server. Approximation Theory and Optimization pp. 167–182 (1997)
11. Liu, D.C., Nocedal, J.: On the limited memory BFGS method for large scale optimization. Technical Report NAM 03, Northwestern University, Evanston, Ill. (1988)
12. Nocedal, J., Wright, S.J.: Numerical Optimization. Springer Series in Operations Research. Springer-Verlag, New York, NY (1999)

Java Automatic Differentiation Tool Using Virtual Operator Overloading

Phuong Pham-Quang and Benoit Delinchant

Abstract AD tools are available and mature for several languages such as C or Fortran, but are just emerging in object oriented language such as Java. In this paper, a Java automatic differentiation tool called JAP is presented which has been defined and developed with specific requirements for the design of engineering systems using optimization. This paper presents JAP requirements and the implementation architecture. It also compares JAP performance to ADOL-C in forward mode on a magnetic MEMS model. JAP has been successfully used on several system optimizations in the field of electromagnetic MEMS.

Keywords Java automatic differentiation • Source transformation • Operator overloading • Forward mode • Optimization

1 Introduction

The main goal of this work is to have a Java code differentiation solution, available in our engineering tools, in order to differentiate physical models "on the fly". These models can be composed of equations, algorithms or/and existing compiled libraries. Not having found such a solution directly usable, we have developed a tool, based on traditional techniques of forward propagation, which is primarily designed to satisfy our needs. The solution is presented here to bring our requirements and approach for Java-based automatic differentiation to the attention of the general AD community.

P. Pham-Quang (✉)
CEDRAT S.A., Meylan Cedex, France
e-mail: phuong.phamquang@cedrat.com

B. Delinchant
Grenoble Electrical Engineering Laboratory, Saint-Martin d'Hères, France
e-mail: benoit.delinchant@grenoble-inp.fr

S. Forth et al. (eds.), *Recent Advances in Algorithmic Differentiation*, Lecture Notes in Computational Science and Engineering 87, DOI 10.1007/978-3-642-30023-3_22,

1.1 Java AD Tool

Based on regularly updated data,[1] one can find a number of tools for AD programming languages such as Fortran, C / C++ or Matlab. But there have been only a few attempts to build tools for the Java language:

- In the Master's work in 2001 by Rune Skjelvik [9], the authors proposed a simulation tool that implements operator overloading. This tool is limited because it only supports certain instructions of the Java language and is limited to scalar functions.
- JavaDiff: This tool was developed in 2004 in our laboratory during the PhD thesis of Fischer [3]. JavaDiff is similar to the tool [9] replacing all the mathematical operators by Java methods. The author used JavaDiff to differentiate models applied to the optimization of electromagnetic devices. However, these uses were not fully automatic and only applied to scalar models.
- ADiJaC: this AD tool has been developed by Slusanschi [10], directed by Prof. C.H. Bischof. ADiJaC generates derivative code by the analysis of the compiled code (bytecode). A joint work has been done using ADiJaC [8], but it has shown the following limits:
  - It can differentiate functions in the presence of tables if we manually intervene in the generated code to initialize the tables. The automation of this step by the authors of ADiJaC is in progress.
  - It can not differentiate functions in the presence of external functions. This property is very important to differentiate complex models using libraries of computation which can be based on other derivatives techniques.
  - The dynamic selection of active variables was not available. This feature is essential to reduce the computation time and also to reduce the memory used.

For these reasons, we decide to make a new version which has to match some requirements raised from the field of semi-analytical model optimization.

1.2 Requirements for System Design Optimization

Given the limits of existing AD tools for Java presented above, we have developed JAP: Java Jacobian Automatic Programming. It has to overcome the limits of current Java AD tools particularly by responding to our needs:

- To support all the complex functions written in Java with data array in one dimension ([]) and two-dimensions ([] []).

---

[1]Community Portal for Automatic Differentiation. http://www.autodiff.org/

- To support external functions provided in the form of libraries (jar extension files: Java ARchive).
- To support runtime selection of the partial derivatives to compute.
- To be automatic and easy to use.

There are some limits that we did not attempt to exceed in this first version of JAP; in particular some object oriented programming functionality such as inheritance.

2 Virtual Operator Overloading

JAP simulates operator overloading by replacing the basic operations and intrinsic functions by Java methods. This is a mixture of the usual methods of operator overloading and source transformation.

2.1 Forward Model Principle

To propagate the derivatives in forward mode, JAP transforms all the active variables of type *double* to *Jdouble1* (1 denotes the order of derivation). Figure 1 shows the *Jdouble1* class definition. This class has three instance variables *(val, nGrad, and dval)* to define its characteristics. The variable *nGrad* is the number of input variables to differentiate.

Figure 2 shows an example of an algorithm to differentiate provided by a JAP user. The example chosen here returns a scalar with scalar *(a1, k)* as well as vector *(av)* arguments. Standard operators, intrinsic functions *(cos, log, abs)*, a loop *(for)* and a conditional branch *(if)* are used.

Figure 3 shows the code generated by JAP after analysis of the class provided. A method working with *Jdouble1* is generated. A prefix *"j"* is added to the method name (the example is *jfunc*). To generate this *"j-method"*, all variables *(double)* are converted into *Jdouble1*, and all operations and intrinsic functions are transformed into static methods of class *Jdouble1*. For instance, function calls *Jdouble1.div, Jdouble1.times, Jdouble1.cos and Jdouble1.abs* correspond in some way to overload operators '/' and '*' and intrinsic functions *cos* and *abs*. We call this technique "virtual operator overloading".

2.2 Using Jacobian and Partial Derivatives

After generating *j-methods* that propagate derivatives along the model using *Jdouble1*, JAP also generates methods in order to get the Jacobian matrix and partial derivatives, as can be done by classical methods.

```
public class Jdouble1 {
    public double val;
    public int nGrad;    // number of variable to derive
    public double[] dval;
    // constructor
    public Jdouble1(int nGrad) {
        this.val = 0;
        this.nGrad = nGrad;
        this.dval = new double[nGrad];      }
    // simulation of method  res=A*B
    public static Jdouble1 times(int nGrad, Jdouble1 A, Jdouble1 B) {
        Jdouble1 res = new Jdouble1(nGrad);
        res.val = A.val * B.val;
        for (int i = 0; i < nGrad; i++)
            res.dval[i] = A.dval[i] * B.val + A.val * B.dval[i];
        return res;
    } … }
```

Fig. 1 Jdouble1: derivative object class

```
public double func(double a1, int k, double[] av) {
    double res;
    int i;
    res = Math.abs(a1/ k);
    for (i = 0; i < av.length; i++)
        res += Math.cos(Math.log(av[i]) + i);
    if (av[0] > a1)
        res *= av[0]- a1;
    else
        res *= a1 / 5;
    return res*2;     }
```

Fig. 2 Sample function to differentiate

```
public Jdouble1 jfunc(Jdouble1 a1, int k, Jdouble1[] av) {
    int jap_nGrad = a1.nGrad;
    Jdouble1 res;
    int i;
    res=Jdouble1.abs(jap_nGrad,Jdouble1.div(jap_nGrad,(a1),(k)));
    for (i = 0; i < av.length; i++)
       res.plusEquals(Jdouble1.cos(jap_nGrad,Jdouble1.plus(
                   jap_nGrad,(Jdouble1.log(jap_nGrad,av[i])),(i))));
    if (av[0].val > a1.val)
        res.timesEquals(Jdouble1.minus(jap_nGrad,(av[0]),(a1)));
    else
        res.timesEquals(Jdouble1.div(jap_nGrad,(a1),(5.0)));
    return Jdouble1.times(jap_nGrad, (2.0),(res)); }
```

Fig. 3 Source transformation of the sample function

```
//external function definition
public double extfunc(double a1,double[] av) { ... }
//Jacobian of the external function
public double[] jacobian_extfunc(double a1,double[] av) { ... }
//partial derivative of external function
public double partialDerivative_extfunc(double a1,double[] av, int index)
```

Fig. 4 Jacobian function naming convention of an external function

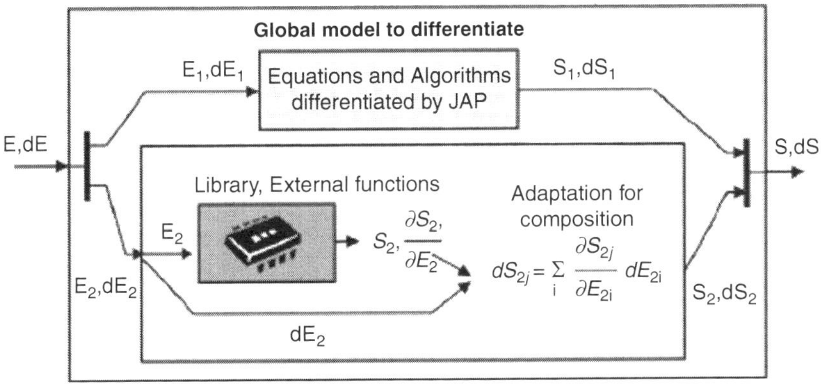

Fig. 5 Model composition, using different propagation techniques

- Full Jacobian matrix: the derivatives of all outputs with respect to all inputs. In this case, *nGrad* is equal to the number of inputs.
- Partial derivative selection: two possibilities
  - The derivatives of all outputs w.r.t. an input which is selected by an index. In this case $nGrad = 1$.
  - The derivatives of all outputs w.r.t. few inputs which are selected by an index array. In this case $nGrad = \textit{index array length}$.

2.3 Management of External Functions

JAP is also capable of supporting external functions provided in the form of computation libraries (Java ARchive). Such a library must provide the computation of the Jacobian or partial derivatives of each function. This computation can be performed by any technique: symbolic, finite differences or automatic differentiation using JAP or any other tools. By convention, the Jacobian is provided by the call of the method whose name is prefixed by "jacobian_" and partial derivatives are provided by the method prefixed by "partialDerivative_". Figure 4 shows an example of the standard external function signature and its corresponding Jacobian and partial derivative.

If the external function is generated by JAP, it provides not only the Jacobian, but also the *j-method*. This *j-method* is used directly for the composition with the global model to improve performance. If the external function is generated by another technique that provides only the Jacobian or some partial derivatives, JAP will make the composition itself by generating a matching method (see Fig. 5). This composition, in particular the transition between these two modes of propagation of sensitivity, must be properly managed to improve computing performance [1].

3 JAP Architecture

JAP is a compiler that processes a text file and automatically generates Java source code. The JAP architecture is divided into three stages (lexical and syntax analysis, source class decomposition, adding forward propagation method). Compared with general code generation techniques [6], JAP does not generate intermediate code, it generates code directly from the source code.

3.1 Lexical and Syntax Analysis

The entry of JAP is a Java source file. This class can contain multiple methods. External libraries (.jar) are placed in the /lib directory at the same level as the class to be differentiated. In this first step, JAP uses the JavaCC[2] library for lexical and syntactic verification of the Java language. If there is a lexical or syntax error (";" missing for example), JAP will report the error and the corresponding line to the user. If no syntax errors are encountered, the process goes to the next step.

3.2 Source Class Decomposition

JAP also uses a parser generated by JavaCC to ignore comments and to split the class into different parts (statement package, import libraries, statement class, declaration of global variables, computation methods). Each method is divided into three parts (method signature, declaration of local variables, core calculation). By convention, local variables must immediately be declared after the method signature.

---

[2] Java Compiler, http://javacc.java.net/

3.3 Adding Forward Propagation Method

JAP adds for each computation method, the associated j-method as shown above.

- Signature of the method: the *double* type is replaced systematically by *Jdouble1*, the method name is prefixed with *"j"*.
- Declaration of local variables: the *double* type is systematically replaced by *Jdouble1*.
- Core computation: all operations and intrinsic functions are transformed into static methods of the *Jdouble1* class.

For each *'public'* computation method, JAP adds three methods in order to compute the full Jacobian, only one partial derivative or several partial derivatives.

3.4 JAP Limits

Given the complexity of the Java language, JAP currently cannot differentiate all kinds of Java code:

- JAP cannot process a file that contains multiple classes. If the functions to differentiate are in multiple classes, the user will have to put them in the same class and in a single file.
- JAP can not handle object inheritance. This mean that there can be no derived classes in the Java program.
- JAP can not handle computing libraries which do not provide their Jacobian.
- JAP only supports the data types: *boolean, double, double[], double[][], int, int[], int[][]*. This means for example, that JAP cannot differentiate the code that it generates.
- JAP conventions must be respected:
  - Forbidden keywords: *"jap_", "jacobian_", "Jdouble", "partialDerivative_"*
  - Overloading a method with the same number of parameter is not allowed, even if its argument types are different.

4 JAP Performance

To evaluate the performance of JAP, it has been compared with ADOL-C [4] on a model which computes a beam deformation under magnetic forces and makes a contact analysis [7]. ADOL-C has been chosen because it is one of the references of the AD community, and it was available in CADES,[3] our optimization

[3] www.cades-solutions.com

Table 1 Beam deformation algorithm specifics

Description	Number
Java source lines	930
Functions	4
Branching structures (if-then-else)	57
Loop structures (for / while)	60

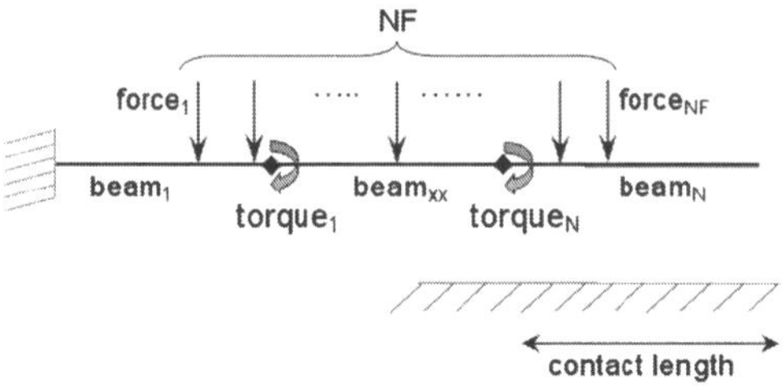

Fig. 6 Forces to compute beam deformation

framework [2]. Table 1 characterizes the contents of the algorithm which computes the beam deformation. This algorithm is quite complex because it contains most control structures offered by programming languages. It may be noted that it is a vector model, method arguments are of type *double, double [], double [][], int.*

The AD tool ADOL-C is used to evaluate the gradient of this algorithm in C++ and JAP for the same algorithm in Java. As the reverse mode is not implemented in JAP, only the forward mode was compared.

The geometry of the beam is shown in Fig. 6. In order to vary the amount of computing time, we vary the number of forces ($NF = 50$, 75, 100) applied to the beam. The model computes two main outputs, *contact length* and *contact force*, which are differentiated with respect to all inputs ($nGrad = 113$, 163, 213).

Table 2 shows the CPU time comparison between ADOL-C and JAP in forward mode. Anyone can observe that the model in C++ is faster than the Java version. That is generally the case between a language compiled directly to machine code and an interpreted language like Java. That is why we do not compare the absolute time, but a relative time (1) which is the time factor between the evaluation of derivatives with respect to an input, and the evaluation time of the model itself.

$$factor = \frac{1}{nGrad} \cdot \frac{Jacobian\ computation\ time}{Model\ computation\ time} \tag{1}$$

It can be observed in Table 2 that the ADOL-C factor and the JAP factor are of the same order of magnitude. This is very comforting about our implementation compared with a mature tool like ADOL-C.

In addition, JAP is able to specify active variables during run time in order to save unnecessary computations. Table 3 shows the CPU time of selective derivative evaluations with respect to some inputs ($nGrad = \{1, 4, 10\}$) compared to the evaluation of the whole Jacobian.

Table 2 CPU performance comparison of forward mode between ADOL-C and JAP

	NF = 50 nGrad = 113	NF = 75 nGrad = 163	NF = 100 nGrad = 213
C++ (model) [s]	0.140	0.188	0.234
Java (model) [s]	0.250	0.343	0.438
ADOL-C (jacobian) [s]	49.03	93.86	147.8
JAP (jacobian) [s]	106.6	251.2	371.3
Factor ADOL-C	3.10	3.06	2.97
Factor JAP	3.77	4.49	3.98

Table 3 Computation time [s] comparison between full and partial Jacobian

	NF = 50	NF = 75	NF = 100
nGrad = 1	3.89	5.77	7.58
nGrad = 4	6.12	9.19	11.73
nGrad = 10	9.75	14.42	19.28
Jacobian	106.6	251.2	371.3

5 Conclusions and Perspectives

In this paper, JAP is presented as a new Java automatic differentiation tool. It has been designed with specific requirements coming from our long experience in the design of engineering systems using optimization. These requirements are mainly based on semi-analytical models of physical devices described in Java language, interoperability with external libraries in which the Jacobian can be produced using symbolic methods, and performance requirements such as run-time selection of active variables.

The JAP architecture has briefly been presented in this paper. Its performance has been compared in this paper with ADOL-C tool in the case of a MEMS deformable beam model, and good results have been obtained. JAP has successfully been used for the design optimization of several devices [5, 8] as well as for DAE simulation. JAP can be improved in two main interesting directions, the first one is Hessian computation which is quite simple with our forward implementation. A *Jdouble2* class has to be defined and associated *j-methods* have to be generated. The second one is the reverse mode which gives sometimes higher performance in our sizing optimization problems.

References

1. Delinchant, B., Wurtz, F., Atienza, E.: Reducing sensitivity analysis time-cost of compound model. IEEE, Transactions on Magnetics **40**(2) (2004)
2. Enciu, P., Wurtz, F., Gerbaud, L., Delinchant, B.: Automatic differentiation for electromagnetic models used in optimization. COMPEL **28**(5) (2009)

3. Fischer, V., Gerbaud, L., Wurtz, F.: Using automatic code differentiation for optimization. IEEE, Transactions on Magnetics **41**(5) (2005)
4. Griewank, A., Juedes, D., Mitev, H., Utke, J., Vogel, O., Walther, A.: ADOL-C: A package for the automatic differentiation of algorithms written in C/C++. Tech. rep., Institute of Scientific Computing, Technical University Dresden (1999). Updated version of the paper published in *ACM Trans. Math. Software* 22, 1996, 131–167
5. Janssen, J., Paulides, J., Lomonova, E., Delinchant, B., Yonnet, J.: Design study on magnetic springs with low resonance frequency. In: LDIA 2011. Eindhoven, The Netherlands (2011)
6. Kowarz, A.: Advanced concepts for automatic differentiation based on operator overloading. Ph.D. thesis, TU Dresden (2007)
7. Pham-Quang, P., Delinchant, B., Coulomb, J.L., du Peloux, B.: Semi-analytical magneto-mechanic coupling with contact analysis for MEMS/NEMS. IEEE, Transactions on Magnetics **47**(5) (2011)
8. Pham-Quang, P., Delinchant, B., Ilie, C., Slusanschi, E., Coulomb, J., du Peloux, B.: Mixing techniques to compute derivatives of semi-numerical models: Application to magnetic nano switch optimization. In: Compumag 2011. Sydney, Australia (2011)
9. Skjelvic, R.: Automatic differentiation in Java. Ph.D. thesis, University of Bergen, Norway (2001)
10. Slusanschi, E.: Algorithmic differentiation of Java programs. Ph.D. thesis, RWTH Aachen University (2008)

High-Order Uncertainty Propagation Enabled by Computational Differentiation

Ahmad Bani Younes, James Turner, Manoranjan Majji, and John Junkins

Abstract Modeling and simulation for complex applications in science and engineering develop behavior predictions based on mechanical loads. Imprecise knowledge of the model parameters or external force laws alters the system response from the assumed nominal model data. As a result, one seeks algorithms for generating insights into the range of variability that can be the expected due to model uncertainty. Two issues complicate approaches for handling model uncertainty. First, most systems are fundamentally nonlinear, which means that closed-form solutions are not available for predicting the response or designing control and/or estimation strategies. Second, series approximations are usually required, which demands that partial derivative models are available. Both of these issues have been significant barriers to previous researchers, who have been forced to invoke computationally intensive Monte-Carlo methods to gain insight into a system's nonlinear behavior through a massive sampling process. These barriers are overcome by introducing three strategies: (1) Computational differentiation that automatically builds exact partial derivative models; (2) Map initial uncertainty models into instantaneous

---

A.B. Younes (✉)
Doctoral Student, Aerospace Engineering, Texas A&M University, College Station, TX, USA
e-mail: olalahmad@gmail.com

J. Turner
Research Professor, Aerospace Engineering, Texas A&M University, College Station, TX, USA
e-mail: turner@aero.tamu.edu

M. Majji
Assistant Professor, Mechanical and Aerospace Engineering, University at Buffalo, Buffalo, NY, USA
e-mail: majjineo@gmail.com

J. Junkins
Distinguished Professor, Aerospace Engineering, Texas A&M University, College Station, TX, USA
e-mail: junkins@tamu.edu

S. Forth et al. (eds.), *Recent Advances in Algorithmic Differentiation*, Lecture Notes in Computational Science and Engineering 87, DOI 10.1007/978-3-642-30023-3_23,

uncertainty models by building a series-based state transition tensor model; and (3) Compute an approximate probability distribution function by solving the Liouville equation using the state transition tensor model. The resulting nonlinear probability distribution function (PDF) represents a Liouville approximation for the stochastic Fokker-Planck equation. Several applications are presented that demonstrate the effectiveness of the proposed mathematical developments. The general modeling methodology is expected to be broadly useful for science and engineering applications in general, as well as grand challenge problems that exist at the frontiers of computational science and mathematics.

Keywords Computational differentiation • Uncertainty propagation • Probability distribution function • Liouville equation • OCEA • State transition tensor

1 Introduction

Complex applications are typically defined by nonlinear mathematical models of the form

$$\dot{x} = f(x, t) \tag{1}$$

where $x \in \Re^n$ and the initial condition is known. The system response is generated by invoking numerical integration methods. For a given initial condition each trajectory is unique. Nevertheless, in the real world, one must address the problem that both initial conditions as well as model parameters may be known only approximately. As a result, the behavior notional systems can only be understood in the context of sampling the system behavior for the anticipated range of uncertain modeling data. Computationally, the solution strategy is straightforward: one attacks this problem by massively sampling the system response to assess the modeling uncertainty. Classically, this problem is handled by introducing Monte-Carlo methods. We seek more efficient methods. To this end, we replace the integration intensive Monte-Carlo method with a state transition tensor series (STTS) approximation. We replace the Monte Carlo model with two steps: (1) numerically integrate the state transition tensor model, and (2) sample the initial uncertainty covariance matrix and use the STTS model where integration is now replaced with algebraic tensor contraction operations. Our current STTS approximation algorithm is limited to fourth order; higher order approximation tools are under development. As with all Taylor-series-like approximations there are legitimate questions regarding the rate of convergence of the series approximations and the solution accuracy provided [8]; future research is required to address all of these issues in a comprehensive way.

Our STTS algorithm strategy consists of three key steps. First, we define an initial n-dimension uncertainty covariance matrix, which is sampled to provide uncertainty propagation vectors. Second, STTS is used to map the uncertainty vectors to points at future times. This calculation provides data for evaluating the non-Gaussian evolution for the initial covariance matrix due to the system's fundamental nonlinear behavior. Third, by selecting sample points on the surface of the initial covariance matrix we are able to identify the n-dimensional volume occupied by the nonlinearly transformed STTS initial condition uncertainty vectors. The determination of the transformed uncertainty volume is important for numerically evaluating expectation values for the non-Gaussian PDF approximation that is generated through our solution for the Liouville equation. Generating the gradient tensors required by the STTS model is only practical by invoking the use of computational differentiation (CD) tools, which are briefly described. As applied in this paper, CD provides a FORTRAN language extension for the automatic generation of mixed partial derivative models, where operator-overloading methodologies embed the chain rule of calculus in the intrinsic and mathematical library functions for generating numerical values for all partial derivatives.

The key contribution of the paper is the presentation of a general-purpose methodology for automatically assembling the tensor gradients and computing the state transition tensor differential equations required for nonlinearly mapping initial condition estimates to future times along a nominal trajectory. The time evolution of the initial condition uncertainty is intimately related to the PDF that characterizes the non-Gaussian statistical behavior of the response. Mathematically, an exact PDF calculation requires the numerical solution for the Fokker-Planck equation, which is defined by an n-dimensional nonlinear partial differential equation. Many expansion methods have been proposed for the solution, however, all are plagued by the problem that the computational domain that defines the solution is unknown and must be recovered dynamically as part of the solution. As a result, exact solutions have been limited to four or five degrees of freedom, even for super-computer-class solution algorithms. The proposed methods bypass this limitation by eliminating diffusion effects, which permits that introduction of a tensor-based series solution for capturing nonlinear behaviors. We transform the underlying nonlinear PDE into an approximate nonlinear ODE which is known as the stochastic Liouville equation. Monte-Carlo simulations are performed to confirm the accuracy of the mean and covariance predictions that are generated from our tensor-based Liouville-equation approximation. The numerical integration of the state equations and the transition tensor ODEs is accomplished by introducing a generalized scalar variable that embeds all of the ODEs. This approach eliminates the annoying task of having to unpack and repack the tensors as numerical approximations are generated. The numerically integrated tensor equations are validated by considering two-body motion predictions from astrodynamics which are well-known to lead to symplectic behaviors. These systems allow the first through fourth order transition tensor calculations to confirm that the resulting partial derivatives satisfy the expected symplectic behavior through fourth order. As shown in the numerical results section, the integrated tensor solutions indeed satisfy the symplectic behavior test, which is a necessary but not sufficient condition to validate the Liouville equation solution.

2 Computational Differentiation

Computational differentiation has existed since the 1960s starting with the seminal works of Wengert [18] and Wilkins [19] and developed by many others [2–5]. Early approaches used an existing coded mathematical model as a template for applying the rules of calculus to write a new code that implements the partial derivatives for the mathematical model. This approach has proven to be very effective for generating first-order sensitivity models. High-order derivative models are generated by using operator-overloading for the standard arithmetic operators, where the transformations necessary to produce derivatives are implicitly handled by the programming language, when the compiler detects a derivative-enhanced data type. In this way, statements like, $x = 5yz + cosh(y/x)$ can be written without being decomposed into elementary operations. The compiler derives the mathematical model, and codes the executable for the partial derivative model. Symbolic manipulation tools are used to derive the tensor index form of each solution, which are then mapped to FORTRAN using symbolic program utilities: no Taylor coefficients are computed, no multi-index array defined, no special addressing scheme is required, all calculations are structured to exploit array processing opportunities. The current version of OCEA supports first through fourth order sensitivity models, extensions for fifth and sixth order have been symbolically derived and coded in FORTRAN but not tested. FORTRAN 2003 and 2008 support the larger array sizes required for extending the results to higher order. All derivative calculations are exact and the resulting numerical calculations are accurate to the working precision of the machine [1, 6, 7, 9–11, 14–17]. The analyst provides only a FORTRAN 95/2003 mathematical model and the compiler automatically generates all partial derivatives without operator intervention.

OCEA operates by identifying the independent variables and defining the following data structure $x := (x, U, 0, 0, 0)$ where $x \in \Re^n$ and *dependent variables* are defined by the more general data structure of the form

$$f := \left(f,\ \nabla f,\ \nabla^2 f,\ \nabla^3 f,\ \nabla^4 f\right) \tag{2}$$

Future versions of **OCEA** will exploit symmetric data structures. Equation 2 assumes that no implicit variable dependence exists. Operational details for using **OCEA** are found in the **OCEA** user manual [13].

3 State Transition Tensor Models

Real-world applications must deal with the reality that both initial conditions as well as system parameters are often uncertain. Monte-Carlo-based sampling methods are often invoked for recovering the desired information. We propose to address this problem by revisiting how uncertainty is propagated in a nonlinear system model by replacing the computationally intensive Monte-Carlo method with a tensor series based approach.

Six algorithmic steps are required:

- Assume a covariance matrix for defining the initial condition uncertainty.
- Sample the initial condition uncertainty covariance ellipse to provide candidate samples to be propagated along the nonlinear system trajectory.
- Develop a series-based nonlinear state-transition tensor model for algebraically propagating the initial condition samples to final trajectory variations.
- Revert the tensor power series to provide transformed initial conditions.
- Compute the gradient of the reverted tensor power series to provide the determinate for the transformed PDF.
- Compute multi-dimensional path integrals for evaluating the nonlinear mean, covariance matrix, and other higher-order statistical measures of the system response.

This approach has several benefits:

- Tensor series replaces potentially millions of state integrations.
- Approximate the nonlinear PDF
- Reverted nonlinear model provides a Liouville equation approximation for Fokker-Planck equation.
- Availability of an approximate PDF makes it possible to re-visit the foundations of estimation theory.

State transition tensor models are developed by computing the sensitivity of the equation of motion defined by (1) w.r.t. the initial conditions for the state vector, in the indicial form

$$\dot{x}_i = f_i \tag{3}$$

from which the first through fourth-order tensor differential equations follow as

$$\begin{aligned}
\dot{x}_{i,j} &= f_{i,r} x_{r,j}; \quad x_{r,j} = \delta_{r,j} \\
\dot{x}_{i,jk} &= f_{i,rs} x_{r,j} x_{s,k} + f_{i,r} x_{r,jk}; \quad x_{r,jk} = 0_{n\times n\times n} \\
\dot{x}_{i,jkl} &= f_{i,rsp} x_{r,j} x_{s,k} x_{p,l} + f_{i,rs}\left(x_{r,jl} x_{s,k} + x_{r,j} x_{s,kl} + x_{r,jk} x_{s,l}\right) \\
&\quad + f_{i,r} x_{r,jkl}; \quad x_{r,jkl} = 0_{n\times n\times n\times n} \\
\dot{x}_{i,jklm} &= f_{i,rspq} x_{r,j} x_{s,k} x_{p,l} x_{q,m} + f_{i,rsp}\left(x_{r,jm} x_{s,k} x_{p,l} + x_{r,j} x_{s,km} x_{s,l} + x_{r,j} x_{s,k} x_{p,lm}\right) \\
&\quad + f_{i,rsp}\left(x_{r,jl} x_{s,k} + x_{r,j} x_{s,kl} + x_{r,jk} x_{s,l}\right) x_{p,m} + f_{i,rs}\Big(x_{r,jlm} x_{s,k} + x_{r,jl} x_{s,km} \\
&\quad + x_{r,jm} x_{s,kl} + x_{r,j} x_{s,klm} + x_{r,jkm} x_{s,l} + x_{r,jk} x_{s,lm} + x_{r,jkl} x_{s,m}\Big) \\
&\quad + f_{i,r} x_{r,jklm}; \quad x_{r,jklm} = 0_{n\times n\times n\times n\times n}
\end{aligned} \tag{4}$$

where Einstein's index notation is employed. As a shorthand notation the state transition tensors are defined by

$$\phi_{1_{i,j}} = \frac{\partial x_i}{\partial x_{0_j}};\ \phi_{2_{i,jk}} = \frac{\partial^2 x_i}{\partial x_{0_j}\partial x_{0_k}};\ \phi_{3_{i,jkl}} = \frac{\partial^3 x_i}{\partial x_{0_j}\partial x_{0_k}\partial x_{0_l}};\ \phi_{4_{i,jklm}} = \frac{\partial^4 x_i}{\partial x_{0_j}\partial x_{0_k}\partial x_{0_l}\partial x_{0_m}}.$$

4 Probability Distribution Function

An exact model of this process requires an investigation into solution strategies for the stochastic Fokker-Planck equation, which is beyond the scope of the approximations considered in this paper. Alternatively, this paper develops a computationally useful PDF approximation by introducing using a tensor series representation.

4.1 Approximate Solutions to the Liouville's Equation

Given a prescribed initial condition for the state, one assumes that (1) is numerically integrated to produce a nominal trajectory, as well as the state transition tensors of (4). Sampling the uncertainty model provides δx_0 which is propagated to the current simulation time by expanding the initial error in the following algebraic series approximation

$$\delta x = \phi_1 \delta x_0 + \frac{1}{2!}\phi_2 \delta x_0 \delta x_0 + \frac{1}{3!}\phi_3 \delta x_0 \delta x_0 \delta x_0 + \frac{1}{4!}\phi_4 \delta x_0 \delta x_0 \delta x_0 \delta x_0 + \cdots \tag{5}$$

where the state transition tensors are previously defined. Equation 5 provides transformed models for the n-dimension covariance surface and volume required for performing the multi-dimensional path integrals for propagating the nonlinear system statistics [12]. Equation 5 defines the forward transformation for the uncertainty model. Assuming that an initial PDF $[P_{\delta x_0}(\delta x_0)]$ is known, (5) defines the transformation $\delta x = g(\delta x_0)$. The nonlinear problem statistics are recovered by assuming that $g(*)$ is an invertible, continuously differentiable mapping, with a differentiable inverse, yielding the transformed PDF

$$P_{\delta x}(\delta x) = P_{\delta x_0}\left(\delta x_0 = g^{-1}(\delta x)\right)\left|det\left(\frac{dg^{-1}(\delta x)}{d(\delta x)}\right)\right| \tag{6}$$

Recognizing that (5) defines the $g(*)$ mapping function, it follows that the transformed PDF is known as g^{-1} as soon as it is recovered.

4.2 Data Structures for Numerical Integration

The calculation of the state and state transition tensors differential equations is handled by introducing a generalized scalar variable, whose substructure components handle all of the uncertainty modeling equations. The assumed data structure for the generalized scalar variable is given by

$$\underbrace{g}_{1\times1} := \Big(\underbrace{t}_{1\times1}, \underbrace{x}_{n\times1}, \underbrace{\phi_1}_{n\times n}, \underbrace{\phi_2}_{n\times n\times n}, \underbrace{\phi_3}_{n\times n\times n\times n}, \underbrace{\phi_4}_{n\times n\times n\times n\times n}\Big) \tag{7}$$

The scalar part is time in the integration process. The vector part is the state vector, which is assumed to be derivative-enhanced as an **OCEA** variable. The state transition tensors are double precision multidimensional arrays. The time derivative of g is assembled at each integration step as

$$\underbrace{\dot{g}}_{1\times1} := \Big(\underbrace{1}_{1\times1}, \underbrace{f}_{n\times1}, \underbrace{\dot{\phi}_1}_{n\times n}, \underbrace{\dot{\phi}_2}_{n\times n\times n}, \underbrace{\dot{\phi}_3}_{n\times n\times n\times n}, \underbrace{\dot{\phi}_4}_{n\times n\times n\times n\times n}\Big) \tag{8}$$

The **OCEA** generated gradients of f are used to build the state transition tensor differential equations.The function multiplications required by numerical integration are handled by defining operator-overloaded capabilities for the generalized scalar data type. The advantage of this approach is that the state transition tensors do not have to be loaded and unloaded at each integration step approximation for carrying out scalar multiplications.

5 Applications

5.1 Unforced Duffing Oscillator

In this example the initial uncertainty in the state vector is propagated in time by employing first-through-fourth order of state transition tensors, see Fig. 1. Initially, the uncertainty forms circular regimes (as $\sigma_1 = \sigma_2 = 0.15$), then it starts to deform based on the trajectory behavior. The state equation is given by

$$\dot{x}_1 = x_2; \qquad \dot{x}_2 = -x_1 - x_1^3 \tag{9}$$

5.2 Two-Body Problem

Unperturbed Keplerian motion is governed by an inverse square gravity field. For object motions near Earth the equation of motion is defined by [17, 20]

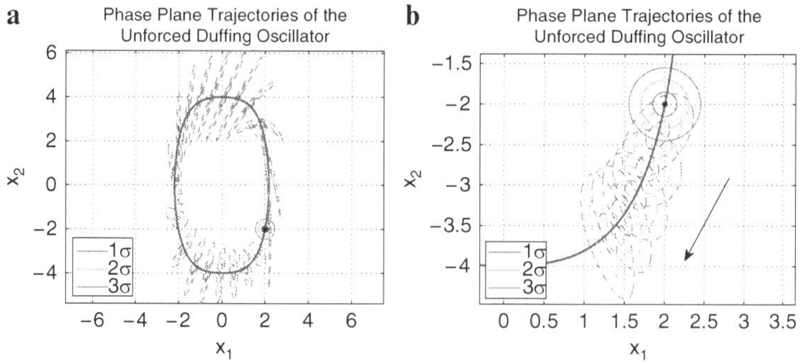

Fig. 1 State uncertainty propagation using first through fourth order state transition tensor

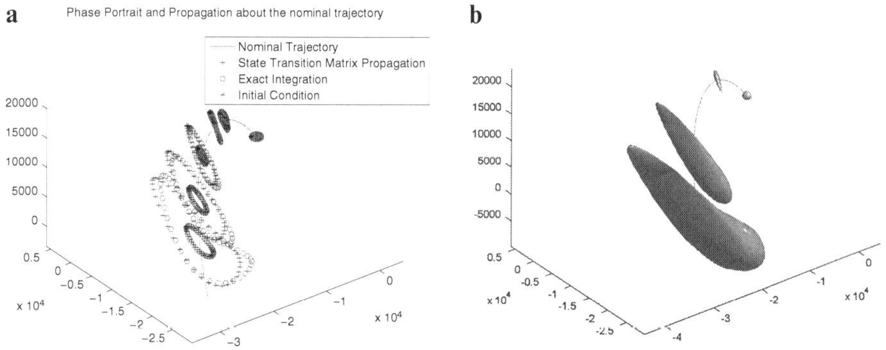

Fig. 2 Phase portrait and uncertainty propagation about the nominal trajectory using first through third order state transition tensor

$$\ddot{\mathbf{r}} = -\frac{\mu_E}{r^3}\mathbf{r} \tag{10}$$

where the inertial vector of position coordinates from the Earth center, $\mathbf{r}$ and μ_E denote the gravitational parameters of the Earth. Bold face letters denote vectors and the corresponding non-bold letters denote the magnitudes of these vectors, that is, $r = |\mathbf{r}|$. The state uncertainty is viewed in a 3D plot, as seen in Fig. 2.

6 Summary and Conclusion

A generalized methodology is presented for achieving three goals: (1) integrating a nonlinear response, (2) propagating an uncertainty envelope, and (3) predicting how the non-Gaussian statistics propagates through the nonlinear system dynamics. Three innovative ideas are presented: (1) a state transition tensor series approach is developed for mapping initial uncertainty vectors into instantaneous uncertainty vectors, (2) a tensor-based reversion of series algorithm is presented for mapping the instantaneous uncertainty vectors into an equivalent initial condition uncertainty vector, (3) the transformed instantaneous uncertainty vectors are processed to evaluate the evolved PDF for the nonlinear system. The resulting nonlinear transformations enable multidimensional integration for analytically computing the nonlinear mean, covariance, and higher-order statistics. Future research will investigate the impact that the availability of the these higher order statistical measures have on generating real-time estimates for the system behavior for system identification and state estimation. The fundamental algorithms represent a Liouville equation approximation for the Fokker-Planck equation. It is anticipated that these results will be broadly useful for knowledge discovery applications in science and engineering.

References

1. Bai, X., Junkins, J.L., Turner, J.D.: Dynamic analysis and adaptive control law of stewart platform using automatic differentiation. AIAA 2006-6286. AIAA/AAS Astrodynamics Specialist Conference and Exhibit, Keystone, Colorado (2006)
2. Bischof, C.H., Carle, A., Corliss, G.F., Griewank, A., Hovland, P.D.: ADIFOR: Generating derivative codes from Fortran programs. Scientific Programming **1**(1), 11–29 (1992)
3. Bischof, C.H., Carle, A., Hovland, P.D., Khademi, P., Mauer, A.: ADIFOR 2.0 user's guide (Revision D). Tech. rep., Mathematics and Computer Science Division Technical Memorandum no. 192 and Center for Research on Parallel Computation Technical Report CRPC-95516-S (1998). URL http://www.mcs.anl.gov/adifor
4. Eberhard, P., Bischof, C.H.: Automatic differentiation of numerical integration algorithms. Mathematics of Computation **68**, 717–731 (1999)
5. Griewank, A., Walther, A.: Evaluating Derivatives: Principles and Techniques of Algorithmic Differentiation, 2nd edn. No. 105 in Other Titles in Applied Mathematics. SIAM, Philadelphia, PA (2008). URL http://www.ec-securehost.com/SIAM/OT105.html
6. Griffith, D.T., Turner, J.D., Junkins, J.L.: Automatic generation and integration of equation of motion for flexible multibody dynamical systems. AAS Journal of the Astronautical Sciences **53**(3), 251–279 (2005)
7. Junkins, J.L., Turner, J.D., Majji, M.: Generalizations and applications of the lagrange implicit function theorem. Special Issue: The F. Landis Markley Astronautics Symposium, The Journal of the Astronautical Sciences **57**(1 and 2), 313–345 (2009)
8. Lohner, R.J.: Enclosing the solution of ordinary initial and boundary value problems. In E. Kaucher, U. Kulisch, and C. Ullrich, Editoisr, Computer Arithmetic: Scientific computation and Programming LnaguagesLanguages pp. 255–286 (1987)
9. Macsyma, Inc: Macsyma, Symbolic/numeric/graphical mathematics software: Mathematics and System Reference Manual, 16th edn. (1996)
10. Majji, M., Junkins, J.L., Turner, J.D.: An investigation of the effects of nonlinearity of algebraic models. AAS-303. Presented at the Terry T. Alfriend Astrodynamics Symposium, Monterey California (2010)
11. Sovinsky, M.C., Hurtado, J.E., Griffith, D.T., Turner, J.D.: The hamel representation: A diagonalized poincare form. ASME Journal of Computational and Nonlinear Dynamics **2**, 316–323 (2007)
12. T., Hahn: Cubaa library for multidimensional numerical integration. Computer Physics Communications **168**(2), 78–95 (2005). DOI 10.1016/j.cpc.2005.01.010. URL http://www.sciencedirect.com/science/article/pii/S0010465505000792

13. Turner, J.: OCEA User Manual. Amdyn System (2006)
14. Turner, J.D.: The application of Clifford algebras for Computing the sensitivity partial derivatives of linked mechanical systems. Nonlinear Dynamics and Control, USNCTAM14: Fourteenth U.S. National Congress Of Theoretical and Applied Mechanics, Blacksburg, Virginia (2002)
15. Turner, J.D.: Automated generation of high-order partial derivative models **41**(8), 1590–1599 (2003)
16. Turner, J.D., Majji, M., Junkins, J.L.: Keynote paper: Fifth-order exact analytic continuation numerical integration algorithm. In: Proceedings of International Conference on Computational and Experimental Engineering and Sciences 2010. Presented to International Conference on Computational and Experimental Engineering and Sciences, Nanjing, China (2010)
17. Turner, J.D., Majji, M., Junkins, J.L.: Keynote paper: High accuracy trajectory and uncertainty propagation algorithm for long-term asteroid motion prediction. In: Proceedings of International Conference on Computational and Experimental Engineering and Sciences 2010. Presented to International Conference on Computational and Experimental Engineering and Sciences, Nanjing, China (2010)
18. Wengert, R.: A simple automatic derivative evaluation program. Communications of the ACM **7**(8), 463–464 (1964)
19. Wilkins, R.D.: Investigation of a new analytical method for numerical derivative evaluation. Communications of the ACM **7**(8), 465–471 (1964)
20. Majji, M., Junkins, J.L., Turner, J.D.: A high order method for estimation of dynamic systems. J. Astronaut. Sci. **56**(3), (2008)

Generative Programming for Automatic Differentiation

Marco Nehmeier

Abstract In this paper we present a concept for a C++ implementation of forward automatic differentiation of an arbitrary order using expression templates and template metaprogramming. In contrast to other expression template implementations, the expression tree in our implementation has only symbolic characteristics. The run-time code is then generated from the tree structure using template metaprogramming functions to apply the rules of symbolic differentiation onto the single operations at compile-time. This generic approach has the advantage that the template metaprogramming functions are replaceable which offers the opportunity to easily generate different specialized algorithms. We tested the functionality, correctness and performance of a prototype in different case studies for floating point as well as interval data types and compared it against other implementations.

Keywords Automatic differentiation • Expression templates • Template metaprogramming • Generative programming • C++ • C++11 • C++0x

1 Introduction

Commonly used implementations of automatic differentiation may be divided into two categories, those using operator overloading to compute the values of the function and the derivative together as well as special tools applying the technique of source transformation to mix in the expressions to compute the derivatives.

In this paper we combined the benefits of both concepts by using template-based C++ techniques like expression templates [13] and template metaprogramming [14] to generate the run-time code for automatic differentiation.

M. Nehmeier (✉)
Institute of Computer Science, University of Würzburg, Am Hubland,
D 97074 Würzburg, Germany
e-mail: nehmeier@informatik.uni-wuerzburg.de

S. Forth et al. (eds.), *Recent Advances in Algorithmic Differentiation*, Lecture Notes in Computational Science and Engineering 87, DOI 10.1007/978-3-642-30023-3_24,

Expression templates is a means for user defined specification of the semantics of expressions. The expression tree is explicitly visible and can be transformed at compile-time.

Template metaprogramming is another powerful programming technique which uses the Turing completeness of C++ templates [15] to realize an interpreter working on types or values of primitive data types at compile-time. This offers the opportunity to write compile-time functions and compile-time control structures.

2 Related Work

Besides automatic differentiation libraries like ADOL-C [8] or FADBAD [4] using operator overloading to compute the values of the function and derivatives together our approach has a strong relation to [6] and especially [2].

In [6], Gil and Gutterman used expression templates and template metaprogramming to perform symbolic differentiation at compile-time. The main concept of this approach is that the symbolic expression or expression tree is assembled out of template classes which use the typedef `tag` for a recursive type composition of template classes to specify the symbolic derivative, see Listing 1. The derivative of a symbolic expression of the type `T` is then easily deduced by the type `T::tag`.

Listing 1 Recursive type composition for symbolic differentiation of the addition used in [6].

```
template<typename T1, typename T2> struct plus_t {
    typedef plus_t<typename T1::tag, typename T2::tag> tag;
};
```

Aubert, Di Césaré and Pironneau used expression templates for forward automatic differentiation in their library FAD [2] to reduce the number of loops, temporaries and copies while computing the first order partial derivatives of an expression. Listing 2 shows the class `Fad` which represents a variable of an expression in this approach. The member `val_` stores the value of the variable whereas the partial derivatives are represented by the vector `dx_`.

Listing 2 Forward automatic differentiation class `Fad` of the FAD library [2].

```
template<typename T> class Fad {
  protected:
    T val_;
    Vector<T> dx_;
    ...
};
```

The call of the member function `diff(int ith, int n)` sets the number of independent variables to `n` and turns the variable of type `Fad` into the `ith` independent variable. This means that the vector `dx_` is initialized with the size n and the ith element of dx_ is set to 1. The member functions val() and dx(int i) are then used to access the value or the partial derivatives, respectively.

Constants are represented by the class FadCst which is almost similar to the class Fad excepted that the vector dx_ as well as the function diff(int ith, int n) are not necessary.

The remaining implementation of the FAD library, like the construction and evaluation of the expression tree, is similar to the classical expression template implementation for vector operations introduced by Veldhuizen [13]. The expression tree is assembled by template classes representing the operations or functions as inner nodes storing references to their child nodes and the classes Fad and FadCst as leaf nodes. The respective template class for an operation or inner node also implements the member functions val() and dx(int i) which apply the operation or the computation of the corresponding derivative onto the values and derivatives of the child nodes.

Note that the FAD library also provides the class TFad<int N, typename T> as an alternative to the class Fad where the number of independent variables is chosen at compile-time. In this implementation the array T dx_[N] is used to represent the partial derivatives.

Sacado is an automatic differentiation library for C++ which supports forward and reverse automatic differentiation. The forward automatic differentiation is based on FAD and uses expression templates for efficiency [12]. Even though it is a complete redesign, the two Sacado implementations DFad and SFad are basically similar to the FAD approaches Fad and TFad, respectively.

3 Implementation

In contrast to the automatic differentiation libraries FAD or Sacado our approach uses expression templates in a completely different way. The expression tree in our implementation has only symbolic characteristics and all the behavior of the represented expression later is associated via trait classes [10] during the generation of the necessary run-time code. This means that the template classes used for the nodes of the expression tree do not provide member functions like val(), dx(int i) or diff(int ith, int n) for the computation of the value or the derivatives. The only exceptions are the classes used for variables and constants of the expression which store the corresponding value and provide the member function val() for the access. All the inner nodes of the tree are almost empty classes which only provide the functionality for building an expression tree at compile-time.

The symbolic characteristic of the expression tree is then used to transform its structure into the required run-time code for the automatic differentiation. This is done by using trait classes which traverse the expression tree, similar to the well known visitor pattern [5], mixing in the necessary run-time code applying the rules of symbolic differentiation onto the single nodes.

Therefore our approach uses expression templates to record the structure of an expression for a later code transformation at compile-time which combine the advantages of the visitor pattern,

- Extensibility: due to the modular design it is easy to extend the implementation for other data types,
- Replaceability: due to the separation of the expression tree and the code generation it is possible to traverse the tree with different "visitor classes" at compile-time to generate different algorithms,

with the efficiency of expression templates and template metaprogramming.

3.1 Implementation Details

Before we start to sketch our approach we want to motivate it with a short example applying our implementation.[1]

Listing 3 Automatic differentiation up to the second order.

```
Var<double, 'x'> x(1.5);
Var<double, 'y'> y(2.3);
Con<double> c(2.0);

// Create expression tree
auto expr = x * x + c * y;
// Perform automatic differentiation (implies code generation)
auto res = expr.df<2>();

// Shortform using res as an intermediate variable
auto res2 = ( sin(res) * y ).df<2>();

cout << "res  d/dx   " << get<'x', 1>(res)  << endl;
cout << "res2 d/dydy " << get<'y', 2>(res2) << endl;
```

As we can see in Listing 3, our approach offers a domain specific language which is intuitive and easy to use. On the other hand the example shows the separation of the expression tree `expr` and the code generation for the automatic differentiation up to the second order which is performed by the call of the template function `df<2>()`.

[1]Note that because of the introduction of type inference with the keyword `auto` in the new C++ standard (C++11, formerly known as C++0x) [3] the usage of expression template and template metaprogramming libraries is extremely simplified for the user.

3.1.1 Expression Tree

As described in Sect. 3 the expression tree in our approach has only symbolic characteristics. Hence the implementation of the main classes for the expression tree is almost straightforward.

As shown in Listing 3 we have a class Var<typename T, int ID> implementing the type for different independent variables. Thereby, the template parameter T specifies the underlying type for the automatic differentiation like double or interval. The parameter ID is an identifier which is used to separate the different independent variables at compile-time using int or char values, e.g Var<double, 1> or Var<double, 'x'>. Hence, the number of independent variables is, like for FAD TFad of Sacado SFad, fixed at compile-time. As described in Sect. 3 the class Var only stores the corresponding value of the independent variable and provides the member function val() for the access.

A type Con<typename T> representing constants of the expression is defined in a similar manner but without the necessity of the template parameter ID.

The inner nodes of the expression tree are represented by the template class ETNode<typename T, typename Policy, typename... Para>, see Listing 4.

Listing 4 Inner node type of an expression tree with an arbitrary number of successors.

```
template<typename T, typename Policy, typename... Para>
class ETNode {
    std::tuple<Para...> s_;
    ...
};
```

This class for the inner nodes uses the C++11 feature variadic templates[7] for the specification of the child nodes or successors. Thereby the ellipsis after the keyword typename of the template identifier Para labels this parameter as a template type parameter pack [7] which can be used with an arbitrary number of arguments offering the feasibility for operations/nodes with an arbitrary number of parameters or successors. The references of the successors of an ETNode are stored in a variadic tuple of the type std::tuple<Para...> which offers the opportunity to address the reference of each element at compile-time [3].

To achieve the symbolic behavior for our expression tree we specify the required operation or function through a policy class [1] associated with the template parameter Policy. The definition of these policy classes is similar to the class in Listing 1 and we use a typedef df to specify the symbolic differentiation. The only difference to [6] is that we perform the symbolic differentiation only on one single operation/node to compute the necessary formula for the automatic differentiation. Hence, we have to introduce an additional policy class RefId<int I, int N>, see Listing 5, to define an abstract or symbolic representation for an argument of an operation which can later be used during the code generation for the "index" of the values and derivatives of the child nodes, see Sect. 3.1.2. Thereby, the template parameter list stands for the N_{th} derivative of the I_{th} argument.

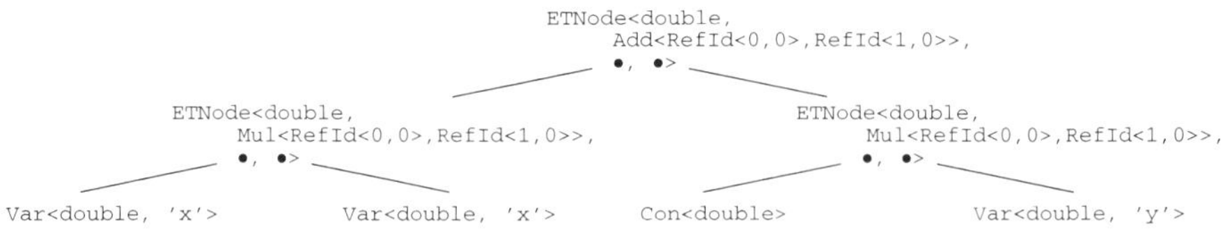

Fig. 1 Expression tree of the type `expr` from Listing 3

Listing 5 Policy class describing the Nth derivative of the Ith argument.

```
template<int I, int N> struct RefId {
    typedef RefId<I, N + 1> df;
};
```

With these descriptive policy classes it is easy to assemble the nested `Policy` describing the operation required by an inner node. The following line defines the policy for an addition

`Add<RefId<0,0>,RefId<1,0>>`

as an addition of the values (0th derivative) of the first (index `0`) and second (index `1`) argument. Now the derivative of this operation

`Add<RefId<0,1>,RefId<1,1>>`

can be easily deduced by using the typedef `df` of the policy at compile-time.

With all these building blocks it is now easy to define the functions or to overload the operators to create an expression tree.[2] Figure 1 shows the detailed expression tree of the type `expr` from Listing 3.

3.1.2 Code Generation

Due to the fact that the code generation of the automatic differentiation is performed by template metaprogramming during compile-time, a special data type to store the results is required. For this purpose we introduce the template class `Derivative<typename T, int N, typename Vars>`. Thereby, the template parameter `T` specifies the underlying type, the parameter `N` represents the desired order for the automatic differentiation and `Vars` is a template structure similar to a type list [1] containing the identifiers[3] of the independent variables.

It is necessary for the template structure `Vars` to offer a list or set functionality at compile-time which can be easily realized in a type `VarList<int... Vars>` using variadic templates and template metaprogramming [7].

[2]Note that we additionally allow instances of the result type `Derivative` from Sect. 3.1.2 as leaf nodes. Hence, it is possible to work with intermediate results, e.g. `res` in Listing 3.

[3]The identifier of an independent variable is specified by the template parameter `ID` of the class `Var`, see Sect. 3.1.1.

The main reason for the template class `Derivative` is the possibility to access the references of the value and partial derivatives at compile-time. To realize this functionality we implemented this class with some similarities to the `tuple` class discussed in [7]. In detail we use the template specialization `Derivative<T, N, VarList<ID, Tail...>>`, see Listing 6, which is derived from the same template class `Derivative<T, N, VarList<Tail...>>` with a reduced `VarList`. The type `Derivative<T, N, VarList<ID, Tail...>>` then stores the `N` partial derivatives for the variable with the identifier `ID` in a member `t_` of the type `std::tuple<T,...,T>`. This tuple structure can be easily generated with a recursive compile-time function `crTuple<int N, typename T>`. Additionally the identifier `ID` is stored in an enum `var`.

Listing 6 Template specialization `Derivative<T, N, VarList<ID, Tail...>>`.

```
template<typename T, int N, int ID, int ... Tail>
class Derivative<T, N, VarList<ID, Tail... >>
  : public Derivative<T, N, VarList<Tail...>> {
    crTuple<N, T>::type t_;
  public:
    enum {var = ID};
    ...
};
```

As a termination condition for the recursive derivation of the class `Derivative` we define the template specialization `Derivative<T, N, VarList<>>` with an empty `VarList` that only defines a member `val_` to store the value of the expression.

With template metaprogramming it is now easy to write template functions like `get<int ID, int N, typename D>(D& d)`, as used in Listing 3, to access the references of the value and derivatives of a `Derivative` structure specified by the parameter `D` at compile-time.[4]

The template function `df<int N>()` from Listing 3, which is used to perform the automatic differentiation by an implicit code generation, now can be defined in the class `ETNode` from Sect. 3.1.1 using an instance of the data structure `Derivative` as the result type.[5] To perform the code generation and evaluation, this function defines an instance of type `Derivative` which will be applied to the "visitor class" generating the run-time code. Note that the necessary `VarList` for the type `Derivative`, which contain the independent variables, can be computed from the expression tree with template metaprogramming.

The "visitor class" that does the code generation for the forward automatic differentiation is realized with the template class `Eval<typename T, typename D, typename Node>` which provides a method `static void eval(D& d, Node const& n)` for the code generation. Thereby, `D` is the template

[4]Note that the compiler can automatically deduce the template parameter `D` from the argument `d`.

[5]For example, the type of `res` from Listing 3 is `Derivative<double, 2, VarList<'x', 'y'>>`.

parameter for the type Derivative and Node is the type of the actual node of the expression tree. Note that the independent variables as well as the required order N are already specified by the type Derivative. The method for the code generation of an inner node performs the following steps using template metaprogramming:

1. Mix in the run-time code to create instances d_1, ..., d_n of the type D for the evaluation of the n successors of the current node. These instances are stored in a tuple t of the type std::tuple<D, ..., D>.
2. Use the class Eval recursively to mix in the evaluation of the child nodes, writing their results (value and derivatives) into the instances $d_1, \ldots, d_n$.
3. Mix in the run-time code for the evaluation of the operation (specified by the Policy of the node) and its partial derivatives of the independent variables $ID_1, \ldots, ID_k$ by applying the tuple t onto a corresponding trait class EvalPol.

- get<ID_1,0>(d) = EvalPol<T,ID_1,Policy>::eval(t)
- get<ID_1,1>(d) = EvalPol<T,ID_1,Policy::df>::eval(t)
- get<ID_1,2>(d) = EvalPol<T,ID_1,Policy::df::df>::eval(t)
- ...

The specialization of the template class Eval for the leaf nodes Var and Con easily fills the provided Derivative structure with the appropriate values.

The trait class EvalPol<typename T, int ID, typename Policy> is a mapping from the symbolic policy class Policy onto the corresponding operations and functions of the type T. As described in Sect. 3.1.1, the policies of an inner node of the expression tree are nested template specifications, e.g. Add<RefId<0,0>,RefId<1,0>>. Hence, the template specialization of EvalPol for an arithmetic operation or function is a trait applying the corresponding operation onto recursive calls of the trait EvalPol for the "child policies".

The specialization EvalPol<T, ID, RefId<I, N>> is used as a termination condition which returns the corresponding reference of the value or partial derivative for the policy RefId<I, N> and variable with index ID from the std::tuple<Derivative, ..., Derivative> structure t in algorithm step 3 at compile-time.

4 Experimental Results

For our test environment we used a Linux 3.0.0 64 Bit system running on an Intel(R) Core(TM)2 Quad CPU Q9550@2.83 GHz with 8 GB random access memory. As a compiler we used the GNU C++ compiler g++ 4.4.6 with the options -O2 as well as -std=c++0x for the required C++0x support.

As a first test case we compared our expression template approach (ET) in double precision against the common automatic differentiation libraries FADBAD++ 2.1, ADOL-C 2.2.0 and Trilinos Sacado 10.4.

Table 1 Performance comparison for the computation of the first derivative in `double` precision

	$x^2y^3 + y\log(x)$	$3x^2y - y^3$	$(1-x)^2 + 100(y-x^2)^2$
ET	8.19 ms	2.22 ms	2.90 ms
By hand	13.38 ms	1.67 ms	1.83 ms
FADBAD++	55.93 ms	40.73 ms	57.77 ms
ADOL-C	206.26 ms	180.74 ms	202.59 ms
ADOL-C reused tape	125.71 ms	114.39 ms	120.69 ms
Sacado DFad	41.42 ms	21.54 ms	22.50 ms
Sacado SFad	17.86 ms	2.67 ms	3.22 ms

Table 2 Performance comparison of the evaluation of interval expressions using automatic differentiation to compute the derivatives up to the first or second order

	$x^3 + x - 1$		$e^x \sin(4x)$	
	$\frac{d}{dx}$	$\frac{d^2}{dx^2}$	$\frac{d}{dx}$	$\frac{d^2}{dx^2}$
ET filib++	6.03 ms	15.23 ms	6.31 ms	19.83 ms
C-XSC	10.03 ms	24.94 ms	8.97 ms	24.02 ms

In this case we measured the time required for the computation of the first derivative for 100,000 randomized input variables x and y. Table 1 shows the measured run-time (the mean of 10 runs).

We have tested two different cases for ADOL-C. The first one records a new *tape* for each computation. For the second one the *tape* is reused for the computation of the other values.

In addition we have measured the time for a differentiation by hand. Surprisingly for expression $x^2y^3 + y\log(x)$, our approach is faster then the hand coded differentiation. Thereby, the symbolic differentiation of the subexpression $y \cdot \log(x)$ requires a second computation of the function $\log(x)$. With an optimized hand coded version which reuses the result of $\log(x)$, the run-time of the hand coded version could be reduced to 7.53 ms.

Additionally we used our approach together with the interval library filib++ 3.0.2 to measure the behavior of our implementation together with interval arithmetic and compared it against the common interval library C-XSC 2.4.0.

The first test cases are similar to the test in `double` precision. We computed the derivatives of the functions for the first and second order for 10,000 different values, see Table 2 (the mean of 10 runs). Due to the fact that C-XSC has implemented the automatic differentiation directly into their operations and functions [9], the speedup is much smaller compared to the speed up measured in `double` precision.

The last test case was a comparison of the iterative `Halley's` method working on intervals [9] to approximate zeros of the non linear functions $x^3 + x - 1$ and $e^x \sin(4x)$ using C-XSC and our filib++ approach. The start interval $x^{(0)}$ was defined as [1.25, 1.25] for both functions and we measured a speed up of 1.76 and 1.39, respectively.

5 Conclusion and Future Work

In this paper we have shown how generative programming using expression templates and template metaprogramming can be used to implement a domain specific language for automatic differentiation of an arbitrary order which is flexible, fast and easy to use.

A comparison of our approach against common automatic differentiation libraries showed a dominating performance for floating point and interval types. Additionally, the flexibility of our approach like the use of different data types for the differentiation or the opportunity to replace the template metaprogramming functions for the code generation is a benefit.

Further investigations to improve the automatic differentiation with generative programming and especially the adaptability of the reverse mode onto our approach are planned. Additionally, we intend to publicly release our prototype in the near future.

For lack of space we sketched only a simplified implementation in this paper to illustrate the idea of our approach. See [11] for a more detailed description.

Acknowledgements We would like to thank the three anonymous referees for their helpful comments and suggestions.

References

1. Alexandrescu, A.: Modern C++ design: generic programming and design patterns applied. Addison-Wesley Longman Publishing Co., Inc., Boston, MA, USA (2001)
2. Aubert, P., Di Césaré, N., Pironneau, O.: Automatic differentiation in C++ using expression templates and application to a flow control problem. Computing and Visualization in Science **3**, 197–208 (2001)
3. Becker, P.: Working Draft, Standard for Programming Language C++. Tech. Rep. N3242=11-0012, ISO/IEC JTC1/SC22/WG21 (2011)
4. Bendtsen, C., Stauning, O.: FADBAD, a flexible C++ package for automatic differentiation. Technical Report IMM–REP–1996–17, Department of Mathematical Modelling, Technical University of Denmark, Lyngby, Denmark (1996)
5. Gamma, E., Helm, R., Johnson, R., Vlissides, J.: Design patterns: elements of reusable object-oriented software. Addison-Wesley Longman Publishing Co., Inc., Boston, MA, USA (1995)
6. Gil, J., Gutterman, Z.: Compile time symbolic derivation with C++ templates. In: Proceedings of the 4th USENIX Conference on Object-Oriented Technologies and Systems, COOTS'98, pp. 249–264. USENIX Association, Berkeley, CA, USA (1998)
7. Gregor, D., Järvi, J., Powell, G.: Variadic templates (revision 3). Tech. Rep. N2080=06-0150, ISO/IEC JTC1/SC22/WG21 (2006)

8. Griewank, A., Juedes, D., Utke, J.: Algorithm 755: ADOL-C: A package for the automatic differentiation of algorithms written in C/C++. ACM Transactions on Mathematical Software **22**(2), 131–167 (1996). URL http://doi.acm.org/10.1145/229473.229474
9. Hammer, R., Ratz, D., Kulisch, U., Hocks, M.: C++ Toolbox for Verified Computing I: Basic Numerical Problems. Springer-Verlag New York, Inc., Secaucus, NJ, USA (1997)
10. Myers, N.: A new and useful template technique: "Traits". C++ Report **7**(5), 32–35 (1995)
11. Nehmeier, M.: Generative programming for automatic differentiation in C++0x. Tech. Rep. 483, University of Würzburg (2011)
12. Phipps, E.T., Bartlett, R.A., Gay, D.M., Hoekstra, R.J.: Large-scale transient sensitivity analysis of a radiation-damaged bipolar junction transistor via automatic differentiation. In: C.H. Bischof, H.M. Bücker, P.D. Hovland, U. Naumann, J. Utke (eds.) Advances in Automatic Differentiation, *Lecture Notes in Computational Science and Engineering*, vol. 64, pp. 351–362. Springer, Berlin (2008). DOI 10.1007/978-3-540-68942-3_31
13. Veldhuizen, T.: Expression templates. C++ Report **7**(5), 26–31 (1995)
14. Veldhuizen, T.: Using C++ template metaprograms. C++ Report **7**(4), 36–43 (1995)
15. Veldhuizen, T.: C++ templates are Turing complete. Tech. rep. (2003)

AD *in* Fortran: Implementation via Prepreprocessor

Alexey Radul, Barak A. Pearlmutter, and Jeffrey Mark Siskind

Abstract We describe an implementation of the Farfel Fortran77 AD extensions (Radul et al. AD in Fortran, Part 1: Design (2012), http://arxiv.org/abs/1203.1448). These extensions integrate forward and reverse AD directly into the programming model, with attendant benefits to flexibility, modularity, and ease of use. The implementation we describe is a "prepreprocessor" that generates input to existing Fortran-based AD tools. In essence, blocks of code which are targeted for AD by Farfel constructs are put into subprograms which capture their lexical variable context, and these are closure-converted into top-level subprograms and specialized to eliminate `EXTERNAL` arguments, rendering them amenable to existing AD preprocessors, which are then invoked, possibly repeatedly if the AD is nested.

Keywords Nesting • Multiple transformation • Forward mode • Reverse mode • Tapenade • ADIFOR • Programming-language implementation

1 Introduction

The *Forward And Reverse Fortran Extension Language* (Farfel) extensions to Fortran77 enable smooth and modular use of AD [7]. A variety of implementation strategies present themselves, ranging from (a) deep integration into a Fortran

A. Radul (✉)
Hamilton Institute, National University of Ireland, Maynooth, Ireland
e-mail: alexey.radul@nuim.ie

B.A. Pearlmutter
Department of Computer Science and Hamilton Institute, National University of Ireland, Maynooth, Ireland
e-mail: barak@cs.nuim.ie

J.M. Siskind
Electrical and Computer Engineering, Purdue University, West Lafayette, IN, USA
e-mail: qobi@purdue.edu

S. Forth et al. (eds.), *Recent Advances in Algorithmic Differentiation*, Lecture Notes in Computational Science and Engineering 87, DOI 10.1007/978-3-642-30023-3_25,

compiler, to (b) a preprocessor that performs the requested AD and generates Fortran or some other high-level language, to (c) a prepreprocessor which itself does no AD but generates input to an existing Fortran AD preprocessor. An earlier attempt to implement a dialect of Fortran with syntactic AD extensions bearing some syntactically similarity to Farfel used strategy (a) [5]. Here we adopt strategy (c), which leverages existing Fortran-based AD tools and compilers, avoiding re-implementation of the AD transformations themselves, at the expense of inheriting some of the limitations of the AD tool it invokes.

Farfallen transforms Farfel input into Fortran77, and invokes an existing AD system [2, 3] to generate the needed derivatives. The process can make use of a variety of existing Fortran-based AD preprocessors, easing the task of switching between them. There is significant semantic mismatch between Farfel and the AD operations allowed by the AD systems used, necessitating rather dramatic code transformations. When the Farfel program involves nested AD, the transformations and staging become even more involved. Viewed as a whole, this tool automates the task of applying AD, including the detailed maneuvers required for nested application of existing tools, thereby extending the reach and utility of AD.

The remainder of the paper is organized as follows: Sect. 2 reviews the Farfel extensions. A complete example program on page 277 illustrates their use. Section 3 describes the implementation in detail using this example program. Section 4 summarizes this work's contributions. Comparison with other ways to offer AD to the practitioner can be found in the design paper [7].

2 Language Extensions

Farfel provides two principal extensions to Fortran77: syntax for AD and for general nested subprograms.

2.1 Extension 1: AD Syntax

Farfel adds the **ADF** construct for forward AD:

```
ADF(TANGENT(var) = expr ...)
statements
END ADF(var = TANGENT(var) ...)
```

Multiple opening and closing assignments are separated by commas. Independent variables are listed in the "calls" to **TANGENT** on the left-hand sides of the opening assignments and are given the specified tangent values. Dependent variables appear in the "calls" to **TANGENT** on the right-hand sides of the closing assignments and the corresponding tangent values are assigned to the indicated destination variables.

The **ADF** construct uses forward AD to compute the directional derivative of the dependent variables at the point specified by the vector of independent variables in the direction specified by the vector of tangent values for the independent variables and assigns it to the destination variables.

An analogous Farfel construct supports reverse AD:

```
ADR(COTANGENT(var) = expr ...)
statements
END ADR(var = COTANGENT(var) ...)
```

Dependent variables are listed in the "calls" to **COTANGENT** on left-hand sides of the opening assignments and are given the specified cotangent values as inputs to the reverse phase. *Independent* variables appear in the "calls" to **COTANGENT** on the right-hand sides of the closing assignments and the corresponding cotangent values at the end of the reverse phase are assigned to the indicated destination variables. The **ADR** construct uses reverse AD to compute the gradient with respect to the independent variables at the point specified by the vector of independent variables induced by the specified gradient with respect to the dependent variables, and assigns it to the destination variables. The expressions used to initialize the cotangent inputs to the reverse phase are evaluated at the end of the forward phase, even though they appear textually prior to the statements specifying the forward phase. This way, the direction input to the reverse phase can depend on the result of the forward phase.

For both **ADF** and **ADR**, implied-**DO** syntax is used to allow arrays in the opening and closing assignments. By special dispensation, the statement **ADF**(*var*) is interpreted as **ADF**(**TANGENT**(*var*)=1) and **ADR**(*var*) as **ADR**(**COTANGENT**(*var*)=1).

2.2 Extension 2: Nested Subprograms

In order to conveniently support distinctions between different variables of differentiation for distinct invocations of AD, as in the example below, we borrow from Algol 60 [1] and generalize the Fortran "statement function" construct by allowing subprograms to be defined inside other subprograms, with lexical scope. As in Algol 60, the scope of parameters and declared variables is the local subprogram, and these may shadow identifiers from the surrounding scope. Implicitly declared variables have the top-level subprogram as their scope.

2.3 Concrete Example

In order to describe our implementation, we employ a concrete example. The task is to find an equilibrium (a^*, b^*) of a two-player game with continuous scalar strategies a and b and given payoff functions A and B. The method is to find roots of

$$a^* = \underset{a}{\operatorname{argmax}}\, A(a, \underset{b}{\operatorname{argmax}}\, B(a^*, b)) \quad (1)$$

The full program is given, for reference, in Listing 7. The heart of the program is the implementation `EQLBRM` of (1). Note that this whole program is only 63 lines of code, with plenty of modularity boundaries. This code is used as a running example for the remainder of the paper.

3 Implementation

Farfel is implemented by the Farfallen preprocessor. The current version is merely a proof of concept, and not production quality: it does not accept the entire Fortran77 language, and does not scale. However, its principles of operation will be unchanged in a forthcoming production-quality implementation. Here we describe the reduction of Farfel constructs to Fortran77, relying on existing Fortran-based AD tools for the actual derivative transformations.

Farfel introduces two new constructs into Fortran77: nested subprograms and syntax for requesting AD. We implement nested subprograms by extracting them to the top level, and communicating the free variables from the enclosing subprogram by passing them as arguments into the new top-level subprogram. This is an instance of *closure conversion*, a standard class of techniques for converting nested subprograms to top-level ones [4]. In order to accommodate passing formerly-free variables as arguments, we must adjust all the call sites of the formerly-nested subprogram; we must specialize all the subprograms that accept that subprogram as an external to also accept the extra closure parameters; and adjust all call sites to all those specialized subprograms to pass those extra parameters.

We implement the AD syntax by constructing new subroutines that correspond to the statements inside each **`ADF`** or **`ADR`** block, arranging to call the AD tool of choice on each of those new subroutines, and transforming the block itself into a call to the appropriate tool-generated subroutine.

3.1 Nested Subprograms in Detail

Let us illustrate closure conversion on our example. Recall `ARGMAX`:

```
      FUNCTION ARGMAX(F, X0, N)
        FUNCTION FPRIME(X)
        FPRIME = DERIV1(F, X)
        END
      ARGMAX = ROOT(FPRIME, X0, N)
      END
```

Listing 7 Complete example Farfel program: equilibria of a continuous-strategy game.

```
C     ASTAR & BSTAR: GUESSES IN, OPTIMIZED VALUES OUT
      SUBROUTINE EQLBRM(BIGA, BIGB, ASTAR, BSTAR, N)
      EXTERNAL BIGA, BIGB
        FUNCTION F(ASTAR)
          FUNCTION G(A)
            FUNCTION H(B)
            H = BIGB(ASTAR, B)
            END
          BSTAR = ARGMAX(H, BSTAR, N)
          G = BIGA(A, BSTAR)
          END
        F = ARGMAX(G, ASTAR, N)-ASTAR
        END
      ASTAR = ROOT(F, ASTAR, N)
      END

      FUNCTION ROOT(F, X0, N)
      X = X0
      DO 1669 I=1,N
      CALL DERIV2(F, X, Y, YPRIME)
 1669 X = X-Y/YPRIME
      ROOT = X
      END

      SUBROUTINE DERIV2(F, X, Y, YPRIME)
      EXTERNAL F
      ADF(X)
      Y = F(X)
      END ADF(YPRIME = TANGENT(Y))
      END

      FUNCTION ARGMAX(F, X0, N)
        FUNCTION FPRIME(X)
        FPRIME = DERIV1(F, X)
        END
      ARGMAX = ROOT(FPRIME, X0, N)
      END

      FUNCTION DERIV1(F, X)
      EXTERNAL F
      ADF(X)
      Y = F(X)
      END ADF(DERIV1 = TANGENT(Y))
      END

      FUNCTION GMBIGA(A, B)
      PRICE = 20-0.1*A-0.1*B
      COSTS = A*(10-0.05*A)
      GMBIGA = A*PRICE-COSTS
      END

      FUNCTION GMBIGB(A, B)
      PRICE = 20-0.1*B-0.0999*A
      COSTS = B*(10.005-0.05*B)
      GMBIGB = B*PRICE-COSTS
      END

      PROGRAM MAIN
      READ *, ASTAR
      READ *, BSTAR
      READ *, N
      CALL EQLBRM(GMBIGA, GMBIGB, ASTAR, BSTAR, N)
      PRINT *, ASTAR, BSTAR
      END
```

This contains the nested function FPRIME. We closure convert this as follows. First, extract FPRIME to the top level:

```
      FUNCTION ARGMAX_FPRIME(X, F)
      ARGMAX_FPRIME = DERIV1(F, X)
      END

      FUNCTION ARGMAX(F, X0, N)
      ARGMAX = ROOT_1(ARGMAX_FPRIME, F, X0, N)
      END
```

Note the addition of a closure argument for F since it is freely referenced in FPRIME, and the addition of the same closure argument at the call site, since FPRIME is passed as an external to ROOT. Then we specialize ROOT to create a version that accepts the needed set of closure arguments (in this case one):

```
      FUNCTION ROOT_1(F, F1, X0, N)
      X = X0
      DO 1669 I=1,N
      CALL DERIV2_1(F, F1, X, Y, YPRIME)
 1669 X = X-Y/YPRIME
      ROOT_1 = X
      END
```

Since ROOT contained a call to DERIV2, passing it the external passed to ROOT, we must also specialize DERIV2:

```
      SUBROUTINE DERIV2_1(F, F1, X, Y, YPRIME)
      EXTERNAL F
      ADF(X)
      Y = F(X, F1)
      END ADF(YPRIME = TANGENT(Y))
      END
```

We must, in general, copy and specialize the portion of the call graph where the nested subprogram travels, which in this case is just two subprograms. During such copying and specialization, we propagate external constants (e.g., FPRIME through the call to ROOT_1, the call to DERIV2_1, and the call site therein) allowing the elimination of the **EXTERNAL** declaration for these values. This supports AD tools that do not allow taking derivatives through calls to external subprograms.

That is the process for handling one nested subprogram. In our example, the same is done for F in EQLBRM, G in F, and H in G. Doing so causes the introduction of a number of closure arguments, and the specialization of a number of subprograms to accept those arguments; including perhaps further specializing things that have already been specialized. The copying also allows a limited form of subprogram reentrancy: even if recursion is disallowed (as in traditional Fortran77) our nested uses of ARGMAX will cause no difficulties because they will end up calling two different specializations of ARGMAX.

Note that we must take care to prevent this process from introducing spurious aliases. For example, in EQLBRM, the internal function F that is passed to ROOT closes over the iteration count N, which is also passed to ROOT separately. When specializing ROOT to accept the closure parameters of F, we must not pass N to the specialization of ROOT twice, lest we run afoul of Fortran's prohibition against assigning to aliased values. Fortunately, such situations are syntactically apparent.

Finally, specialization leaves behind unspecialized (or underspecialized) versions of subprograms, which may now be unused, and if so can be eliminated. In this case, that includes ROOT, DERIV2, ARGMAX, FPRIME, and DERIV1 from the original program, as well as some intermediate specializations thereof.

3.2 AD Syntax in Detail

We implement the AD syntax by first canonicalizing each **ADF** or **ADR** block to be a single call to a (new, internal) subroutine, then extracting those subroutines to the top level, then rewriting the block to be a call to an AD-transformed version of the subroutine, and then arranging to call the AD tool of choice on each of those new subroutines to generate the needed derivatives.

Returning to our example program, closure conversion of nested subprograms produced the following specialization of DERIV1:

```
FUNCTION DERIV1_1(ASTAR, BIGA, BIGB, BSTAR, N, X)
ADF(X)
Y = EQLBRM_F_G(X, ASTAR, BIGA, BIGB, BSTAR, N)
END ADF(DERIV1_1 = TANGENT(Y))
END
```

which contains an **ADF** block. We seek to convert this into a form suitable for invoking the AD preprocessor. We first canonicalize by introducing a new subroutine to capture the statements in the **ADF** block, producing the following:

```
FUNCTION DERIV1_1(ASTAR, BIGA, BIGB, BSTAR, N, X)
  SUBROUTINE ADF1()
  Y = EQLBRM_F_G(X, ASTAR, BIGA, BIGB, BSTAR, N)
  END
ADF(X)
CALL ADF1()
END ADF(DERIV1_1 = TANGENT(Y))
END
```

Extracting the subroutine ADF1 to the top level as before yields the following:

```
SUBROUTINE DERIV1_1_ADF1(X, ASTAR, BIGA, BIGB, BSTAR, N, Y)
Y = G(X, ASTAR, BIGA, BIGB, BSTAR, N)
END

FUNCTION DERIV1_1(ASTAR, BIGA, BIGB, BSTAR, N, X)
ADF(X)
CALL DERIV1_1_ADF1(X, ASTAR, BIGA, BIGB, BSTAR, N, Y)
END ADF(DERIV1_1 = TANGENT(Y))
END
```

Now we are properly set up to rewrite the **ADF** block into a subroutine call—specifically, to a subroutine that will be generated from DERIV1_1_ADF1 by AD. The exact result depends on the AD tool that will be used to construct the derivative of DERIV1_1_ADF1; for Tapenade, the generated code looks like this:

```
 FUNCTION DERIV1_1(ASTAR, BIGA, BIGB, BSTAR, N, X)
 X_G1 = 1
 ASTAR_G1 = 0
 BSTAR_G1 = 0
 CALL DERIV1_1_ADF1_G1(X, X_G1, ASTAR, ASTAR_G1, BIGA, BIGB,
+BSTAR, BSTAR_G1, N, Y, DERIV1_1)
 END
```

Different naming conventions are used for DERIV1_1_ADF1_G1 when generating code for ADIFOR; the parameter passing conventions of Tapenade and ADIFOR agree in this case. Farfallen maintains the types of variables in order to know whether to generate variables to hold tangents and cotangents (which are initialized to zero if they were not declared in the relevant opening assignments.)

The same must be repeated for each **ADF** and **ADR** block; in our example there are five in all: two in specializations of DERIV1 and three in specializations of DERIV2. We must also specialize EQLBRM and its descendants in the call graph, by the process already illustrated, to remove external calls to the objective functions.

Finally, we must invoke the user's preferred AD tool to generate all the needed derivatives. Here, Farfallen might invoke Tapenade as follows:

```
#! /bin/sh
tapenade -root deriv1_2_adf2 -d -o eqlbrm42 -diffvarname "_g2"\
    -difffuncname "_g2" eqlbrm42.f
tapenade -root deriv2_1_2_adf4 -d -o eqlbrm42 -diffvarname "_g4"\
    -difffuncname "_g4" eqlbrm42{,_g2}.f
tapenade -root deriv1_1_adf1 -d -o eqlbrm42 -diffvarname "_g1"\
    -difffuncname "_g1" eqlbrm42{,_g2,_g4}.f
tapenade -root deriv2_1_1_adf3 -d -o eqlbrm42 -diffvarname "_g3"\
    -difffuncname "_g3" eqlbrm42{,_g2,_g4,_g1}.f
tapenade -root deriv2_2_adf5 -d -o eqlbrm42 -diffvarname "_g5"\
    -difffuncname "_g5" eqlbrm42{,_g2,_g4,_g1,_g3}.f
```

We must take care that multiple invocations of the AD tool to generate the various derivatives occur in the proper order, which is computable from the call graph of the program, to ensure that generated derivative codes that are used in subprograms to be differentiated are available to be transformed. For example, all the derivatives of `EQLBRM_F_G_H` needed for its optimizations must be generated before generating derivatives of (generated subprograms that call) `EQLBRM_F_G`. We must also take care to invoke the AD tool with different prefixes/suffixes, so that variables and subprograms created by one differentiation do not clash with those made by another.

3.3 Performance

We tested the performance of code generated by Farfallen in conjunction with Tapenade and with ADIFOR. We executed Farfallen once to generate Fortran77 source using Tapenade for the automatic differentiation, and separately targeting ADIFOR. In each case, we compiled the resulting program (`gfortran` 4.6.2-9, 64-bit Debian sid, `-Ofast -fwhole-program`, single precision) with $N = 1,000$ iterations at each level and timed the execution of the binary on a 2.93 GHz Intel i7 870. For comparison, we translated the same computation into vlad [6] (Fig. 1), compiled it with Stalingrad [8], and ran it on the same machine. Stalingrad has the perhaps unfair advantage of being an optimizing compiler with integrated support for AD, so we are pleased that Farfallen was able to achieve performance that was nearly competitive.

Tapenade	ADIFOR	Stalingrad
6.97	8.92	5.83

The vlad code in Fig. 1 was written with the same organization, variable names, subprogram names, and parameter names and order as the corresponding Farfel code in Listing 7 to help a reader unfamiliar with Scheme, the language on which vlad is based, understand how Farfel and vlad both represent the same essential notions of nested subprograms and the AD discipline of requesting derivatives precisely where they are needed. Because functional-programming languages, like Scheme and vlad, prefer higher-order functions (i.e., operators) over the block-based style that is prevalent in Fortran, AD is invoked via the $\overrightarrow{\mathcal{J}}$ and $\overleftarrow{\mathcal{J}}$ operators rather than via the **`ADF`** and **`ADR`** statements. However, there is a one-to-one correspondence between

$$\overrightarrow{\mathcal{J}} : f \times x \times \acute{x} \rightarrow y \times \acute{y}$$

an operator that takes a function f as input, along with a primal argument x and tangent argument $\acute{x}$ and returns both a primal result y and a tangent result $\acute{y}$, and:

```
ADF(TANGENT(x) = x́)
y = f(x)
END ADF(ý = TANGENT(y))
```

$$
\begin{array}{l}
\text{EQLBRM}(A,B,a_0,b_0,n) \triangleq \textbf{let } f(a') \triangleq \textbf{let } g(a) \triangleq \textbf{let } h(b) \triangleq B(a',b) \\
\qquad\qquad\qquad\qquad\qquad\qquad\qquad\qquad\qquad\qquad \textbf{in } A(a, \text{ARGMAX}(h,b_0,n)) \\
\qquad\qquad\qquad\qquad\qquad\qquad\qquad\quad \textbf{in } \text{ARGMAX}(g,a',n) - a' \\
\qquad\qquad\qquad\qquad\qquad\qquad\quad a^* = \text{ROOT}(f,a_0,n) \\
\qquad\qquad\qquad\qquad\qquad\quad \textbf{in let } h(b) \triangleq B(a^*,b) \\
\qquad\qquad\qquad\qquad\qquad\qquad\quad b^* = \text{ARGMAX}(h,b_0,n) \\
\qquad\qquad\qquad\qquad\qquad\qquad \textbf{in } a^*, b^*
\end{array}
$$

$$
\begin{array}{l}
\text{ROOT}(f,x,n) \triangleq \textbf{if } n = 0 \\
\qquad\qquad\qquad\quad \textbf{then } x \\
\qquad\qquad\qquad\quad \textbf{else let } y, y' = \text{DERIV2}(f,x) \\
\qquad\qquad\qquad\qquad\quad \textbf{in } \text{ROOT}(f, x - \frac{y}{y'}, n-1)
\end{array}
$$

$$\text{DERIV2}(f,x) \triangleq \overrightarrow{\mathbf{J}}(f,x,1)$$

$$
\begin{array}{l}
\text{ARGMAX}(f,x_0,n) \triangleq \textbf{let } f'(x) \triangleq \text{DERIV1}(f,x) \\
\qquad\qquad\qquad\qquad\quad \textbf{in } \text{ROOT}(f',x_0,n)
\end{array}
$$

$$
\begin{array}{l}
\text{DERIV1}(f,x) \triangleq \textbf{let } x, \acute{y} = \overrightarrow{\mathbf{J}}(f,x,1) \\
\qquad\qquad\qquad\quad \textbf{in } \acute{y}
\end{array}
$$

$$
\begin{array}{l}
\text{A}(a,b) \triangleq \textbf{let } \textit{price} = 20 - 0.1 \times a - 0.1 \times b \\
\qquad\qquad\qquad\quad \textit{costs} = a \times (10 - 0.05 \times a) \\
\qquad\qquad \textbf{in } a \times \textit{price} - \textit{costs}
\end{array}
$$

$$
\begin{array}{l}
\text{B}(a,b) \triangleq \textbf{let } \textit{price} = 20 - 0.1 \times b - 0.0999 \times a \\
\qquad\qquad\qquad\quad \textit{costs} = b \times (10.005 - 0.05 \times b) \\
\qquad\qquad \textbf{in } b \times \textit{price} - \textit{costs}
\end{array}
$$

$$
\begin{array}{l}
\textbf{let } a_0 = \text{READREAL}() \\
\quad b_0 = \text{READREAL}() \\
\quad n = \text{READREAL}() \\
\quad a^*, b^* = \text{EQLBRM}(\text{A}, \text{B}, a_0, b_0, n) \\
\textbf{in } \text{WRITEREAL}(a^*) \\
\quad \text{WRITEREAL}(b^*)
\end{array}
$$

Figure 1 Complete vlad program for our concrete example with the same organization and functionality as the Farfel program in Listing 7

Similarly, there is a one-to-one correspondence between

$$\overleftarrow{\mathbf{J}} : f \times x \times \grave{y} \to y \times \grave{x}$$

an operator that takes a function f as input, along with a primal argument x and cotangent $\grave{y}$ and returns both a primal result y and a cotangent $\grave{x}$, and

```
ADR(COTANGENT(y) = ẏ)
y = f(x)
END ADR(ẋ = COTANGENT(x))
```

The strong analogy between how the callee-derives AD discipline is represented in both Farfel and vlad serves two purposes: it enables the use of the Stalingrad compiler technology for compiling Farfel and facilitates migration from legacy imperative languages to more modern, and more easily optimized, pure languages.

4 Conclusion

We have illustrated an implementation of the Farfel extensions to Fortran—nested subprograms and syntax for AD [7]. These extensions enable convenient, modular programming using a callee-derives paradigm of automatic differentiation. Our implementation is a preprocessor that translates Farfel Fortran77 extensions into input suitable for an existing AD tool. This strategy enables modular, flexible use of AD in the context of an existing legacy language and tool chain, without sacrificing the desirable performance characteristics of these tools: in the concrete example, only 20–50% slower than a dedicated AD-enabled compiler, depending on which Fortran AD system is used.

Acknowledgements This work was supported, in part, by Science Foundation Ireland grant 09/IN.1/I2637, National Science Foundation grant CCF-0438806, Naval Research Laboratory Contract Number N00173-10-1-G023, and Army Research Laboratory Cooperative Agreement Number W911NF-10-2-0060. Any views, opinions, findings, conclusions, or recommendations contained or expressed in this document or material are those of the authors and do not necessarily reflect or represent the views or official policies, either expressed or implied, of SFI, NSF, NRL, ONR, ARL, or the Irish or U.S. Governments. The U.S. Government is authorized to reproduce and distribute reprints for Government purposes, notwithstanding any copyright notation herein.

References

1. Backus, J.W., Bauer, F.L., Green, J., Katz, C., McCarthy, J., Naur, P., Perlis, A.J., Rutishauser, H., Samelson, K., Vauquois, B., Wegstein, J.H., van Wijngaarden, A., Woodger, M.: Revised report on the algorithmic language ALGOL 60. The Computer Journal **5**(4), 349–367 (1963). DOI 10.1093/comjnl/5.4.349
2. Bischof, C.H., Carle, A., Corliss, G.F., Griewank, A., Hovland, P.D.: ADIFOR: Generating derivative codes from Fortran programs. Scientific Programming **1**(1), 11–29 (1992)
3. Hascoët, L., Pascual, V.: TAPENADE 2.1 user's guide. Rapport technique 300, INRIA, Sophia Antipolis (2004). URL http://www.inria.fr/rrrt/rt-0300.html
4. Johnsson, T.: Lambda lifting: Transforming programs to recursive equations. In: Functional Programming Languages and Computer Architecture. Springer Verlag, Nancy, France (1985)

5. Naumann, U., Riehme, J.: A differentiation-enabled Fortran 95 compiler. ACM Transactions on Mathematical Software **31**(4), 458–474 (2005). URL http://doi.acm.org/10.1145/1114268.1114270
6. Pearlmutter, B.A., Siskind, J.M.: Using programming language theory to make automatic differentiation sound and efficient. In: C.H. Bischof, H.M. Bücker, P.D. Hovland, U. Naumann, J. Utke (eds.) Advances in Automatic Differentiation, *Lecture Notes in Computational Science and Engineering*, vol. 64, pp. 79–90. Springer, Berlin (2008). DOI 10.1007/978-3-540-68942-3_8
7. Radul, A., Pearlmutter, B.A., Siskind, J.M.: AD in Fortran, Part 1: Design (2012). URL http://arxiv.org/abs/1203.1448
8. Siskind, J.M., Pearlmutter, B.A.: Using polyvariant union-free flow analysis to compile a higher-order functional-programming language with a first-class derivative operator to efficient Fortran-like code. Tech. Rep. TR-ECE-08-01, School of Electrical and Computer Engineering, Purdue University, West Lafayette, IN, USA (2008). URL http://docs.lib.purdue.edu/ecetr/367

An AD-Enabled Optimization ToolBox in LabVIEW^{TM}

Abhishek Kr. Gupta and Shaun A. Forth

Abstract LabVIEW^{TM} is a visual programming environment for data acquisition, instrument control and industrial automation. This article presents LVAD, a graphically programmed implementation of forward mode Automatic Differentiation for LabVIEW. Our results show that the overhead of using overloaded AD in LabVIEW is sufficiently low as to warrant further investigation and that, within the graphical programming environment, AD may be made reasonably user friendly. We further introduce a prototype LabVIEW Optimization Toolbox which utilizes LVAD's derivative information. Our toolbox presently contains two main LabVIEW procedures `fzero` and `fmin` for calculating roots and minima respectively of an objective function in a single variable. Two algorithms, Newton and Secant, have been implemented in each case. Our optimization package may be applied to graphically coded objective functions, not the simple string definition of functions used by many of the optimizers of LabVIEW's own optimization package.

Keywords Forward mode AD • LabVIEW • Graphical programming • Optimization

A.Kr. Gupta (✉)
Department of Electrical Engineering, IIT Kanpur, Kanpur, UP 208016, India
e-mail: g.kr.abhishek@gmail.com

S.A. Forth
Applied Mathematics and Scientific Computing, Cranfield University, Shrivenham, Swindon, SN6 8LA, UK
e-mail: S.A.Forth@cranfield.ac.uk

S. Forth et al. (eds.), *Recent Advances in Algorithmic Differentiation*, Lecture Notes in Computational Science and Engineering 87, DOI 10.1007/978-3-642-30023-3_26,

1 Introduction

LabVIEW[1] is a programming environment for data acquisition, instrument control and industrial automation [7]. LabVIEW programs are written in the visual programming language *G* [7]. Visual programming languages facilitate writing complex codes as flow diagrams by dragging and dropping inbuilt graphical icons representing an instrument, module or a subprogram and wiring icons together to establish the program's data flow. Visual programming eliminates the writing of programs as a collection of text commands making it popular among non-programming engineers and scientists. LabVIEW programs are called virtual instruments (VI) as they typically represent actual laboratory equipment. Each LabVIEW VI has two components: the *Front Panel* containing the VI's controls (inputs) and displays (outputs); and the *Block Diagram* which defines the VI's data flow.

Many scientific and engineering applications involve optimization so necessitating an optimization package in LabVIEW. LabVIEW provides a handful of optimization algorithms but many of these are limited to optimizing objective functions coded as simple equation expressions within a string [7]. Further, there appears to be no way for the user to provide derivative information.

AD tools exist for a wide number of programming languages e.g., C, C++, Fortran, MATLAB, Python.[2] However, AD of *visual programming languages*, such as LabVIEW, appears under-researched. The equation-based simulation language Modelica[3] is frequently programmed via a visual programming environment. Elsheikh et al. [2] considered AD of Modelica by source transformation of the model's representation in the Modelica programming language and also by a symbolic approach [1]; differentiation of the visual program was not considered.

Our LVAD package implements forward mode AD using operator overloading in LabVIEW's visual programming language G as described in Sect. 2. This is the first presentation of LVAD outside of the student competition paper [6]. Section 3 describes the implementation of our Optimization toolbox with results presented in Sect. 4 and conclusions in Sect. 5.

2 Implementation of AD in LabVIEW: The LVAD Package

Our LVAD package's forward mode AD [5, Chap. 3] differs from that for standard operator overloading [5, Chap. 6] since it is implemented in LabVIEW's visual programming paradigm. In Sect. 2.1 we define an LVAD class whose objects

[1]LabVIEW™ is a trademark of National Instruments. This publication is independent of National Instruments, which is not affiliated with the publisher or the author, and does not authorize, sponsor, endorse or otherwise approve this publication.

[2]See www.autodiff.org for a list of such tools.

[3]https://modelica.org/

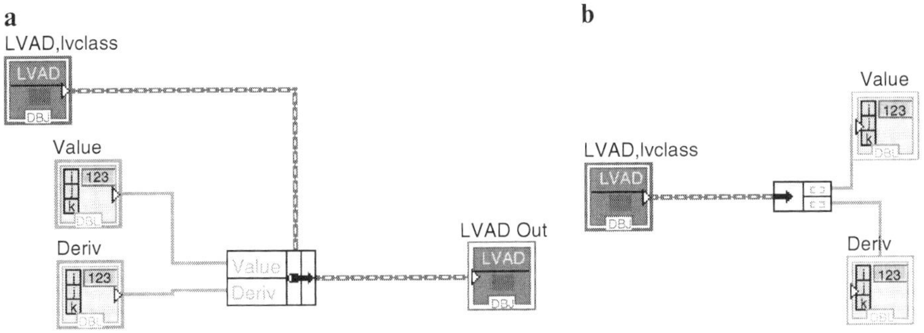

Fig. 1 Visual programming of an LVAD object's `set` and `get` methods. (**a**) The LVAD set method. (**b**) The LVAD get method

posses value and derivative components, in Sect. 2.2 we use visual programming to overload arithmetic operations and we present an example of use in Sect. 2.3.

2.1 *LVAD Class*

Figure 1a shows the visual programming of a LVAD object's `set` operation that takes as input the `Value` and `Deriv` supplied by a front panel, say, and assigns them to the appropriate components of an LVAD object. The attributes of `Value` and `Deriv` indicate that these are both arrays (the `i,j,k` attribute) and contain numeric data (the `1 2 3` attribute) of type double (the `DBL` attribute). Figure 1b shows the `get` method for extracting an object's two components.

2.2 *Operator Overloading*

The usual arithmetic operations for forward mode AD [5, Sect. 3.1] are defined by overloading LVAD objects. Figure 2 implements the product rule for multiplication of two LVAD objects. Note how the `get` operation of Fig. 1a is used to access the `Value` and `Deriv` components of `X` and `Y` and then the intrinsic addition and multiplication operations form the product's value and derivatives before `set` assigns them to the product object's components.

Similarly, other arithmetic operations and intrinsic functions (e.g., sin) may be overloaded [6].

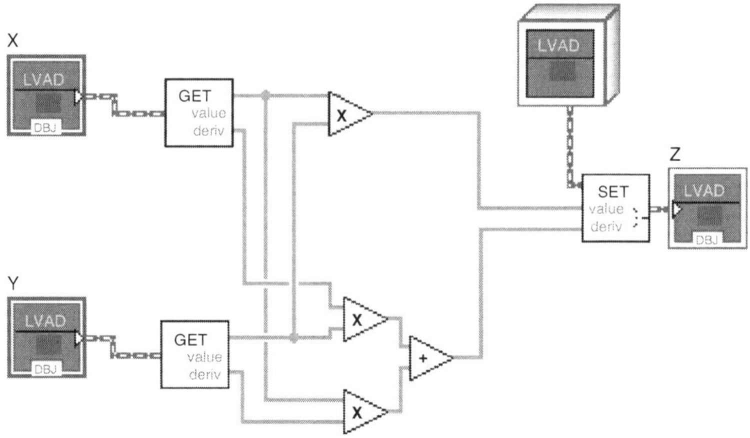

Fig. 2 Overloaded multiplication of two LVAD objects $Z = X * Y$

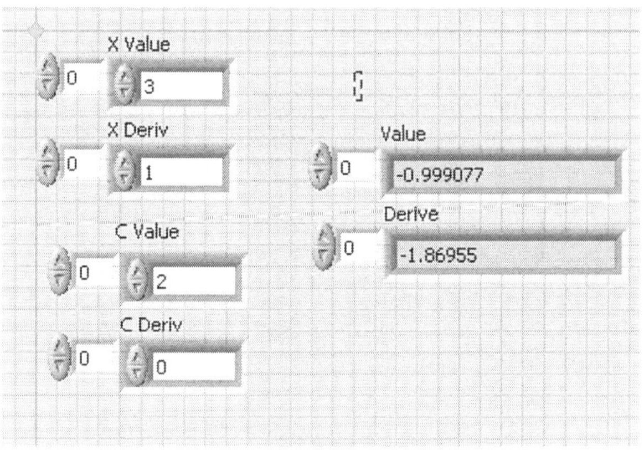

Fig. 3 *Front panel* to enable differentiation of the function (1)

2.3 Examples

Consider obtaining the derivatives $\partial f/\partial x$ and $\partial f/\partial c$ of the scalar function,

$$f(x) = (x - c)^2 \sin \frac{4\pi e^{\cos(x)}}{\sqrt{(x-a)}}, \tag{1}$$

with constant $a = 2$ and variable $c = 2$. The front panel of Fig. 3 permits the user to set the value and derivative of both x and c and observe the computed function's value and derivative. The function value and derivative are correct for $\partial f/\partial x$ at

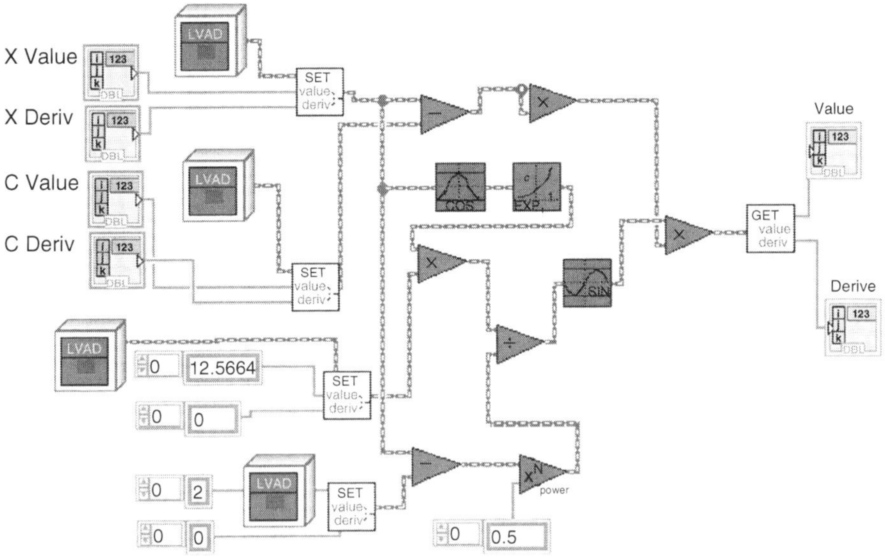

Fig. 4 Visual programming to differentiate function (1)

$x = 3$ and $c = 2$. To perform this overloaded AD computation the function was programmed using the arithmetic operations and functions of the LVAD class by standard LabVIEW techniques as seen in Fig. 4.

3 Implementation of a LabVIEW Optimization Toolbox

The two main Subroutine Virtual Instruments (subVIs) of our optimization toolbox are Sect. 3.1 `fzero` (1-D root finding) and Sect. 3.2 `fmin` (1-D minimization).

3.1 Root Finding: The `fzero` subVI

Figure 5a depicts the `fzero` front panel and Fig. 5b the corresponding subVI in the case of Newton iteration; derivative-free Secant iteration may also be employed. The front panel allows the user to: nominate an objective function VI which must have a single LVAD input and single LVAD output; set an initial value for the iteration, or two such values for the Secant method; set solver options (tolerances, maximum iterations, method) within the LabVIEW equivalent of a structure termed

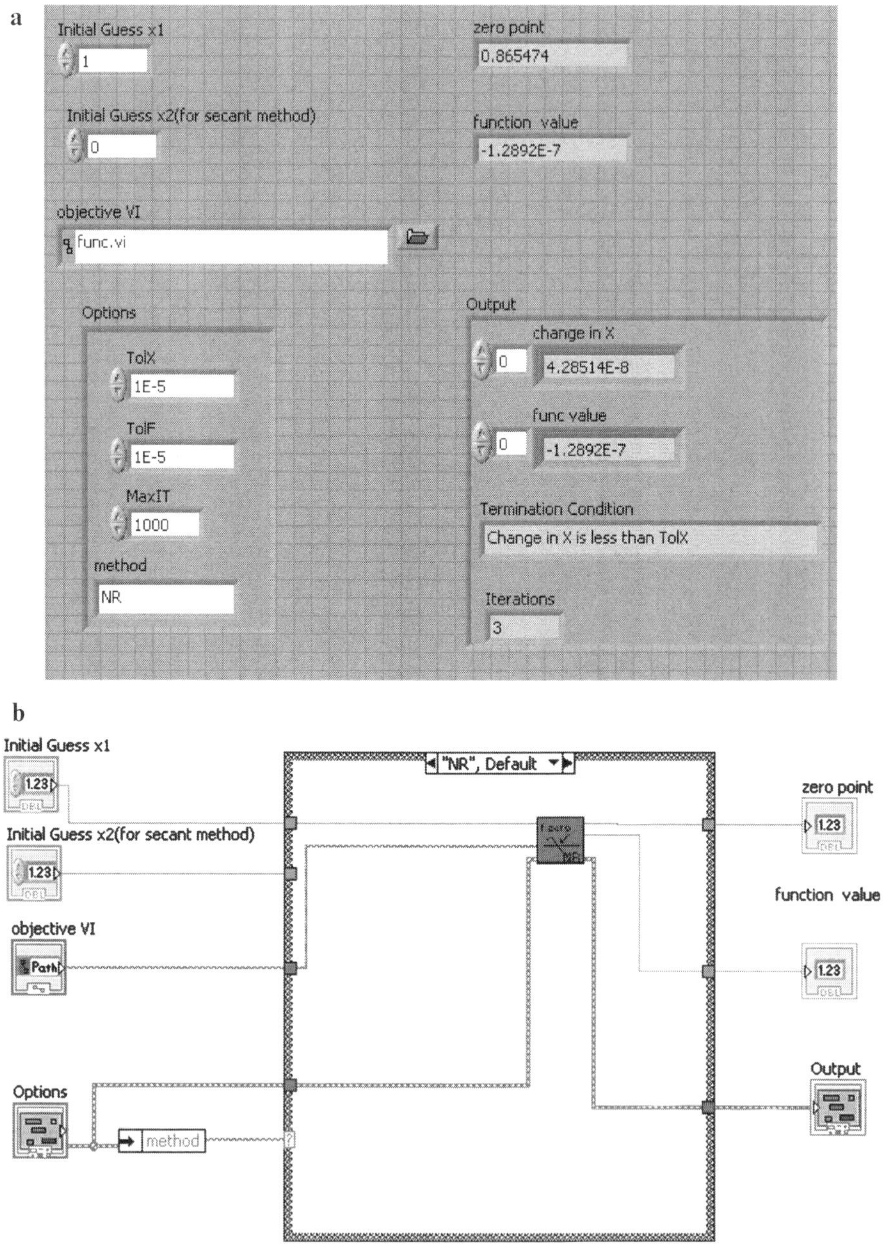

Fig. 5 The `fzero` subVI. (**a**) `fzero` front panel. (**b**) `fzero` block diagram when Newton-Raphson selected

a *cluster*.[4] On completion the calculated root x^*, function value $f(x^*)$ and, via a cluster, iteration summary outputs are displayed.

The subVI of Fig. 5b shows that, depending on the method selected, `fzero` calls either the `fzeroNR` or the `fzerosecant` VI.[5] For this proof-of-concept work we adopt iteration without any global convergence enhancements [8]. Iteration continues until simple convergence conditions are met ($|\Delta x| <$ `tolX`, $|f(x)| <$`tolF`, or maximum iterations exceeded, with Δx the change in x over one iteration).

3.2 *Minimization: The* `fmin` *subVI*

Minimizing a function in a single variable is performed by Newton or Secant iteration on the stationary equation $f'(x) = 0$ by the `fmin` VI. The user must provide the objective function $f(x)$ in the form of a subVI. The program calculates $f'(x)$ from $f(x)$ by overloaded AD for both the Secant and Newton methods. For the Newton method, the second derivative $f''(x)$ is calculated by one-sided differencing of $f'(x)$.

4 Results

We present a simple example of our package use for root finding in Sect. 4.1 and then present performance testing in Sect. 4.2.

4.1 *Simple Example*

The objective VI of Fig. 6 corresponds to the objective function

$$f(x) = \cos(x) - x^3, \tag{2}$$

The input, output and all arithmetic operations and function calls are of LVAD class.

Figure 7 shows a suitable front panel and block diagram for use of Fig. 5 `fzero` to find a zero of (2); the zero is found in five iterations at $x = 0.865474$. The function may similarly be minimized using `fmin` of Sect. 3.2.

---

[4] A further VI (details omitted for brevity) is supplied to set these options within the cluster.

[5] Both block diagrams omitted for brevity.

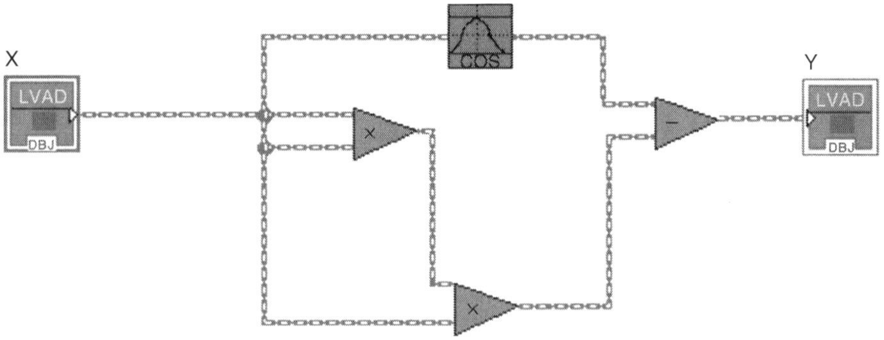

Fig. 6 Objective function VI for $f(x) = \cos(x) - x^3$

4.2 Performance Testing

For objective function (2) and a tolerance of 1.0×10^{-6}, Table 1 compares performance of: our Secant `fzero` method without overloading the objective; our Secant and Newton Raphson methods when overloaded with LVAD; and the inbuilt LabVIEW Newton Raphson zero finder [7]. Our Newton method used an initial $x = 1$ and all others used the initial pair $x = 1, -3$. As the root is located at $x \approx 0.865$ this avoids giving an *unfair advantage* to the latter three methods.

The timing difference between the two Secant methods is due to the overhead of overloaded AD (n.b., the derivatives computed are not used). The improved convergence rate of Newton makes up some of this overhead by using one less iteration. The inbuilt method has least execution time owing to its optimized implementation as an executable.

Table 2 gives CPU times to minimize $f(x) = x^2 - \sin(x)$ to a tolerance of 1.0×10^{-6} with: our package's Secant and Newton methods (with overloaded AD); and LabVIEW's inbuilt Quasi-Newton and Brent's methods. We indicate whether the objective is supplued as a VI or string. For the (Quasi) Newton methods an initial value of $x = 1$ was used; for Secant the pair $x = -1.2, 1$; and for Brent's the triplet $x = -1.2, 1, 2.2$ (n.b., the minimum is located at $x \approx 0.45$). The inbuilt functions have least execution time owing to their optimized implementation as executables; the large number of function evaluations is possibly due to poor derivative accuracy preventing asymptotic superlinear convergence. Our LVAD:Secant method is, encouragingly for a one-dimensional optimization, only some 30% slower. We are currently unable to explain the high CPU time for LVAD:Newton.

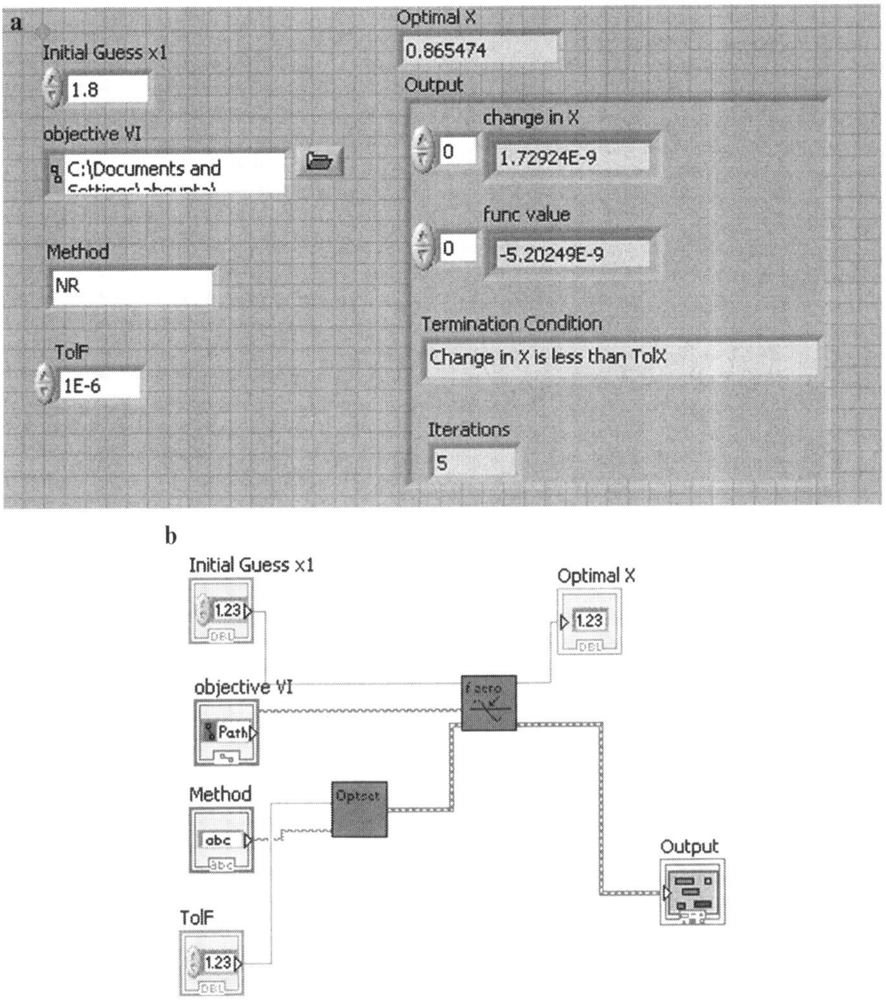

Fig. 7 *Front panel* and subVI for function zero example. (**a**) Front panel. (**b**) subVI

5 Conclusions

Following Sect. 2 description of our overloaded forward mode AD package, Sect. 3 detailed how we utilized AD in Newton methods for single variable root finding and minimization in a prototypical LabVIEW optimisation package. Our package accepts graphically coded definitions of the objective function, an advantage over the restrictive string definitions of many of LabVIEW's inbuilt optimization functions. Section 4 performance testing showed that overloading overheads, though noticeable, are not sufficiently large to warrant discarding our approach.

Table 1 Performance testing for `fzero`: variation in number of iterations per solution and mean CPU time (ms) per solution with method for 1,000 repeated solutions. A dash indicates unavailable information

Method	AD	Iterations	CPU time (ms)
LVAD:Secant	No	4	2.232
LVAD:Secant	Yes	4	14.52
LVAD:Newton	Yes	3	9.670
Inbuilt:Newton	No	–	0.199

Table 2 Performance testing for `fmin`: variation in number of iterations per solution and mean CPU time (ms) per solution with method for 1,000 repeated solutions. A dash indicates unavailable information

Method	AD	Objective supplied as	Function evaluations	CPU time (ms)
LVAD:Secant	Yes	VI	19	0.531
LVAD:Newton	Yes	VI	6	3.620
Inbuilt:Quasi-Newton	No	String	47	0.447
Inbuilt:Quasi-Newton	No	VI	47	0.400
Inbuilt:Brent's	No	String	–	0.433

A disadvantage of our LabVIEW approach, compared to AD in compiled languages, is that the objective function must be re-coded by replacing all the subVI's inbuilt class function calls and arithmetic operations by those of the LVAD class. In compiled languages one simply changes, perhaps automated by scripting or templating, the class or type of the objects [4, 9]. This task is unnecessary in MATLAB as objects acquire the class of the result of the assignment that creates them [3].

The LVAD class might be extended to vector forward mode, perhaps by utilizing a specialized vector derivative storage and linear combination class [3]. Then we might extend our optimization toolbox for functions with $x \in \mathbb{R}^n$.

Finally, we note that LabVIEW saves VIs in a propriety format making source-transformation AD approaches almost impossible without the cooperation of LabVIEW's owner National Instruments.

Acknowledgements The authors thank National Instruments for permission to include LabVIEW screenshots.

References

1. Elsheikh, A., Noack, S., Wiechert, W.: Sensitivity analysis of Modelica applications via automatic differentiation. In: 6th International Modelica Conference, vol. 2, pp. 669–675. Bielefeld, Germany (2008)

2. Elsheikh, A., Wiechert, W.: Automatic sensitivity analysis of DAE-systems generated from equation-based modeling languages. In: C.H. Bischof, H.M. Bücker, P.D. Hovland, U. Naumann, J. Utke (eds.) Advances in Automatic Differentiation, *Lecture Notes in Computational Science and Engineering*, vol. 64, pp. 235–246. Springer, Berlin (2008). DOI 10.1007/978-3-540-68942-3_21
3. Forth, S.A.: An efficient overloaded implementation of forward mode automatic differentiation in MATLAB. ACM Transactions on Mathematical Software **32**(2), 195–222 (2006). URL http://doi.acm.org/10.1145/1141885.1141888
4. Griewank, A., Juedes, D., Utke, J.: Algorithm 755: ADOL-C: A package for the automatic differentiation of algorithms written in C/C++. ACM Transactions on Mathematical Software **22**(2), 131–167 (1996). URL http://doi.acm.org/10.1145/229473.229474
5. Griewank, A., Walther, A.: Evaluating Derivatives: Principles and Techniques of Algorithmic Differentiation, 2nd edn. No. 105 in Other Titles in Applied Mathematics. SIAM, Philadelphia, PA (2008). URL http://www.ec-securehost.com/SIAM/OT105.html
6. Gupta, A.K., Agrahari, A.: LVAD package: Implementation of forward mode automatic differentiation in LabVIEW using operator overloading. VI Mantra Technical Paper Writing Contest, National Instuments (2008). URL https://sites.google.com/site/gkrabhishek/projects/publications
7. LabVIEW 2011 Help (2011). URL http://digital.ni.com/manuals.nsf/websearch/7C3F895E4B50A03D862578D400575C01
8. Nocedal, J., Wright, S.J.: Numerical Optimization, 2nd edn. Springer, New York (2006)
9. Pryce, J.D., Reid, J.K.: ADO1, a Fortran 90 code for automatic differentiation. Tech. Rep. RAL-TR-1998-057, Rutherford Appleton Laboratory, Chilton, Didcot, Oxfordshire, OX11 OQX, England (1998). URL ftp://ftp.numerical.rl.ac.uk/pub/reports/prRAL98057.pdf

CasADi: A Symbolic Package for Automatic Differentiation and Optimal Control

Joel Andersson, Johan Åkesson, and Moritz Diehl

Abstract We present CasADi, a free, open-source software tool for fast, yet efficient solution of nonlinear optimization problems in general and dynamic optimization problems in particular. To the developer of algorithms for numerical optimization and to the advanced user of such algorithms, it offers a level of abstraction which is notably lower, and hence more flexible, than that of algebraic modeling languages such as AMPL or GAMS, but higher than working with a conventional automatic differentiation (AD) tool.

CasADi is best described as a minimalistic computer algebra system (CAS) implementing automatic differentiation in eight different flavors. Similar to algebraic modeling languages, it includes high-level interfaces to state-of-the-art numerical codes for nonlinear programming, quadratic programming and integration of differential-algebraic equations. CasADi is implemented in self-contained C++ code and contains full-featured front-ends to Python and Octave for rapid prototyping. In this paper, we present the AD framework of CasADi and benchmark the tool against AMPL for a set of nonlinear programming problems from the CUTEr test suite.

Keywords Automatic differentiation • Dynamic optimization • Optimal control • Nonlinear programming • Source code transformation • Operator overloading • C++ • Python • Octave

J. Andersson (✉) · M. Diehl
Electrical Engineering Department (ESAT) and Optimization in Engineering Center (OPTEC), K.U. Leuven, Kasteelpark Arenberg 10, 3001, Heverlee, Belgium
e-mail: joel.andersson@esat.kuleuven.be; moritz.diehl@esat.kuleuven.be

J. Åkesson
Department of Automatic Control, Faculty of Engineering, Lund University, BOX 118, 21100, Lund, Sweden
e-mail: johan.akesson@control.lth.se

S. Forth et al. (eds.), *Recent Advances in Algorithmic Differentiation*, Lecture Notes in Computational Science and Engineering 87, DOI 10.1007/978-3-642-30023-3_27,

1 Introduction and Motivation

In *dynamic optimization* or *optimal control*, solutions are sought for decision-making problems constrained by dynamic equations in the form of ordinary differential equations (ODE) or differential-algebraic equations (DAE). Optimal control problems (OCPs) can cover many industry-relevant features, such as multi-stage problem formulations, problems with integer valued decision variables, problems with multi-point constraints and problems with uncertainty. The fact that the problem formulation easily becomes so general, along with a large set of different solution algorithms, makes it difficult to implement software tools that treat optimal control problems with some generality. While tools do exist that deal with a broad range of OCPs – examples include ACADO Toolkit [21], DIRCOL [24], DyOS [1] and MUSCOD-II [8, 23] – the usage of these tools, in particular for a user not directly involved in their actual development, is typically limited to comparably restricted problem formulations.

The CasADi project addresses optimal control problems using a different approach than the above packages. Rather than providing the end user with a "black box" OCP solver, the focus is a framework that allows advanced users to implement their method of choice, with any complexity. The framework takes the form of a minimalistic computer algebra system (CAS) with a general implementation of automatic differentiation (AD) on expression graphs made up of sparse and matrix-valued atomic operations. Around this *symbolic core* is a set of interfaces to numerical codes for nonlinear programming (NLP), ODE/DAE integration, quadratic programming (QP) and solution of linear systems as well as front-ends to the scripting languages Python and Octave for fast prototyping and easy access to facilities for visualization and user interaction.

The remainder of the paper is organized as follows: Sect. 2 introduces some AD concepts needed for the presentation of the structure of CasADi in Sect. 3. We benchmark CasADi against AMPL in Sect. 4 before ending the paper with some concluding remarks in Sect. 5.

2 Background: Eight Flavors of Automatic Differentiation

In this work, we consider eight different *flavors* of AD, each flavor being a combination of the following three mutually orthogonal dimensions:

- Forward versus reverse mode
- Operator overloading versus source code transformation
- Scalar versus matrix-valued atomic operations

We assume that the reader is familiar with the forward and adjoint modes of AD. For an introductory discussion, we refer to Griewank and Walther [19]. The other two dimensions are briefly explained next.

2.1 *Operator Overloading versus Source Code Transformation*

The two classical ways to implement AD are through operator overloading (OO) and source code transformation (SCT). For our purposes, we shall not use the notion of OO and SCT for the implementation approach, but for the method: Assume that we have an expression graph constructed in some manner, be it from directly parsing source code, by recording an *operation trace* (as in e.g. ADOL-C [17]) or through a symbolic modeling process (the approach used in CasADi). The OO approach to AD can then be seen as a method that loops over the operations of the expression graph of the *nondifferentiated function* while also propagating directional derivatives. In the SCT approach, on the other hand, a new expression graph is constructed. For more discussion about OO and SCT and their pros and cons, we refer to [6].

2.2 *Scalar versus Matrix-Valued Atomic Operations*

While most AD tools break down structurally complex functions into sequences of a small set of *scalar* atomic operations, AD can also be applied to an algorithm consisting of a sequence of *vector-* or *matrix-valued* atomic operations. This has advantages both in terms of memory usage and speed, which can be seen from considering the reverse mode automatic differentiation of a function performing a matrix-matrix multiplication. While using the matrix chain rule directly gives an expression for the derivative that only involves matrix products and transposes (see e.g. [25]), a breakdown into scalar-valued operations would cause the number of stored elements to grow rapidly with the matrix dimensions and would not make use of computationally efficient routines for matrix products.

The first popularized implementation of AD with matrix-valued operations was done in MATLAB by Verma and Coleman through their ADMAT/ADMIT packages [10, 25], using an OO approach. More recent implementations, also in MATLAB, include the ADiMat package [7], MAD [13] and MSAD [22]. Implementations in other programming languages include the Python-based Theano package [5].

3 Structure of CasADi

CasADi is written in self-contained C++ and is designed to have the *look and feel* of a traditional computer algebra system, but borrows its representation of symbolic expressions from the field of automatic differentiation. This can be illustrated by considering the following piece of code for MATLAB's Symbolic Math Toolbox:

```
x = sym('x');
for i=1:100
  x = x*sin(x);
```

```
end
disp x
```

This example causes a memory overflow in MATLAB (version 7.9) and equivalent scripts in Maple (version 15) and SymPy (version 0.6.6) give the same result. Because subexpressions are copied in these tools, the 100 consecutive copying operations would result in a graph with some $2^{100} \approx 10^{30}$ nodes, explaining the overflow.

After replacing the function `sym` above with either `ssym` or `msym`, corresponding to expression graphs with scalar- and matrix-valued atomic operations respectively (see Sect. 3.2), the syntax is that of CasADi's Octave front-end. Since CasADi references rather than copies subexpressions, the script will execute in a fraction of a second as the expression graph contains only 200 nodes (100 sines and 100 multiplications).

CasADi contains a growing set of operations normally associated with computer algebra tools, such as symbolic evaluations, (rudimentary) expression simplifications and solution of linear systems of equations that contain symbolic expressions, but the capabilities are still far from state-of-the-art computer algebra systems. For concrete examples on how to use these and other features, in particular in the context of optimal control, we refer to CasADi's user's guide and examples collection.

The main focus of CasADi is automatic differentiation, where a multiple paradigm approach is employed to implement all of the eight flavors of AD as presented in Sect. 2. This is described in the following.

3.1 Two Different Expression Graph Representations

While an AD implementation that supports vector- or matrix-valued atomic operations is both economical and fast for algorithms consisting mainly of vector-valued operations, this extra generality can come at the price of more overhead both in terms of memory and in terms of computations when used for algorithms consisting solely of scalar-valued operations. To avoid this, CasADi uses a combination of two expression graph representations with scalar and matrix-valued atomic operations, respectively.

Using either graph representation, the user uses a CAS-like syntax to define multiple-input, multiple-output functions. The nodes of these expressions are then topologically sorted (using variants of *depth-first search* or *breadth-first search*) to give sequences of operations that can be evaluated in two different virtual machines, one for each graph representation. Both virtual machines support forward and reverse mode directional derivatives using both OO and SCT.

3.1.1 Scalar-Valued Atomic Operations

The first graph representation used by CasADi is designed for minimal overhead and consists of atomic operations that are either unary or binary scalar-valued operations. These restrictions make it possible to represent expressions with millions of nodes and numerically evaluate the corresponding functions with an overhead comparable to an operator overloading tool such as ADOL-C [17] and CppAD [4] or an algebraic modeling language such as AMPL [14], as shown in the comparison in Sect. 4.

By maintaining lists of *live variables*, it is possible to reuse the memory locations of intermediate variables that go out of scope. This decreases the overall memory footprint, but it also makes it possible to identify "inplace" binary operations of the form $x := x+y \Leftrightarrow x\mathrel{+}=y$ or $x := x*y \Leftrightarrow x\mathrel{*}=y$. While the savings made by using the inplace operators as such are small (compared to the overall memory overhead of the operation), a feature of CasADi is to *combine* these inplace operations with other elementary operations, giving such atomic operations as "inplace addition and multiplication", $x\mathrel{+}=y * z$ or "inplace multiplication and sine", $x\mathrel{*}=sin(y)$. This increases the number of atomic operations by a factor of five, since every operation can be combined with ":=", "+=", "-=", "*=" and "/=", respectively.

We note that this feature is particularly interesting for the expression of a Jacobian or Hessian calculated through SCT, where a large proportion of the atomic operations will be either additions or multiplications. Tests on the `det_minor` speed benchmark (in e.g. CppAD [4]) have shown a decrease in overhead from around 6 ns to less than 4 ns per elementary operation (with 49,055 operations in total).

Even faster code (by another factor 5) is possible through generation of C-code, which can be compiled and dynamically linked with CasADi without halting the execution of the program. The faster execution times are, however, often offset by long compilation times for the generated code.

3.1.2 Sparse, Matrix-Valued Atomic Operations

The second graph representation used by CasADi uses atomic operations that are multiple-input, multiple-output operations, with inputs and outputs being sparse[1] matrices. To represent matrix sparsity, a *compressed row storage* format [2] is used. The sparsity patterns are always fixed in the virtual machines and can be shared between multiple expressions (the ownership of the sparsity pattern objects are controlled through *reference counting* in C++).

The atomic operations in the matrix-valued graph representations have been selected carefully to give a small set of operations that are general enough to be able to express very general symbolic expressions, but at the same time few enough

[1] A *sparse input* in this context is an input that is never used by the operation.

to be maintainable (especially in the context of AD by SCT) and allow automatic simplification of expressions, for example during element assignment operations.

3.2 *Three Different Symbolic Types*

While having two different expression graph representations does not per se dictate the need to expose both formulations to the user, we believe that doing so increases the transparency of the tool and allows users to design symbolic expressions that use the two graph representations in an optimal way.

CasADi has, from a user's point of view, three different, but related symbolic types. Firstly, there is a scalar type, `SX`, which uses the scalar graph representation and can be used in similar way to e.g. `adouble` in ADOL-C [17], though in contrast to this type, `SX` does *not* contain any numerical value that can be used for determining the control flow. Switches (currently unsupported for `SX`) must be explicitly defined by the user. The second symbolic type is a sparse matrix type, `Matrix<SX>`,[2] whose elements are of the above `SX` type. `Matrix<SX>` uses a MATLAB-inspired *everything-is-a-matrix* syntax, with scalars represented as 1-by-1 matrices and vectors as n-by-1 matrices. The third symbolic type is the matrix expression type `MX`, which corresponds to the matrix graph representation and uses the same syntax as the `Matrix<SX>` type.

We point out that, although different, there are several common features in the above classes. The `SX` and `MX` share the same code for functionalities such as topological sorting and sparsity pattern calculation, `Matrix<SX>` and `MX` share code corresponding to the syntax and sparsity pattern and all the three classes use the same code for differentiation rules. This is achieved by employing C++ idioms such as reference counting as well as static polymorphism and other template metaprogramming techniques.

3.3 *Function Objects*

CasADi defines a base class for function objects (or *functors*), the two virtual machines being two examples. These functions objects correspond to multiple-input, multiple-output functions where inputs as well as outputs are allowed to be sparse (in the above sense), hence matching the matrix graph representation. When a function is defined by a symbolic expression, the sparsity patterns of the output will be the same as the sparsity patterns of the expressions defining the function.

---

[2]The template class `Matrix<>` can also be instantiated with other types, including numerical types and (though it has not been tested) the scalar types of other AD tools.

The function objects define a uniform way for calculating directional derivatives by either OO or SCT as well as complete Jacobians. With matrix-valued inputs and outputs, one would expect a (usually very sparse) fourth order tensor to be the Jacobian corresponding to one input–output pair. Vectorizing both the input and output matrices, including either *all* matrix elements or just the *structurally nonzero* matrix entries depending on context, such a *Jacobian block* can be represented by a (sparse) two dimensional matrix.

3.4 Sparsity Pattern Detection and Graph Coloring for Jacobians

CasADi calculates Jacobian blocks by either OO or SCT in three steps: Firstly, the sparsities of the different blocks are calculated by (forward or adjoint mode) propagation of dependencies (see e.g. [18]). Secondly, graph coloring is used to find a small set of seed directions that can be used to uniquely determine the Jacobian. Two algorithms are implemented; a direct unidirectional algorithm which can be applied to either the rows or the columns of the Jacobian and a direct star coloring algorithm for Hessians. Both methods are described by Gebremedhin et al. in [15]. The third and last step of the Jacobian calculation is to assemble the complete Jacobian.

For SCT using the matrix graph representation, we first allow the expression graph for the Jacobian to contain *references* to the Jacobian blocks of function objects called from the graph. Then, in a second step, which takes place during the topological sorting of the derivative expression, all the Jacobian references are sorted by (nondifferentiated) function and input arguments and replaced in the expression by a new expression that calculates a set of Jacobian blocks along with the nondifferentiated function.

Note that when substituting the Jacobian references, we can choose to form expressions only for the Jacobian blocks (or directional derivatives) that are actually needed. A planned extension of CasADi is to use *partial coloring* (see [15]) to generate this set of Jacobian blocks efficiently.

3.5 External Solver Interfaces

For the development of optimization code, several solver packages have been written or interfaced. These solvers all take the form of function objects (Sect. 3.3), allowing them to be referenced from expression graphs.

Interfaced tools include ODE/DAE integrators, in particular the Sundials suite [20], where directional derivatives are calculated through automatic formulation and solution of the forward and adjoint sensitivity equations. Other interfaced solvers are quadratic programming (QP) solvers such as qpOASES [12], OOQP [16] and CPLEX [11] and nonlinear programming (NLP) solvers such as Ipopt [26] and KNITRO [9]. Whenever derivative information is needed (e.g. in ODE/DAE integrators or NLP solvers, this is automatically generated.

3.6 Python and Octave Front-Ends

While C++ is an excellent language for writing high-performance code, scripting languages such as Matlab, Python or Octave are often preferred when working interactively and the user wishes to make use of packages for scientific computing or visualization. CasADi uses the open-source tool SWIG [3] to parse CasADi's C++ header files, and then automatically generate full-featured and easily maintained front-ends to Python and Octave. These front-ends make rapid prototyping possible without any compiler in the loop and can typically be used with small differences in performance, since the computationally heavy calculations take place in CasADi's virtual machines and in interfaced code.

4 Numerical Results

To assess the performance of CasADi in a real world setting, we use Bob Vanderbei's AMPL version of the CUTEr test problems suite. We parse the problems using AMPL and solve them using Ipopt [26] (Version 3.10 using MA27 as a linear solver) in two different ways: By Ipopt's interface to the AMPL Solver Library (ASL) and by converting the expressions for the objective function and constraints to CasADi scalar expression graphs and using CasADi to generate derivatives and its interface to Ipopt. From the 111 nonlinear problems successfully solved (out of a total of 135), we select the problems where the iterates are identical for CasADi and ASL – a different number in the iterates can often be explained by the finite precision floating point arithmetics. In Table 1, we present the six problems where Ipopt reports more than 0.200 s solution time for either CasADi or ASL. All calculations have been performed on a Dell Latitude E6400 laptop with an Intel Core Duo processor (only one core was used) of 2.4 GHz, with 4 GB of RAM, 3072 KB of L2 Cache and 128 KB of L1 cache, running Ubuntu 10.4.

In the table, we show the total solution time as reported by Ipopt and in parentheses the part of this time actually spent in ASL or CasADi (most of the remainder is spent in the linear solver). For five of the six benchmarks, the time spent in CasADi is half or less than that of ASL. Note that there is no compiler in the loop for neither CasADi nor ASL. Through C-code generation, significantly shorter runtimes would be possible.

Table 1 Benchmarking against AMPL solver library (ASL)

Benchmark	Variables	Constraints	NLP iterations	Time ASL [s]	Time CasADi [s]
gpp	250	498	22	0.492 **(0.272)**	0.500 **(0.264)**
Reading1	10,001	5,000	22	0.712 **(0.408)**	0.306 **(0.104)**
Orthrgds	10,003	5,000	16	0.949 **(0.568)**	0.512 **(0.164)**
Svanberg	5,000	5,000	30	2.492 **(0.520)**	2.300 **(0.272)**
Orthregd	10,003	5,000	6	0.332 **(0.208)**	0.160 **(0.060)**
Orthrgdm	10,003	5,000	6	0.328 **(0.208)**	0.156 **(0.068)**

5 Conclusion

We have introduced the free, open-source optimization framework CasADi, which is a symbolic tool aimed at giving users an efficient, yet transparent approach to automatic differentiation and to shortening the development times for algorithms for large-scale nonlinear optimization in general and optimal control in particular.

At the core of the tool is a general implementation of automatic differentiation using two different expression graph representations: one with scalar atomic operations for minimal overhead and one with sparse matrix-valued atomic operations for maximum generality. The combination of these two representations makes it possible to design code that is both fast *and* generic. Using either graph formulation, CasADi implements automatic differentiation in forward as well as adjoint mode using both an operator overloading and a source code transformation approach.

Acknowledgements Joel Andersson and Moritz Diehl were supported by the Research Council KUL: GOA/11/05 Ambiorics, GOA/10/09 MaNet , CoE EF/05/006 Optimization in Engineering (OPTEC) en PFV/10/002 (OPTEC), IOF-SCORES4CHEM, several PhD/postdoc & fellow grants; Flemish Government: FWO: PhD/postdoc grants, projects: G0226.06 (cooperative systems and optimization), G0321.06 (Tensors), G.0302.07 (SVM/ Kernel), G.0320.08 (convex MPC), G.0558.08 (Robust MHE), G.0557.08 (Glycemia2), G.0588.09 (Brain-machine) research communities (WOG: ICCoS, ANMMM, MLDM); G.0377.09 (Mechatronics MPC) IWT: PhD Grants, Eureka-Flite+, SBO LeCoPro, SBO Climaqs, SBO POM, O&O-Dsquare Belgian Federal Science Policy Office: IUAP P6/04 (DYSCO, Dynamical systems, control and optimization, 2007–2011) ; EU: ERNSI; FP7-HD-MPC (INFSO - ICT - 223854), COST intelliCIS, FP7-EMBOCON (ICT-248940), FP7-SADCO (MC ITN-264735), ERC HIGHWIND (259 166) Contract Research: AMINAL Helmholtz: vICERP ACCM

Johan Åkesson was supported by the Swedish Research Council in the framework the Lund Center for Control of Complex Engineering Systems.

The authors also would like thank the anonymous reviewers who helped to improve the original manuscript and everyone who has contributed to the CasADi project, in particular Joris Gillis, Attila Kozma and Carlo Savorgnan.

References

1. DyOS User Manual. RWTH Aachen University, Germany, 2.1 edn. (2002)
2. Anderson, E., et al.: Lapack Users' Guide, 2 edn. SIAM, Philadelphia (1995)

3. Beazley, D.M.: Automated scientific software scripting with SWIG. Future Gener. Comput. Syst. **19**, 599–609 (2003)
4. Bell, B.: CppAD: a package for C++ algorithmic differentiation. Computational Infrastructure for Operations Research **coin-or**(http://www.coin-or.org/CppAD) (2012)
5. Bergstra, J., Breuleux, O., Bastien, F., Lamblin, P., Pascanu, R., Desjardins, G., Turian, J., Bengio, Y.: Theano: a CPU and GPU math expression compiler. In: Proceedings of the Python for Scientific Computing Conference (SciPy) (2010). Oral
6. Bischof, C.H., Bücker, H.M.: Computing derivatives of computer programs. In: J. Grotendorst (ed.) Modern Methods and Algorithms of Quantum Chemistry: Proceedings, Second Edition, *NIC Series*, vol. 3, pp. 315–327. NIC-Directors, Jülich (2000). URL http://www.fz-juelich.de/nic-series/Volume3/bischof.pdf
7. Bischof, C.H., Bücker, H.M., Lang, B., Rasch, A., Vehreschild, A.: Combining source transformation and operator overloading techniques to compute derivatives for MATLAB programs. In: Proceedings of the Second IEEE International Workshop on Source Code Analysis and Manipulation (SCAM 2002), pp. 65–72. IEEE Computer Society, Los Alamitos, CA, USA (2002). DOI 10.1109/SCAM.2002.1134106
8. Bock, H., Plitt, K.: A multiple shooting algorithm for direct solution of optimal control problems. In: Proceedings 9th IFAC World Congress Budapest, pp. 243–247. Pergamon Press (1984). URL http://www.iwr.uni-heidelberg.de/groups/agbock/FILES/Bock1984.pdf
9. Byrd, R.H., Nocedal, J., Waltz, R.A.: KNITRO: An integrated package for nonlinear optimization. In: G. Pillo, M. Roma (eds.) Large Scale Nonlinear Optimization, pp. 35–59. Springer Verlag (2006)
10. Coleman, T.F., Verma, A.: ADMAT: An automatic differentiation toolbox for MATLAB. Tech. rep., Computer Science Department, Cornell University (1998)
11. Corp., I.: IBM ILOG CPLEX V12.1, User's Manual for CPLEX (2009)
12. Ferreau, H.: qpOASES – An Open-Source Implementation of the Online Active Set Strategy for Fast Model Predictive Control. In: Proceedings of the Workshop on Nonlinear Model Based Control – Software and Applications, Loughborough, pp. 29–30 (2007)
13. Forth, S.A.: An efficient overloaded implementation of forward mode automatic differentiation in MATLAB. ACM Transactions on Mathematical Software **32**(2), 195–222 (2006). URL http://doi.acm.org/10.1145/1141885.1141888
14. Gay, D.M.: Automatic differentiation of nonlinear AMPL models. In: A. Griewank, G.F. Corliss (eds.) Automatic Differentiation of Algorithms: Theory, Implementation, and Application, pp. 61–73. SIAM, Philadelphia, PA (1991)
15. Gebremedhin, A.H., Manne, F., Pothen, A.: What color is your Jacobian? graph coloring for computing derivatives. SIAM Review **47**(4), 629–705 (2005). DOI 10.1137/S0036144504444711. URL http://link.aip.org/link/?SIR/47/629/1
16. Gertz, E., Wright, S.: Object-Oriented Software for Quadratic Programming. ACM Transactions on Mathematical Software **29**(1), 58–81 (2003)
17. Griewank, A., Juedes, D., Mitev, H., Utke, J., Vogel, O., Walther, A.: ADOL-C: A package for the automatic differentiation of algorithms written in C/C++. Tech. rep., Institute of Scientific Computing, Technical University Dresden (1999). Updated version of the paper published in *ACM Trans. Math. Software* 22, 1996, 131–167
18. Griewank, A., Mitev, C.: Detecting Jacobian sparsity patterns by Bayesian probing. Mathematical Programming, Ser. A **93**(1), 1–25 (2002). DOI 10.1007/s101070100281
19. Griewank, A., Walther, A.: Evaluating Derivatives: Principles and Techniques of Algorithmic Differentiation, 2nd edn. No. 105 in Other Titles in Applied Mathematics. SIAM, Philadelphia, PA (2008). URL http://www.ec-securehost.com/SIAM/OT105.html
20. Hindmarsh, A., et al.: SUNDIALS: Suite of nonlinear and differential/algebraic equation solvers. ACM Transactions on Mathematical Software **31**, 363–396 (2005)
21. Houska, B., Ferreau, H., Diehl, M.: An Auto-Generated Real-Time Iteration Algorithm for Nonlinear MPC in the Microsecond Range. Automatica **47**(10), 2279–2285 (2011)

22. Kharche, R.V., Forth, S.A.: Source transformation for MATLAB automatic differentiation. In: V.N. Alexandrov, G.D. van Albada, P.M.A. Sloot, J. Dongarra (eds.) Computational Science – ICCS 2006, *Lecture Notes in Computer Science*, vol. 3994, pp. 558–565. Springer, Heidelberg (2006). DOI 10.1007/11758549_77
23. Leineweber, D.: Efficient reduced SQP methods for the optimization of chemical processes described by large sparse DAE models, *Fortschritt-Berichte VDI Reihe 3, Verfahrenstechnik*, vol. 613. VDI Verlag, Düsseldorf (1999)
24. von Stryk, O.: User's Guide for DIRCOL. Technische Universität Darmstadt (2000)
25. Verma, A.: Structured automatic differentiation. Ph.D. thesis, Department of Computer Science, Cornell University, Ithaca, NY (1998)
26. Wächter, A., Biegler, L.: On the Implementation of a Primal-Dual Interior Point Filter Line Search Algorithm for Large-Scale Nonlinear Programming. Mathematical Programming **106**(1), 25–57 (2006)

Efficient Expression Templates for Operator Overloading-Based Automatic Differentiation

Eric Phipps and Roger Pawlowski

Abstract Expression templates are a well-known set of techniques for improving the efficiency of operator overloading-based forward mode automatic differentiation schemes in the C++ programming language by translating the differentiation from individual operators to whole expressions. However standard expression template approaches result in a large amount of duplicate computation, particularly for large expression trees, degrading their performance. In this paper we describe several techniques for improving the efficiency of expression templates and their implementation in the automatic differentiation package Sacado (Phipps et al., Advances in automatic differentiation, Lecture notes in computational science and engineering, Springer, Berlin, 2008; Phipps and Gay, Sacado automatic differentiation package. http://trilinos.sandia.gov/packages/sacado/, 2011). We demonstrate their improved efficiency through test functions as well as their application to differentiation of a large-scale fluid dynamics simulation code.

Keywords Forward mode • Operator overloading • Expression templates • C++

E. Phipps (✉)
Optimization and Uncertainty Quantification Department, Sandia National Laboratories,† Albuquerque, NM, USA
e-mail: etphipp@sandia.gov

R. Pawlowski
Multiphysics Simulation Technologies Department, Sandia National Laboratories,† Albuquerque, NM, USA
e-mail: rppawlo@sandia.gov

† Sandia National Laboratories is a multi-program laboratory managed and operated by Sandia Corporation, a wholly owned subsidiary of Lockheed Martin Corporation, for the U.S. Department of Energy's National Nuclear Security Administration under contract DE-AC04-94AL85000.

S. Forth et al. (eds.), *Recent Advances in Algorithmic Differentiation*, Lecture Notes in Computational Science and Engineering 87, DOI 10.1007/978-3-642-30023-3_28,

1 Introduction

Automatic differentiation (AD) techniques for compiled languages such as C++ and Fortran fall generally into two basic categories: source transformation and operator overloading. Source transformation involves a preprocessor that reads and parses the code to be differentiated, applies differentiation rules to this code, and generates new source code for the resulting derivative calculation that can be compiled along with the rest of the undifferentiated source code. This approach is quite popular for simpler languages such as Fortran and C, however is challenging for C++ due to the complexity of the language. An alternative approach for C++ (and many other languages) is operator overloading. Here new derived types storing derivative values and corresponding overloaded operators are created so that when the fundamental scalar type in the calculation (float or double) is replaced by these new types and evaluated, the relevant derivatives are computed as a side-effect. This approach is attractive in that it uses native features of the language, making operator overloading-based AD tools simple to implement and use. There are two basic challenges for operator overloading schemes however: run-time efficiency and facilitating the necessary type change from the floating point type to AD types. For forward mode AD, expression templates can be used to partially address the first of these challenges. However achieving the full performance benefits of expression templates is challenging and is the subject of this paper. For the second challenge, we advocate a templating-based approach which has been described elsewhere[4, 13–15].

This paper is organized as follows. We first describe standard expression template techniques and their application to forward mode automatic differentiation in Sect. 2. For concreteness, we describe the simple implementation of these techniques in the AD package Sacado [15, 16]. Then in Sect. 3 we describe two techniques for improving the performance of expression templates: caching and expression-level reverse mode. We demonstrate significantly improved performance for these techniques, particularly for large expressions, by applying them to two test functions. We then briefly describe applying all of these techniques to a large-scale fluid dynamics simulation in Sect. 4, again demonstrating improved performance on a real-world application. We then close with several concluding remarks in Sect. 5.

2 Expression Templates for Forward Mode AD

As described above, operator overloading-based AD schemes work by first creating a new derived type and corresponding overloaded operators. While operator overloading can be used for any AD mode, and there are many ways of implementing the overloaded operators for any given AD mode, we will restrict this discussion to tapeless implementations of the first-order vector forward mode. Here the AD type typically contains a floating point value to represent the value of an intermediate variable, and an array to store the derivatives of that intermediate variable with respect to the independent variables (see [11] for an introduction to basic AD implementations). The implementation of each overloaded operator then involves calculation of the value of that operation from the values of the arguments and stored in the value of the result, and a loop over the derivative components using the corresponding derivative rule from basic differential calculus.

There are two basic problems with this approach. First, each intermediate operation within an expression requires creation of at least one temporary object, and creating and destroying this object adds significant run-time overhead. Second, the AD implementation is limited to differentiating one operation at a time, each involving a loop over derivative components. Together these problems often result in inefficient derivative code. Expression templates [19] are a technique that can address these issues. They were first used for AD in the Fad package [2, 3], and later incorporated into Sacado. Here the AD type is fundamentally the same, however the operators return an object encoding the operation type and a handle to their arguments, instead of directly evaluating the derivative. As each term in the expression is evaluated, a tree is created encoding the structure of the whole expression. Then the assignment operator for the AD type loops through this tree recursively applying the chain rule. An implementation of these ideas for the $\times$ operator is shown below.

Listing 1 Partial expression template-based operator overloading implementation

```
// Expression template—based Forward AD type
template <typename T> class Expr {};
class ETFadTag {};
class ETFad : public Expr<ETFadTag> {
  double val_;                  // value
  std::vector<double> dx_;      // derivatives
public:
  explicit ETFad(int N) : val_(0), dx_(N) {} // Constructor
  int size() const { return dx_.size(); }
  double  val() const { return val_; } // Return value
  double& val()       { return val_; } // Return value
  double  dx(int i) const { return dx_[i]; } // Return derivative
  double& dx(int i)       { return dx_[i]; } // Return derivative

  // Expression template assignment operator
  template <typename T> ETFad& operator=(const Expr<T>& x)  {
    val_ = x.val();
    dx_.resize(x_.size());
    for (int i=0; i<x.size(); i++)
      dx_[i] = x.dx(i);
  }
};

// Specialization of Expr to multiplication
template <typename ExprT1, typename ExprT2>  class MultTag {};
template <typename T1, typename T2> class Expr<MultTag<Expr<T1>, Expr<T2> > > {
  const Expr<T1>& a_; const Expr<T2>& b_;
public:
  Expr(const Expr<T1>& a, const Expr<T2>& b) : a_(a), b_(b) {}
  int size() const { return a.size(); }
  double val() const { return a.val() * b.val(); }
  double dx(int i) const { return a.val()*b.dx(i)+a.dx(i)*b.val(); }
};

// Expression template implementation of a*b
template <typename T1, typename T2> Expr< MultTag< Expr<T1>, Expr<T2> > >
operator * (const Expr<T1>& a, const Expr<T2>& b) {
  return Expr< MultTag< Expr<T1>, Expr<T2> > >(a,b);
}
```

Each overloaded operator (operator $*$()) in this case) returns a simple expression object that stores just references to its arguments with the kind of operation encoded in the type of the expression (though MultTag in this case). Notice that the overloaded multiplication operator takes general expressions as arguments, and thus an *expression tree* is generated by each term in any given expression. The class ETFad stores the value and derivatives of any given intermediate variable, and also being derived from a specialization of the Expr, is a leaf node in an expression tree. The tree is then traversed to actually compute the value of the expression and its derivatives through the template assignment operator (ETFad::operator=()) when the expression is assigned to an ETFad intermediate variable. Since the nodes in the expression tree (such as Expr< MultTag< > >) just contain references, a good optimizing compiler can often eliminate them all together and generate code that is functionally equivalent to that shown below when applied to $d = a \times b \times c$.

Listing 2 Equivalent derivative code resulting from differentiation of $a \times b \times c$

```
d.val() = a.val()*b.val()*c.val();
for (int i=0; i<d.size(); i++)
  d.dx(i) = (a.val()*b.val())*c.dx(i)+(a.val()*b.dx(i)+a.dx(i)*b.val())*c.val();
```

Thus all of the intermediate temporary AD objects have been removed and the loops have been fused into a single loop over the derivative components for d. This often removes much of the overhead associated with a simple operator overloading approach. We note that constants and passive variables introduce additional complexity into the implementation which is not shown or discussed here. Also, a dynamically allocated derivative array was chosen to allow the number of derivative components to be determined at run time. In cases when this is known at compile time, a fixed-length array can be chosen to improve performance.

3 Improving Performance of Expression Templates

While the expression template approach can significantly reduce the overhead associated with operator overloading, there is still room for improvement in reducing the cost of the differentiation. Careful examination of Listing 2 reveals a basic problem: the calculation of the value portion of intermediate terms in the expression can be repeated multiple times. This is particularly troublesome for large expressions involving many terms or expressions involving transcendental functions whose values are expensive to compute. In theory, a good optimizing compiler should be able to remove these redundant calculations through common sub-expression elimination, however our experience has been that no compiler actually does (which is supported by the numerical experiments below).

To remedy this, the compiler must be coerced into computing any needed values just once for all of the intermediate operations in the expression tree. We have investigated overcoming this problem by caching the value and/or partial derivatives of each intermediate operation in the expression objects themselves. The small modifications to the multiplication expression template from Listing 1 are shown below where the cache() method in this case stores the values of the operator arguments a and b (which in this special case is both the values and partial derivatives of the arguments). These cached values are then used in any subsequent calls to val() or dx(). Since an expression class may be copied many times during the creation of a full expression-template, the caching computation is deferred until the expression template construction is complete by modifying the top-level expression-template assignment operator to call cache() before any calls to val() or dx() (as opposed to caching within the expression constructors). This approach eliminates the duplicate computation of intermediate values, at the expense of more complicated expression objects that the compiler may not be able to optimize away (including the now non-trivial copy constructors). Nonetheless we have found this approach more efficient for recent compilers that support aggressive C++ optimization.

Listing 3 Modifications from Listing 1 for caching expression template-based operator overloading. Only the modifications from Listing 1 are presented. For brevity the simple modification to the ETFad assignment operator is suppressed

```
template <typename T1, typename T2> class Expr<MultTag<Expr<T1>, Expr<T2> > > {
  const Expr<T1>& a_; const Expr<T2>& b_;
  mutable double a_val_, b_val_;
public:
  Expr(const Expr<T1>& a, const Expr<T2>& b) : a_(a), b_(b) {}
  void cache() const {
    a_.cache(); b_.cache(); a_val_ = a_.val(); b_val_ = b_.val();
  }
  double val() const { return a_val_*b_val_; }
  double dx(int i) const { return a_val_*b_.dx(i)+a_.dx(i)*b_val_; }
};
```

A second technique that can be used to generally improve the performance of forward mode AD is expression-level reverse mode [7]. This results from the recognition that while derivatives are generally being propagated forward through the calculation, any given expression likely has multiple inputs and only one output. Thus it should be more efficient to compute the derivative of the expression outputs with respect to its inputs using reverse mode AD and then combine those derivatives with the derivatives of the inputs using the chain rule. This technique is common in source transformation tools such as ADIFOR [6], and the ADTAGEO [17] tool implemented an instant graph elimination technique in an operator overloading fashion that is equivalent to expression-level reverse mode. However we are unaware of any use of this approach in expression template-based approaches. The challenge is implementing the technique in a way that can be effectively optimized by the compiler.

We have implemented expression-level reverse mode within the forward mode classes in our tool Sacado using template meta-programming techniques [1] similar to those found in the Boost MPL library [9]. Referring to Listing 4, the total number of arguments to the expression is accumulated in the num_args member[1] of each expression class as the expression tree is built up. Leaves in the tree (objects of type ELRFad) are treated as single argument identity functions. Each binary operation

[1]Note that num_args is a compile-time constant that is uniquely determined by each expression in the code, and thus while it is "static", it isn't a static variable in the traditional sense.

such as operator $*()$ from Listing 4 treats its full set of expression arguments as the union of its two arguments, thus the operation $a \times a$ would be treated as having two arguments. The computePartials () method for each expression class computes the partial derivatives of the result of that operation with respect to the expression arguments using reverse mode AD (bar stores the derivative of the expression result with respect to that intermediate variable/operation). These are stored in the partials array, whose length is determined by num_args, with the partials arising from the first argument in a binary operation stored in the first num_args1 locations and those from the second argument in the remaining num_args2 locations. Then the arguments of the expression tree are returned by the getArg() method allowing extraction of their derivative components.

Listing 4 Additional expression template interface incorporating expression-level reverse mode

```
class ETFad : public Expr<ETFadTag> {
public:
  static const int num_args = 1; // Number of expression args

  // Return partials w.r.t. arguments
  void computePartials(double bar, double partials[]) const {
    partials[0] = bar;  }

  // Return argument Arg of expression
  template <typename Arg> const ETFad& getArg() const { return *this; }
};

template <typename T1, typename T2> class Expr<MultTag<Expr<T1>, Expr<T2> > > {
public:
  // Number of arguments to expression
  static const int num_args1 = Expr<T1>::num_args;
  static const int num_args2 = Expr<T2>::num_args;
  static const int num_args = num_args1 + num_args2;

  // Compute partial derivatives w.r.t. arguments
  void computePartials(double bar, double partials[]) const {
    a_.computePartials(bar*b_.val(), partials);
    b_.computePartials(bar*a_.val(), partials+num_args1);
  }

  // Return argument Arg for expression
  template <int Arg> const ETFad& getArg() const {
    if (Arg < num_args1) return a_.template getArg<Arg>();
    else return b_.template getArg<Arg-num_args1>();
  }
};
```

These methods are then used to combine the expression-level reverse mode with the overall forward AD propagation through the new implementation of the assignment operator shown in Listing 5. First the derivatives with respect to the expression arguments are computed using reverse mode AD. These are then combined with the derivative components of the expression arguments using the functor LocalAccumOp and the MPL function for_each. The overloaded operator () of LocalAccumOp computes the contribution of expression argument Arg to final derivative component i using the chain rule. The MPL function for_each then iterates over all of the expression arguments by iterating through the integral range $[0, M)$ where M is the number of expression arguments. Since M is a compile-time constant and for_each uses template recursion to perform the iteration, this effectively an unrolled loop.

Listing 5 Expression template forward AD propagation using expression-level reverse mode

```
// Functor for mpl::for_each to multiply partials and tangents
template <typename ExprT> struct LocalAccumOp {
  const ExprT& x_;
  mutable double t_;
  double partials_[Expr<T>::num_args];
  int i;
  LocalAccumOp(const ExprT& x_) : x(x_) {}
  template <typename ArgT> void operator () (ArgT arg) const {
    const int Arg = ArgT::value;
    t += partials[Arg] * x.template getArg<Arg>().dx(i);
  }
};

class ETFad : public Expr<ETFadTag> {
public:
  // ELR expression template assignment operator
  template <typename T> ELRFad& operator=(const Expr<T>& x) {
    val_ = x.val();
    dx_.resize(x.size());

    // Compute partials w.r.t. expression arguments
    LocalAccumOp< Expr<T> > op(x);
    x.computePartials(1.0, op.partials);

    // Multiply partials with derivatives of arguments
    const int M = Expr<T>::num_args;
    for(op.i=0; op.i<x.size(); ++op.i) {
      op.t = 0.0;
      mpl::for_each< mpl::range_c< int, 0, M > > f(op);
      dx_[i] = op.t;
    }
    return *this;
  }
};
```

Note that as in the simple expression template implementation above, the value of each intermediate operation in the expression tree may be computed multiple times. However the values are only computed in the `computePartials()` and `val()` methods, which are each only called once per expression, and thus the amount of re-computation only depends on the expression size, not the number of derivative components. Clearly the caching approach discussed above can also be incorporated with the expression-level reverse mode approach, which will not be shown here.

To test the performance of the various approaches, we apply them to

$$y = \prod_{i=1}^{M} x_i \quad (1) \quad \text{and} \quad y = \overbrace{\sin(\sin(\ldots\sin(x)\ldots))}^{M \text{ times}} \quad (2)$$

for $M = 1, 2, 3, 4, 5, 10, 15, 20$. Test function (1) tests wide but shallow expressions, whereas function (2) tests deep but narrow expressions, and together they are the extremes for expressions seen in any given computation. In Fig. 1 we show the scaled run time (average wall clock time divided by the average undifferentiated expression evaluation time times the number of derivative components N) of propagating $N = 5$ and $N = 50$ derivative components through these expressions for each value of M using the standard expression template, expression-level reverse mode, caching, and caching expression-level reverse mode approaches implemented in Sacado. Also included in these plots is a simple (tapeless) forward

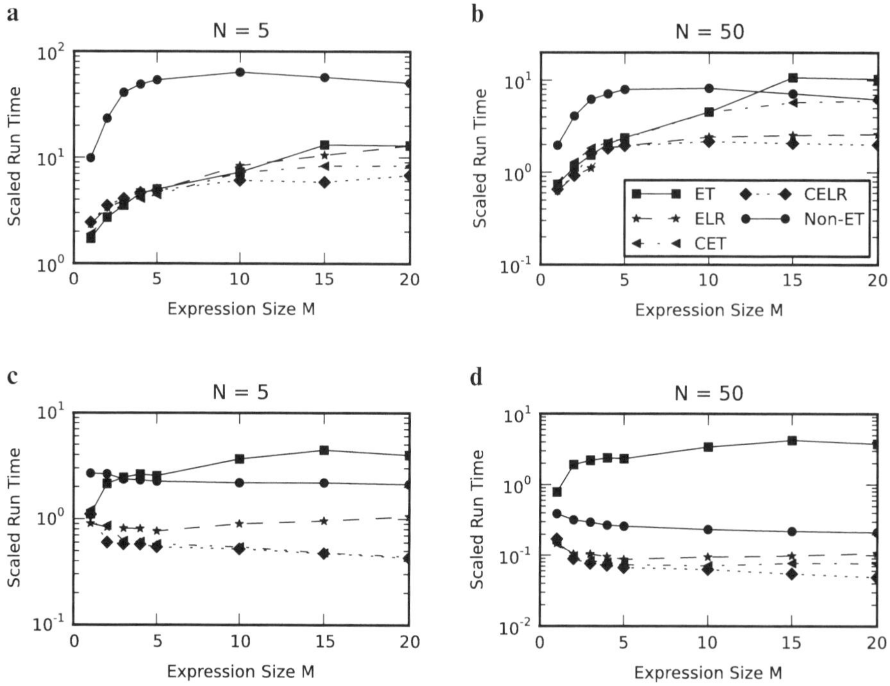

Fig. 1 Scaled derivative propagation time for expressions of various sizes. Here ET refers to standard expression templates, ELR to expression-level reverse mode, CET/CELR to caching versions of these approaches, and Non-ET to an implementation without expression templates. (**a**) Multiply function (1) for $N = 5$. (**b**) Multiply function (1) for $N = 50$. (**c**) Nested function (2) for $N = 5$. (**d**) Nested function (2) for $N = 50$

AD implementation without expression templates. These tests were conducted using Intel 12.0 and GNU 4.5.3 compilers using standard aggressive optimization options (-O3), run on a single core of an Intel quad-core processor. The GNU and Intel results were qualitatively similar with the GNU results shown here. One can see that for a larger number of derivative components or expressions with transcendental terms, the standard expression-template approach performs quite poorly due to the large amount of re-computation. All three of caching, expression-level reverse mode, and caching expression-level reverse mode are significant improvements, with the latter generally being the most efficient. Moreover, even for expressions with one transcendental term but many derivative components, these approaches are a significant improvement. For small expression sizes with no transcendental terms and few derivative components however, the differences are not significant. Since most applications would likely consist primarily of small expressions with a mixture of algebraic and transcendental terms, we would still expect to see some improvement.

Table 1 Scaled Jacobian evaluation time for reaction/transport problem

Implementation	Scaled Jacobian evaluation time
Standard expression template	0.187
Expression-level reverse	0.121
Caching expression template	0.129
Caching expression-level reverse	0.120

4 Application to Differentiation of a Fluid Dynamics Simulation

To demonstrate the impact of these approaches to problems of practical interest, we apply them to the problem of computing a steady-state solution to the decomposition of dilute species in a duct flow. The problem is modeled by a system of coupled differential algebraic equations that enforce the conservation of momentum, energy, and mass under non-equilibrium chemical reaction. The complete set of equations, the discretization technique and the solution algorithms are described in detail in [18]. The system is discretized using a stabilized Galerkin finite element approach on an unstructured hexahedral mesh of 8,000 cells with a linear Lagrange basis. We solve three momentum equations, a total continuity equation, an energy equation and five species conservation equations resulting in ten total equations. Due to the strongly coupled nonlinear nature of the problem, a fully coupled, implicit, globalized inexact Newton-based solve [10] is applied. This requires the evaluation of the Jacobian sensitivity matrix for the nonlinear system. An element-based automatic differentiation approach [4, 15] is applied via template-based generic programming [13, 14] and Sacado, resulting in 80 derivative components in each element computation. The five species decomposition mechanism uses the Arrhenius equation for the temperature dependent kinetic rate, thus introducing transcendental functions via the source terms for the species conservation equations.

Table 1 shows the evaluation times for the global Jacobian required for each Newton step, scaled by the product of the residual evaluation time and the number of derivative components per element. The calculation was run on 16 processor cores using MPI parallelism and version 4.5.3 of the GNU compilers and -O3 optimization flags. As would be expected, both caching and expression-level reverse mode approaches are significant improvements.

5 Concluding Remarks

In this paper we described challenges for using expression template techniques in operator overloading-based implementations of forward mode AD in the C++ programming language, and two approaches for overcoming them: caching and expression-level reverse mode. While expression-level reverse mode is not a new idea, we believe our use of it in expression template approaches, and its implementation using template meta-programming is unique. Together, these techniques significantly improve the performance of expression template approaches on a wide range of expressions, demonstrated through small test problems and application to a reacting flow fluid dynamics simulation. In the future we are interested in applying the approach in [12] for accumulating the expression gradient even more efficiently, which should be feasible with general meta-programming techniques.

References

1. Abrahams, D., Gurtovoy, A.: C++ Template Metaprogramming: Concepts, Tools, and Techniques from Boost and Beyond. Addison-Wesley (2004)
2. Aubert, P., Di Césaré, N.: Expression templates and forward mode automatic differentiation. In: Corliss et al. [8], chap. 37, pp. 311–315
3. Aubert, P., Di Césaré, N., Pironneau, O.: Automatic differentiation in C++ using expression templates and application to a flow control problem. Computing and Visualization in Science **3**, 197–208 (2001)
4. Bartlett, R.A., Gay, D.M., Phipps, E.T.: Automatic differentiation of C++ codes for large-scale scientific computing. In: V.N. Alexandrov, G.D. van Albada, P.M.A. Sloot, J. Dongarra (eds.) Computational Science – ICCS 2006, *Lecture Notes in Computer Science*, vol. 3994, pp. 525–532. Springer, Heidelberg (2006). DOI 10.1007/11758549_73
5. Bischof, C.H., Bücker, H.M., Hovland, P.D., Naumann, U., Utke, J. (eds.): Advances in Automatic Differentiation, *Lecture Notes in Computational Science and Engineering*, vol. 64. Springer, Berlin (2008). DOI 10.1007/978-3-540-68942-3
6. Bischof, C.H., Carle, A., Khademi, P., Mauer, A.: ADIFOR 2.0: Automatic differentiation of Fortran 77 programs. IEEE Computational Science & Engineering **3**(3), 18–32 (1996)
7. Bischof, C.H., Haghighat, M.R.: Hierarchical approaches to automatic differentiation. In: M. Berz, C. Bischof, G. Corliss, A. Griewank (eds.) Computational Differentiation: Techniques, Applications, and Tools, pp. 83–94. SIAM, Philadelphia, PA (1996)
8. Corliss, G., Faure, C., Griewank, A., Hascoët, L., Naumann, U. (eds.): Automatic Differentiation of Algorithms: From Simulation to Optimization, Computer and Information Science. Springer, New York, NY (2002)
9. Dawes, B., Abrahams, D.: Boost C++ Libraries. http://www.boost.org (2011)
10. Eisenstat, S.C., Walker, H.F.: Globally convergent inexact Newton methods. SIAM J. Optim. **4**, 393–422 (1994)
11. Griewank, A.: Evaluating Derivatives: Principles and Techniques of Algorithmic Differentiation. No. 19 in Frontiers in Appl. Math. SIAM, Philadelphia, PA (2000)
12. Naumann, U., Hu, Y.: Optimal vertex elimination in single-expression-use graphs. ACM Transactions on Mathematical Software **35**(1), 1–20 (2008). DOI 10.1145/1377603.1377605
13. Pawlowski, R.P., Phipps, E.T., Salinger, A.G.: Automating embedded analysis capabilities using template-based generic programming. Scientific Programming (2012). In press.
14. Pawlowski, R.P., Phipps, E.T., Salinger, A.G., Owen, S.J., Siefert, C., Staten, M.L.: Applying template-based generic programming to the simulation and analysis of partial differential equations. Scientific Programming (2012). In press.
15. Phipps, E.T., Bartlett, R.A., Gay, D.M., Hoekstra, R.J.: Large-scale transient sensitivity analysis of a radiation-damaged bipolar junction transistor via automatic differentiation. In: Bischof et al. [5], pp. 351–362. DOI 10.1007/978-3-540-68942-3_31

16. Phipps, E.T., Gay, D.M.: Sacado Automatic Differentiation Package. http://trilinos.sandia.gov/packages/sacado/ (2011)
17. Riehme, J., Griewank, A.: Algorithmic differentiation through automatic graph elimination ordering (adtageo). In: U. Naumann, O. Schenk, H.D. Simon, S. Toledo (eds.) Combinatorial Scientific Computing, no. 09061 in Dagstuhl Seminar Proceedings. Schloss Dagstuhl - Leibniz-Zentrum fuer Informatik, Germany, Dagstuhl, Germany (2009)
18. Shadid, J.N., Salinger, A.G., Pawlowski, R.P., Lin, P.T., Hennigan, G.L., Tuminaro, R.S., Lehoucq, R.B.: Large-scale stabilized FE computational analysis of nonlinear steady-state transport/reaction systems. Computer methods in applied mechanics and engineering **195**, 1846–1871 (2006)
19. Veldhuizen, T.: Expression templates. C++ Report **7**(5), 26–31 (1995)

Computing Derivatives in a Meshless Simulation Using Permutations in ADOL-C

Kshitij Kulshreshtha and Jan Marburger

Abstract In order to compute derivatives in a meshless simulation one needs to take into account the ever changing neighborhood relationships in a point-cloud that describes the domain. This may be implemented using permutations of the independent and dependent variables during the assembly of the discretized system. Such branchings are difficult to handle for operator overloading AD tools using traces. In this paper, we propose a new approach that allows the derivative computations for an even larger class of specific branches without retracing.

Keywords Meshless simulation • Finite pointset • ADOL-C • Permutations

1 Introduction

In general, branching in code is difficult to handle for algorithmic differentiation tools based on operator overloading. Using a recent extension of ADOL-C [8], functions containing a certain class of branches can be differentiated quite easily. As one possible application, we consider meshless simulation methods that have become popular for problems with moving boundaries or changing domains and are often applied for the optimal control of fluid flows or in the context of shape optimization problems. In order to efficiently solve such optimization problems, it is imperative that derivatives be available cheaply. One meshless method is the Finite Pointset method [4, 5]. The method relies on approximating the partial differential

---

K. Kulshreshtha (✉)
Institut für Mathematik, Universität Paderborn, Warburger Str. 100, 33098 Paderborn, Germany
e-mail: kshitij@math.upb.de

J. Marburger
Fraunhofer-Institut für Techno- und Wirtschaftsmathematik, Fraunhofer-Platz 1, 67663 Kaiserslautern, Germany
e-mail: jan.marburger@itwm.fhg.de

S. Forth et al. (eds.), *Recent Advances in Algorithmic Differentiation*, Lecture Notes in Computational Science and Engineering 87, DOI 10.1007/978-3-642-30023-3_29,

operators in small neighborhoods on a point cloud. Using the lagrangian formulation in space, as the optimizer iterates, the points change their positions, thereby changing the neighborhood relationships as well. The approximated partial differential operators in these neighborhoods are assembled into system of equations. The discrete solution is then obtained by solving this system.

In order to compute derivatives of the solution with respect to the positions in the point-cloud one needs to track the changing neighborhood relationships correctly. An algorithmic differentiation tool like ADOL-C that traces the computation to form an internal function representation, from which all required derivatives may be computed [8], would fail as soon as the neighborhood information has changed, since then ADOL-C will require retracing the function evaluation. This is a particularly expensive process and would result in expensive computation of derivatives.

The simulation handles these ever changing neighborhood relationships by permuting the variables as required for the current state. So the obvious alternative for ADOL-C is to be able to use permutations without the need for retracing.

In Sect. 2 we give a short introduction of the Finite Pointset Method. Then in Sect. 3 we shall describe, how permutations can be handled in ADOL-C. Section 4 reports on a numerical simulation. In Sect. 5 we discuss some other applications of this implementation of permutations before we conclude in Sect. 6.

2 Finite Pointset Method

The basic idea of this method is exemplified by the Laplacian. Let $\Omega \subset \mathbb{R}^2$ be a bounded domain and $f : \Omega \to \mathbb{R}$ a sufficiently smooth function. Moreover, let $P = \{\mathbf{x}_1, \dots, \mathbf{x}_N\}$, $\mathbf{x}_i = (x_i, y_i) \in \Omega$ denote a given point set. Then we approximate f by $f_h(\mathbf{x}) = \sum_{j=1}^{N} c_j(\mathbf{x}) f_j$ where c_j are some, yet unknown, functions used as approximation weights and $f_j = f(\mathbf{x}_j)$ supporting values. For the Laplacian we get

$$\Delta f(\mathbf{x}) \simeq \Delta\Big(\sum_{j=1}^{N} c_j(\mathbf{x}) f_j\Big) = \sum_{j=1}^{N} (\Delta c_j(\mathbf{x})) f_j =: \sum_{j=1}^{N} \tilde{c}_j(\mathbf{x}) f_j .$$

To obtain the weights $\tilde{c}_j$ we use the following properties of the continuous Laplacian

$$\Delta\text{const} = \Delta x = \Delta y = \Delta(xy) = 0 \qquad \text{and} \qquad \Delta(x^2) = \Delta(y^2) = 2.$$

Hence, for each point $\mathbf{x} \neq \mathbf{x}_j \in P$ and weighting functions $\omega_j(\mathbf{x})$ depending on the distance from $\mathbf{x}$ to $\mathbf{x}_j$, e.g., a Gaussian function as shown in Fig. 1, the weights $\tilde{c}_j(\mathbf{x})$ have to satisfy

Fig. 1 Weight function

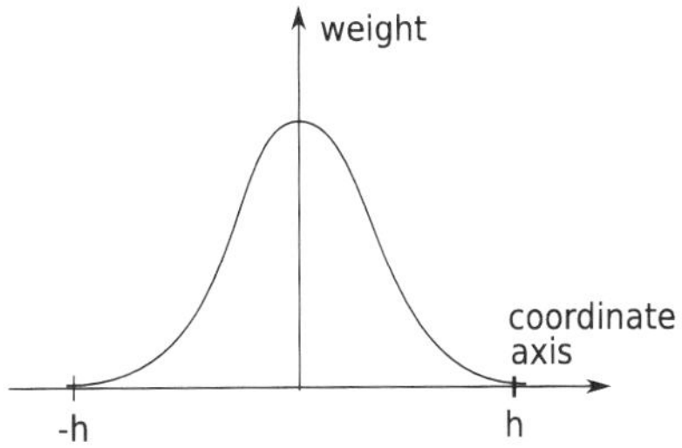

Fig. 2 Point set and smoothing length h

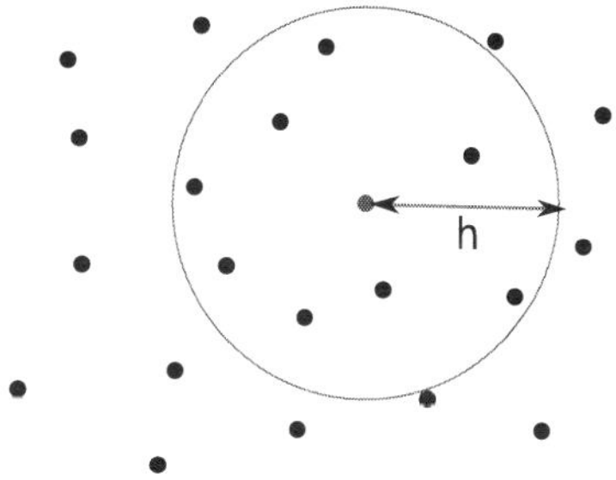

$$\sum_{j=1}^{N} \omega_j(\mathbf{x})\tilde{c}_j(\mathbf{x}) = 0, \qquad \sum_{j=1}^{N} \omega_j(\mathbf{x})\tilde{c}_j(\mathbf{x})(x - x_j)(y - y_j) = 0,$$

$$\sum_{j=1}^{N} \omega_j(\mathbf{x})\tilde{c}_j(\mathbf{x})(x - x_j) = 0 \qquad \sum_{j=1}^{N} \omega_j(\mathbf{x})\tilde{c}_j(\mathbf{x})(y - y_j) = 0,$$

$$\sum_{j=1}^{N} \omega_j(\mathbf{x})\tilde{c}_j(\mathbf{x})(x - x_j)^2 = 2, \qquad \sum_{j=1}^{N} \omega_j(\mathbf{x})\tilde{c}_j(\mathbf{x})(y - y_j)^2 = 2$$

Since the weighting function tends to zero for distances greater than the smoothing length h, cf. Fig. 2, it suffices to consider only particles with a distance less than h for the above approximation conditions. This finally yields an underdetermined linear system, as we use more supporting points than approximation conditions. The resulting system is solved by a QR factorization, for instance.

In this fashion, all spatial derivatives are approximated. Also complex boundary conditions can be implemented in that way. For example, the derivative in normal direction, i.e. $\nabla f \cdot \mathbf{n}$, can be approximated by the conditions

$$\nabla \text{const} \cdot \mathbf{n} = 0, \qquad \nabla x \cdot \mathbf{n} = n_x \qquad \text{and} \qquad \nabla y \cdot \mathbf{n} = n_y$$

for $\mathbf{n} = (n_x, n_y)^T$. Moreover, the extension of the above approach to 3D is straight forward, i.e. appropriate conditions for the z-direction are added. For more details we refer to [4].

3 Permutations and Selections

The simplest implementation of a permutation is via indices of the elements in an array or vector, i.e., a list of indices that point to the permuted elements of the original vector. In linear algebra a permutation matrix may be computed by permuting the columns of the identity matrix via such an index list. In order to incorporate such a permutation in ADOL-C we need to look closely at the internal function representation of ADOL-C, that is traced and evaluated for computing derivatives.

While a function evaluation is traced in ADOL-C various overloaded operators and functions are called that create an internal function representation. The arguments to these operators and functions are the basic active type of ADOL-C `badouble`, which is the parent type of both named active variables `adouble` and unnamed active temporaries `adub`. This representation records the operations that are performed, the constants required, and the locations of the variable arguments in the internal representation for each operation. The operations are recorded in the form of an opcode and the locations of its arguments, similar to the way a machine language does. This trace can then be evaluated at any given set of arguments to the original function by performing the operations recorded using a trace interpreter.

The hurdle in this representation for permutations is that indices in an array translate to locations of the variables in the internal representation. These locations are fixed as soon as the trace is created and cannot be changed for different points of evaluation. A permutation can however be computed differently on different points and thus translate to different locations in the internal representation.

We thus need a way to carry out this translation of the index to the location of the variable in internal representation on the fly during the evaluation instead of doing it while the trace is being created. This operation is what we call active subscripting.

3.1 *Active* rvalue *and* lvalue *Subscripting*

In order to implement active subscripting one has to distinguish between the usage as *rvalue* and *lvalue*. This usage nomenclature is standard in programming language theory and refers to the appearance of a certain value on the right hand side and left hand side of an assignment-type operator respectively.

In order to implement *rvalue* active subscripting in forward mode of AD, it is enough to copy the required taylor buffers of the indexed element of a vector to a new location that is then used in further calculations. In reverse mode this means adding the adjoint buffer at the temporary location to the ones of the indexed location. This can be done easily by defining a class in ADOL-C containing a `std::vector<adouble>`. We call this class `advector`. For *rvalue* active subscripting we define a overloaded operator

```
adub advector::operator [] (const adouble& index) const
```

This operator uses the computed value `index` to pick out the correct element out of the vector and copies it to a temporary location, which is then returned as an `adub`.

The operation is traced using a new opcode and the arguments are the locations of the vector elements (**adouble**s stored in the vector) and the location of the index and the result is the location of the new temporary variable.

Although `index` is a floating point number, for its use as an index into the underlying **std::vector** the sign and the fractional part are disregarded. This only leaves a non-negative integer to be used for indexing. For computed indices the user must be careful during initialization and computation to ensure correct results. Despite this seeming difficulty we decided to use a floating point based implementation since this allows for coverage of other applications as described in Sect. 5.

In early versions of ADOL-C as detailed in [3] there were classes called **avector** and **amatrix** that represented vectors and matrices. Due to significant bugs these classes were removed from the ADOL-C code with version 1.9. The current implementation of **advector** is not based on the old one and does not use any part of the old code. One major reason for this is that the internal memory manager of ADOL-C has completely changed with version 2.2 to allow also non-LIFO allocations as done, for example, when using STL classes.

Implementing *lvalue* active subscripting is a little more tricky, since *lvalues* should be able to support various assignment-type operators of C++. The crux of the matter is the fact, that the returned object from the subscripting operator behaves like a C++ reference type, which is distinct from value types. So this object must not belong to the established class hierarchy in ADOL-C, but should still be able to perform some internal operations, which are not seen by the user.

To this end we devised a new class **adubref** to represent an active temporary reference variable, just as **adub** represents an active temporary value. In this class one must overload all the ADOL-C relevant assignment-type operators. Also a way to convert from a reference to a value is needed, which is analogous to *rvalue* active subscripting, in that we just copy the correct values to a temporary location. The following is an essential signature of the class **adubref**.

```
class adubref {
  friend class adub;      // this one should be symmetric
  friend class advector;  // advector is also a symmetric
                          // friend of adub
protected:
  locint location;
  locint refloc;
  explicit adubref(locint lo, locint ref);
  // various other constructors each protected, as they should only
  // be called by friend classes
public:
  adub operator++(int);   // postfix increment operator does not
  adub operator--(int);   // return an lvalue
  adubref& operator++();
  adubref& operator--();
  adubref& operator = ( double );
  adubref& operator = ( const badouble& );
  adubref& operator = ( const adubref& );
  adubref& operator +=  ( double );
  adubref& operator +=  ( const badouble& ); // similarly for -= *= and /=
  adubref& operator <<= ( double );  void declareIndependent();
  adubref& operator >>= ( double& );  void declareDependent();
  friend void condassign(adubref, const badouble&, const badouble&,
                         const badouble&);
  friend void condassign(adubref, const badouble&, const badouble&);
  operator adub() const;  // used to convert to rvalue
};
```

All constructors except the copy constructor must be protected, so that the only way to construct a **adubref** object in a program is to call the active subscripting operator. The copy assignment operator must be overloaded to carry out the internal tracing operations too. This is required to handle cases such as the following situation.

```
{   advector a, b; adouble index1, index2;
    // ...
    a[index1] = b[index2];
    // ...
}
```

Now we can define the *lvalue* active subscripting operator in the **advector** class as

```
adubref advector::operator [] (const badouble& index)
```

As seen from the above signature, an **adubref** object has a location in the internal ADOL-C representation and it stores the referenced location in `refloc` during tracing as well as on the taylor buffer (zeroth order coefficient). The opcodes recorded in the internal function representation are therefore different from those for basic active value types. As arguments to these opcodes only the location of the **adubref** object is recorded, which does not change whenever the computed index changes. What changes is the referenced location that is stored on the taylor buffer, which is already transient information. During forward and reverse mode evaluations for these operations the location of the referenced element is read from the taylor buffer and then the corresponding assignment-type operation is performed on the referenced location directly.

In expressions where the basic active type of ADOL-C **badouble** are required, the **adubref** object is converted to an **adub** object by copying out data at the referenced location.

This is significantly different from the older implementation in [3], as the referenced location in that case was stored in the locations buffer as an argument to the subscript operation and the subscript output type **asub** was itself a derived type from **badouble**. Thus they used the same opcodes in the recorded internal function representation for all operations as the basic active variables.

3.2 Permutations in Finite Pointset Method

The data structure containing the domain information for the Finite Pointset method consists of two parts. Firstly there are the coordinates of the points in the point cloud. Secondly there is a list of indices for each neighborhood in this point cloud, such that the index of coordinates of the central point of a certain neighborhood is the first element in the list followed by the rest of the indices. The local spacial derivative operator is then approximated and forms a row in the system matrix corresponding to the central point.

In the implementation we select the correct coordinates from the coordinate vector using the indices in the neighborhood list. This is *rvalue* usage of the active coordinates. Then the local derivative operator is approximated in an external function of ADOL-C by solving underdetermined linear systems as described in Sect. 2. An external function is a function pointer stored on the ADOL-C trace, which is called with appropriate arguments, whenever the trace is evaluated by ADOL-C (cf. [8] and the documentation in ADOL-C sources). This returns one particular row of the system matrix. This row vector is then to be placed in the correct elements of the system matrix. The derivatives for the approximate local operator involve the derivative of a pseudoinverse and are also obtained in this external function as described in [2], when ADOL-C evaluates the trace in forward or reverse mode. At the end we solve the system of equations. This may be done using an LU factorization with pivoting or an iterative solver like GMRES.

4 Numerical Simulation with Finite Pointset Method

Let us consider the following shape optimization problem

$$\min_{\substack{x\in\Omega\\ \Omega\in\mathcal{M}}} \frac{1}{2}\|u(x)\|_2^2 \quad \text{such that} \quad L(x,u(x)) = f(x) \text{ in } \Omega$$

where Ω is a domain in $\mathbb{R}^2$ out of a some suitable set $\mathcal{M}$ of domains, L is a differential operator on this domain and f is a constant source term. In case L is nonlinear, one can solve for a feasible solution for any given domain using the Newton's method. If the operator $L(x, u(x))$ is approximated via the Finite Pointset method, the Jacobian $\partial_u L$ is straightforward. The discrete spacial derivatives approximated by Finite Pointset only occur as constants in this Jacobian. In order to compute the gradient of the objective function at the solution point of the constraint, one can use the following relationship

$$\nabla_x u = -(\partial_u L)^{-1}(\partial_x L)$$

Permutations and selections are required to compute the Jacobian $\partial_x L$. We therefore create two traces, the first one where only the dependence of L on u is encoded and a second one where the dependence of L on both x and u is encoded. We first solve the constraint system using the Newton's method and the Jacobian $\partial_u L$. Then we compute the Jacobian $\partial_x L$ from the second trace at the solution u of the constraint. Then the Jacobian of the solution w.r.t. the domain is obtained by solving linear systems for $(\partial_u L)^{-1}(\partial_x L)$. The gradient of the objective w.r.t. the domain discretization is thus $\langle u, \nabla_x u\rangle$.

Figure 3 shows the result $u(x)$ for the solution of $L(x, u(x)) = f(x)$ on a coarsely discretized domain with $x \in [0, 1] \times [0, 1]$ in 2D. Here $L(x, u(x)) =$

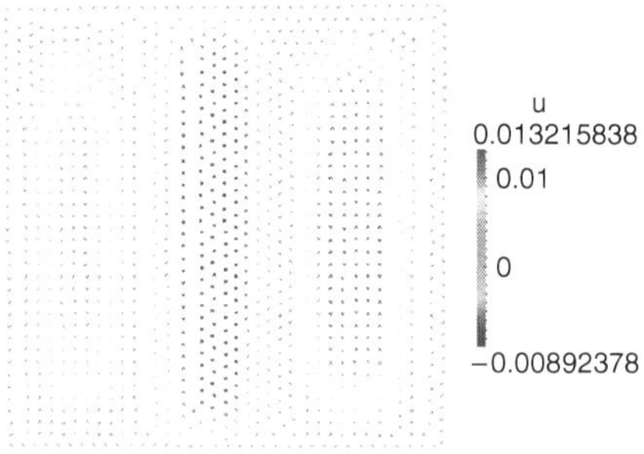

Fig. 3 Solution $u(x)$ for $L(x, u(x)) = f(x)$

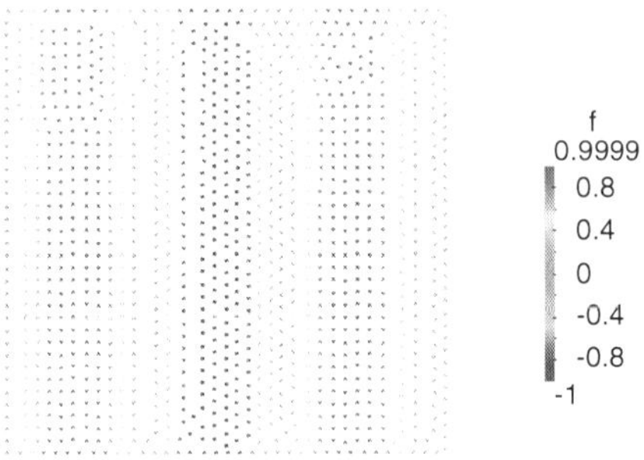

Fig. 4 Right hand side $f(x)$

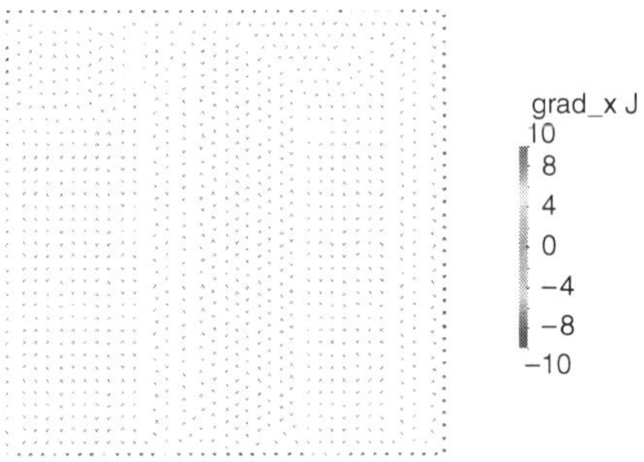

Fig. 5 Gradient of the objective in x_1 components

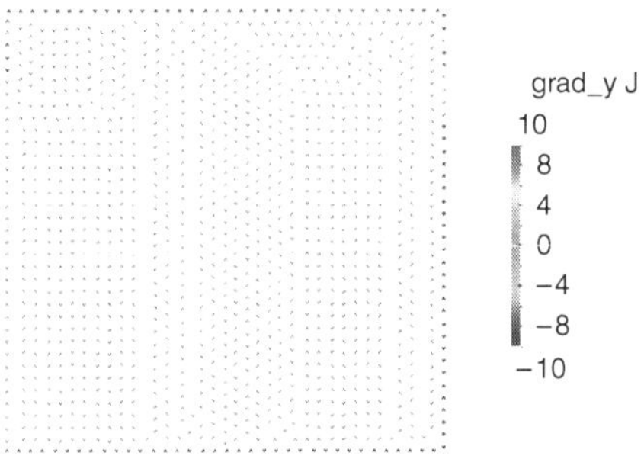

Fig. 6 Gradient of the objective in x_2 components

$u(x)^2 - \Delta u(x)$ and $f(x) = \sin(10x_1)$ (illustrated in Fig. 4) with Dirichlet boundary conditions. Figures 5 and 6 show $\langle u, \nabla_{x_1} u \rangle$ and $\langle u, \nabla_{x_2} u \rangle$ at the primal solution. The gradient components are zero in this case except at the Dirichlet boundary, where there are discontinuities. The same quantities at for a perturbation of the

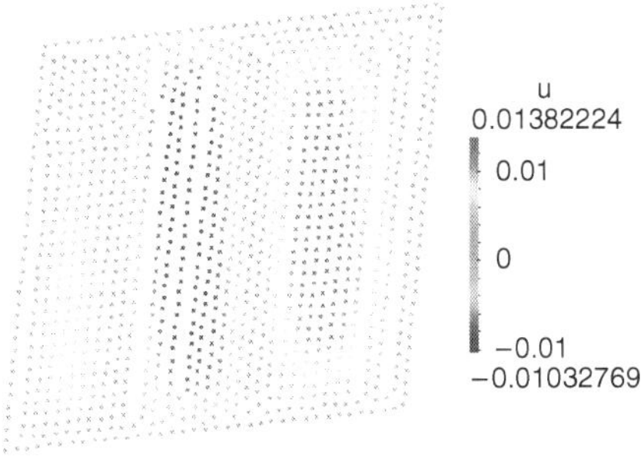

Fig. 7 Solution $u(x)$ for $L(x, u(x)) = f(x)$ in perturbed domain

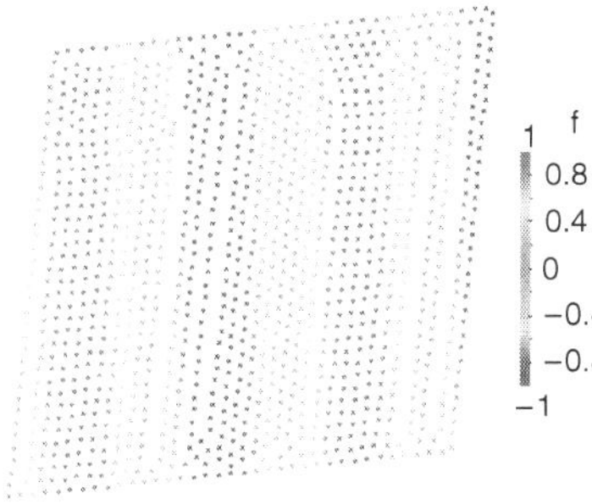

Fig. 8 Right hand side $f(x)$ in perturbed domain

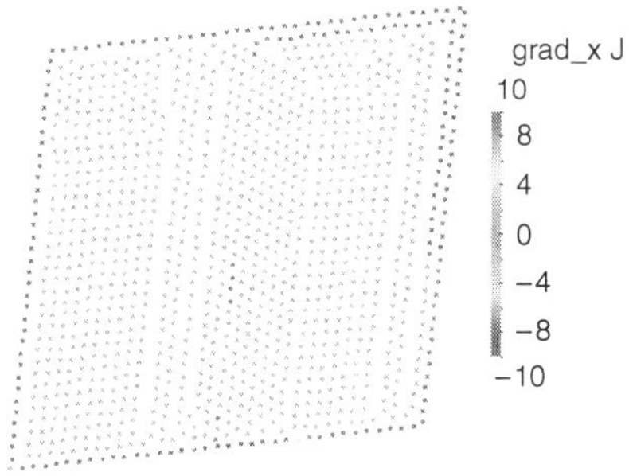

Fig. 9 Gradient of the objective in x_1 components in perturbed domain

domain, where some points have moved closer and some others away from each other, obtained from the same ADOL-C trace as before are illustrated in Figs. 7–10 respectively. The derivative obtained using this method is, in fact, just an algebraic derivative of the discretization strategy. It is unclear if the derivative obtained in this manner can be directly used for optimization purposes. The interior gradient being zero does not provide a direction for an optimization algorithm to proceed in this setting. The disconitinuous derivative at the boundary may be a cause for concern. Further analysis is required to ascertain that the derivative conforms to theoretical considerations for PDE constrained optimization problems or not. Since this problem with discrete derivatives is common to a wide class of optimization tasks, smoothing techniques for these discrete derivatives similar to the ones used in multigrid methods [9] are required. Alternatively, regularisation

Fig. 10 Gradient of the objective in x_2 components in perturbed domain

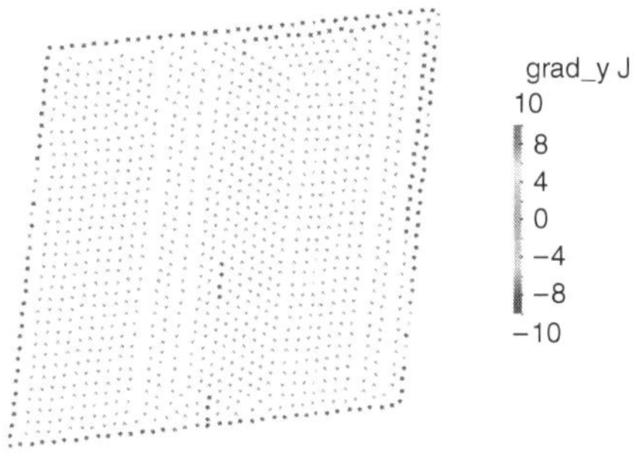

techniques for such optimisation problems as proposed by Tröltsch [7] may have to be applied.

5 Other Applications

The implementation of permutations and selections in ADOL-C as described in Sect. 3 has various other applications. One of the simplest application are matrix factorizations with pivoting. Such factorizations may be necessary when linear systems need to be solved during a function evaluation and the system is not a priori positive definite. The derivatives of such a function may now be computed by tracing the factorization using `advector` objects in ADOL-C. The permutation required for pivoting will be handled internally and implicitly in such an implementation.

Whereas similar functionality for pivoting was already available with the former active subscripting as described in [3] the new implementation allows a handling of much more general branches as, for example, it allows the appropriate taping of piecewise defined functions. That is, functions that are defined as a composition of smaller functions, each of which may be defined piecewise on their respective domains. Each such piecewise function may be implemented by evaluating all pieces and storing the results in an `advector` and then conditionally assigning the result. When used in this manner, `advector` is a generalization of the conditional assignment already implemented in ADOL-C with a choice between two values. Griewank and coauthors are currently investigating the computation of piecewise defined functions in the context of generalized differentiation (Bouligand subdifferentials [1,6]) with the help of this generalized conditional assignment.

Although the name of the class newly implemented is `advector`, it is unsuitable for implementing vector valued computations in $\mathbb{R}^n$ as atomic operations, since the implementation is based on the C++ STL class `std::vector`. Thus the class is only a container for simpler types and the operations are performed on the elements themselves.

6 Conclusions

We have demonstrated the use of permutations and selections in the computation of derivatives in a meshless simulation using the Finite Pointset method. The implementation of active subscripting has many other applications like in the differentiation of piecewise functions.

The major benefit of this implementation is that it avoids retracing the function evaluation, whenever a branch switch occurs. This was a major drawback of ADOL-C until now for any function evaluation with different branches. This still exists if the evaluation code consists of general if-then-else branches. However for simpler cases, like out-of-order evaluation of intermediates or dependents, selection of independents on runtime, piecewise definition of subfunctions etc. we can now evaluate the function and its derivatives without having to retrace the evaluation at the new point of evaluation.

In the context of shape optimization, being able to compute the gradient of the objective function with respect to the domain as well as the Jacobian of the constraints numerically allows one to compute adjoint states for problems, where the adjoint is hard to compute theoretically. The derivatives and the adjoint state may then be used in a suitable optimization procedure to compute optimal shapes. Actual implementation of such an optimization strategy with derivatives obtained from ADOL-C for shape optimization or optimal control of fluid flows (Euler or Navier-Stokes) equations as constraints remains a subject for further work.

References

1. Clarke, F.: Optimization and Nonsmooth Analysis. Wiley, New York (1983). Reprinted by SIAM, Philadelphia, 1990
2. Golub, G.H., Pereyra, V.: The differentiation of pseudo-inverses and nonlinear least squares problems whose variables separate. SIAM Journal on Numerical Analysis **10**(2), pp. 413–432 (1973). URL http://www.jstor.org/stable/2156365
3. Griewank, A., Juedes, D., Mitev, H., Utke, J., Vogel, O., Walther, A.: ADOL-C: A package for the automatic differentiation of algorithms written in C/C++. Tech. rep., Institute of Scientific Computing, Technical University Dresden (1999). Updated version of the paper published in *ACM Trans. Math. Software* 22, 1996, 131–167
4. Kuhnert, J.: Finite pointset method based on the projection method for simulations of the incompressible Navier-Stokes equations . Springer LNCSE: Meshfree methods for Partial Differential Equations **26**, 243–324 (2002)
5. Marburger, J.: Optimal Control based on meshfree approximations. Ph.D. thesis, Technische Universität Kaiserslautern (2011). Verlag Dr. Hut München
6. Mifflin, R.: Semismooth and semiconvex functions in constrained optimization. SIAM Journal of Control and Optimization **15**, 952–972 (1977)
7. Tröltsch, F.: Regular lagrange multipliers for control problems with mixed pointwise control-state constraints. SIAM J. on Optimization **15**(2), 615–634 (2005)
8. Walther, A., Griewank, A.: Getting started with ADOL-C. In: U. Naumann, O. Schenk (eds.) Combinatorial Scientific Computing. Chapman-Hall (2012). See also http://www.coin-or.org/projects/ADOL-C.xml
9. Wesseling, P.: An Introduction to Multigrid Methods. Wiley (1992)

Lazy K-Way Linear Combination Kernels for Efficient Runtime Sparse Jacobian Matrix Evaluations in C++

Rami M. Younis and Hamdi A. Tchelepi

Abstract The most notoriously expensive component to develop, extend, and maintain within implicit PDAE-based predictive simulation software is the Jacobian evaluation component. While the Jacobian is invariably sparse, its structure and dimensionality are functions of the point of evaluation. The application of Automatic Differentiation to develop these tools is highly desirable. The challenge presented is in providing implementations that treat dynamic sparsity efficiently without requiring the developer to have any a priori knowledge of sparsity structure. Under the context of dynamic sparse Operator Overloading implementations, we develop a direct sparse lazy evaluation approach. In this approach, an efficient runtime variant of the classic Expression Templates technique is proposed to support sparsity. The second aspect is the development of two alternate multi-way Sparse Vector Linear Combination kernels that yield efficient runtime sparsity detection and evaluation.

Keywords Implicit • Simulation • Sparsity • Jacobian • Thread-safety

1 Introduction

A focal area of scientific computing is the predictive simulation of complex physical processes. Implicit simulation methods require the evaluation and solution of large systems of nonlinear residual equations and their Jacobian matrices. In the context of emerging simulation applications, the Jacobian matrix is invariably large and sparse. Moreover the actual sparsity structure and dimensionality may both be functions of the point of evaluation. Additionally, owing to the model complexity,

R.M. Younis (✉) · H.A. Tchelepi
Department of Energy Resources Engineering, Stanford University, Stanford, CA, USA
e-mail: ryounis@stanford.edu; tchelepi@stanford.edu

S. Forth et al. (eds.), *Recent Advances in Algorithmic Differentiation*, Lecture Notes in Computational Science and Engineering 87, DOI 10.1007/978-3-642-30023-3_30,

the evaluation of the Jacobian matrix typically occurs over numerous modules and stages, requiring the storage of resultants from a wide range of intermediate calculations. The resultants of such calculations vary dramatically in terms of their level of sparsity, ranging from univariate variables to dense and block sparse multivariates. Finally, given an interest in rapidly modifiable codes to include new physics or alternate sets of independent unknowns, the most notoriously expensive software component to develop, extend, and maintain is the Jacobian matrix evaluation component. Dynamic, sparse Automatic Differentiation (AD) offers a clearly recognized potential solution to the design and development challenges faced by implicit simulator developers. Several comprehensive introductions to AD are available [7, 9, 15]. The efficient runtime computation of dynamic sparse Jacobian matrices is the topic of several recent contributions. There are two broad approaches to dynamic sparse AD.

The first approach uses results from sparsity pattern analysis by graph coloring techniques in order to obtain the Jacobian from a compressed intermediate dense matrix [8, 13]. This is accomplished by inferring the sparsity pattern of the Jacobian and analyzing it to determine an appropriate compression operator that is referred to as the *seed matrix*. The dense intermediate matrix is computed using AD, and the target sparse Jacobian is backed-out from it using the seed matrix. Since the AD operations are performed in a dense format, they can be implemented efficiently. Advances in efficient dense AD implementations include the Operator Overloading (OO) tools as described in [1, 14]. These approaches report the use of a lazy evaluation generic metaprogramming technique known as Expression Templates (ET) [10, 11] in order to attain close to optimal dense AD operation efficiencies. The computational costs of the compression and de-compression however can be significant and can involve heavy sparse memory bandwidth limited operations. In situations where the sparsity pattern is constant or is known a priori, this cost may be amortized since the seed matrix remains unchanged. In the context of general purpose predictive simulation this is not the case.

The second approach is intrinsically dynamic, and it uses sparse vector data structures to represent derivatives. The core computational kernel of direct runtime sparse AD is a SParse-vector Linear Combination (SPLC). This is because the derivative of any expression with $k > 0$ arguments can be expressed as a linear combination of the k sparse vector derivatives of the expression arguments; $c_1\mathbf{f}_1' + \ldots + c_k\mathbf{f}_k'$. SPLC operations perform sparsity structure inference along with the computation of the sparse Jacobian entries. Examples of implementations with such a capability include the SparsLinC module [4] within the Source Transformation tools ADIFOR [3] and ADIC [5]. Direct sparse treatment offers complete runtime flexibility with transparent semantics. On the other hand, since the computational kernel consists of a set of sparse vector operations, it is a challenge to attain reasonable computational efficiency on modern computer architectures. Sparse algebra operations involve a heavy memory bandwidth requirement leading to notoriously memory-bound contexts [12]. Existing codes such as the SparsLinC module provide various sparse vector data structures that attempt to hide the costs of dynamic memory and memory latency to some extent.

1.1 This Work

This work extends the lazy evaluation performance optimization techniques that are applied in dense AD approaches to direct dynamic sparsity OO implementations. The extension requires two advances. The first, is to extend the data structures and construction mechanism of the ET technique to suit sparsity. The second is to develop single pass algorithms to execute SPLC expressions more efficiently.

Section 2 introduces the challenges of extending the classic compile-time ET technique to support sparse arguments directly. A run-time alternative form of ET is developed. In particular, the run-time variant is designed to deliver competitive levels of efficiency compared to static approaches while directly supporting sparsity.

Section 3 reviews current SPLC algorithms and develops an alternate class of single-pass SPLC evaluation algorithms. The algorithms execute SPLC expressions involving k sparse vectors in one go while improving the ratio of memory and branch instructions to floating point operations over current alternatives.

Finally, Sect. 4 presents computational results using a large-scale Enhanced Oil Recovery simulation.

2 Lazy Evaluation Techniques for Dynamic Sparsity

The compile-time (static) ET technique is a lazy evaluation OO implementation that overcomes the well-recognized performance shortcoming of plain OO [6]. Along the *pairwise evaluation* process of OO, the ET technique generates expression graph nodes instead of dense vector intermediate resultants. The expression nodes are allocated on the stack, and they are meant to be completely optimized away by the compiler. The execution of the expression is delayed until the expression is assigned to a cached dense vector variable. At that point, the expression is executed with a single fused loop.

Since dense linear combinations involve vectors of the same dimension, the single-pass loop is performed entry-by-entry. Each iteration produces the appropriate scalar resultant by recursively querying parent nodes in the ET graph. The recursion terminates at the leaf dense vectors which simply return their corresponding entry. On the way out of the recursion, intermediate nodes perform scalar operations on the returned values and pass the resultant down along the recursion stack.

The extension of the classic ET technique to SPLCs requires a different ET data structure. In sparse operations, non-zero entries do not always coincide, and subsequently the depth of the fused loop is not known until the entire SPLC is executed. Moreover, every node within the ET data structure would need to maintain its own intermediate scalar state. This implies that the ET nodes for a sparse expression grow recursively in size on the stack with no constraints on the recursion depth. This is exacerbated by the fact that OO intermediates have a temporary life-cycle, and so ET nodes need to store parent nodes by value to avoid undefined

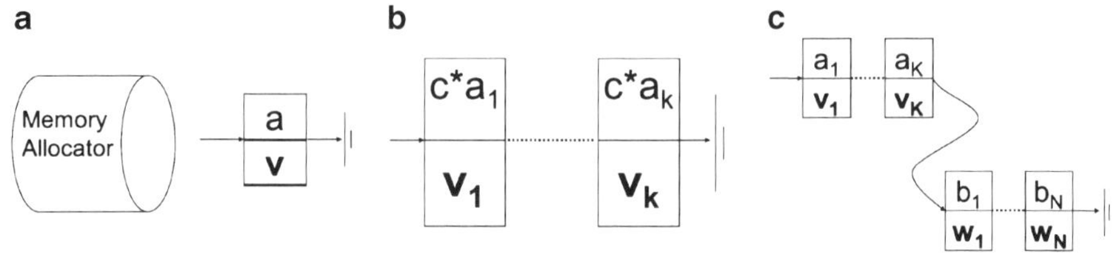

Fig. 1 SPLC expressions are represented by a one-directional linked list. The list is built by the OO pairwise evaluation process involving three fundamental operations only. (**a**) $c.\mathbf{V}$. (**b**) $c.\sum_{i=1}^{k} a_i.\mathbf{V}_i$. (**c**) $\sum_{i=1}^{K} a_i.\mathbf{V}_i + \sum_{j=1}^{N} b_j.\mathbf{W}_j$

behavior. The exception to this is the leaf nodes since they refer to vector arguments that are persistent in memory. This costly situation suggests a value in dynamic SPLC expression data structures that are inexpensive to build at runtime.

Once they are multiplied through, forward mode derivative expressions become vector linear combinations. The SPLCs can be represented by a list where each entry is a pair of a scalar weight and a sparse vector argument. Owing to their efficiency of concatenation, singly linked list data structures can be used to efficiently store and represent runtime SPLC expressions. In the proposed approach, the scalar operators are overloaded to generate an SPLC list through three fundamental building blocks. These three operations are illustrated Fig. 1.

The first operation depicted in Fig. 1a involves the multiplication of a scalar weight and a sparse vector argument. This operation would be used for example whenever the chain rule is applied. Only in this operation is it necessary to allocate memory dynamically. Since the elements of SPLC expressions are allocated dynamically, their lifespan can be controlled explicitly. Subsequently, nodes can be made to persist beyond a statement's scope and it is only after the evaluation stage that the SPLC expressions need to be freed.

The second operation is the multiplication of a scalar and a SPLC sub-expression. As illustrated in Fig. 1b, this is accomplished most efficiently by multiplying the weights in the linked list through, leaving the dynamic memory intact as is it returned by the operator.

Finally, Fig. 1c illustrates the third building block; the addition of two SPLC sub-expressions, each containing one or more terms. The addition simply involves the re-assignment of the tail pointer of one sub-list to the head node of the other. In total, using dynamic memory pools, the run-time lists require $O(k)$ operations.

3 K-Way SPLC Evaluation Kernels

Upon assignment to a resultant, the SPLC needs to be evaluated. In this section, we develop two evaluation algorithms that exploit the fact that all k arguments are already available.

The first algorithm employs a caching implicit binary tree to generalize the SparsLinC algorithms. The cached intermediate nodes store non-zero elemental information, thereby substantially reducing the number of non-zero index comparisons and the associated memory bandwidth requirements. The second algorithm is inspired by the seed matrix mappings used in other forms of sparse AD. Before presenting the two algorithms, we review the current approach to SPLC evaluation in AD tools.

The SparsLinC module generalizes a 2-way algorithm in a static manner to accommodate more arguments. The 2-way SPLC uses a running pointer to each of the two vector arguments. Initially both of the two pointers are bound to the first non-zero entry of its respective sparse vector argument. While both running pointers have not completely traversed their respective vector, the following sequence of operations is performed. The column indices of the two running pointers are compared. If the nonzero entry column indices are equal, the two entries are linearly combined and inserted into the resultant. Both running pointers are advanced. On the other hand, if they are not equal, then the entry with the smaller column index is inserted, and only its running pointer is advanced. At the end of the iteration, if one of the two sparse arrays involves any remaining untraversed entries, they are simply inserted into the resultant. The SparsLinC module executes K-Way combinations by repeating this 2-way algorithm in a pairwise process.

3.1 K-Way SPLC Kernel 1: Caching Nodal Binary Tree

This approach generalizes the pairwise evaluation process used by SparsLinC in order to perform the evaluation in one pass while minimizing the number of index comparisons that are necessary. A binary tree is designed to maintain non-zero elemental state information at each node. This state information consists of a single nonzero entry (a pair of an integer column index and a value), as well as a logical index that maintains a node's *activation*. There are two types of nodes that make-up the tree.

1. Terminal leaves are the parent terminal nodes, and each leaf refers to a SPLC argument. A running pointer to the argument's sparse vector is maintained. The pointer is initialized to the sparse vector's first nonzero entry. Terminal leaves are *active* provided that their running pointer has not traversed the entire sparse vector.
2. Internal nodes, including the terminal root node, have two parents. Such nodes maintain the linear combination nonzero resultant of the parent nodes. Internal nodes also maintain a coded activation variable that distinguishes between each of the following four scenarios:
   (a) Internal nodes are inactive if both parents are.
   (b) The left parent entry has a smaller column index than the right parent.
   (c) The right parent's column index is smaller than the left's.

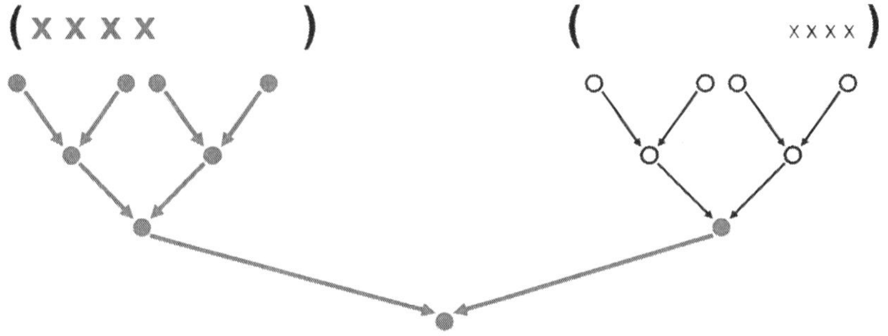

Fig. 2 A hypothetical SPLC expression and caching binary tree. The *left sub-tree* resultant is a sparse vector with nonzero entries with low column indices. The *right sub-tree* resultant has nonzero entries with large column indices. At the initial stages of the evaluation process (the first four iterations), only the *left sub-tree* is queried for column index comparisons

(d) The column indices of both parents are equal.

The evaluation is performed in a single pass process starting from the root internal node. At each step in the fused evaluation loop, two reverse topological sweeps are executed by recursion. The first sweep is an *Analyze Phase* that labels the activation codes. The second sweep is an *Advance Phase* where all advanceable nodes are visited to evaluate their nonzero entry value and to update the running pointers of the active leaf nodes. The iteration continues so long as the root node remains active.

Consider the hypothetical SPLC scenario illustrated in Fig. 2. The proposed algorithm requires at most half of the number of comparisons and associated reads and writes as would be required by a SparsLinC kernel.

3.2 K-Way SPLC Kernel 2: Prolong-and-Restrict Dense Accumulation

This algorithm is inspired by the seed matrix approaches to sparse AD. As illustrated in Fig. 3, the algorithm proceeds in a two stage process. In the prolong phase (Fig. 3a), each of the k sparse array arguments is added into a zero-initialized dense buffer. In the restrict phase (Fig. 3b), the entire dense buffer is traversed to deduce the non-zero entries producing a sparse resultant. This algorithm performs poorly whenever the dimension of the required enclosing dense buffer is very large compared to the number of non-zero entries in the resultant. On the other hand, when that is not the case, this algorithm is very effective as it uses a more favorable memory locality profile and involves no branching in the prolong phase.

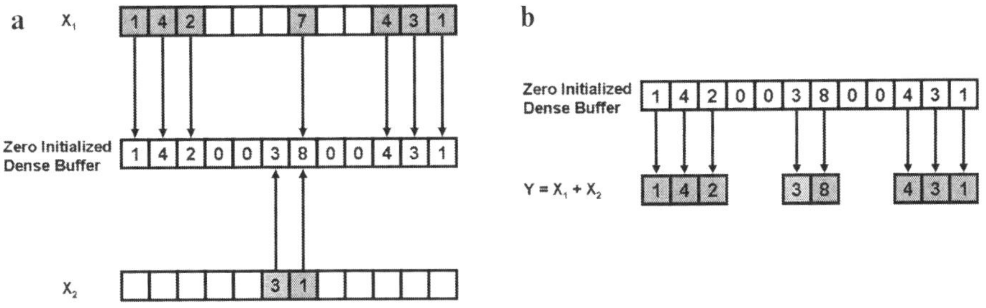

Fig. 3 An illustration of the two stages of the prolong-and-restrict k-way SPLC kernel. In this example, there are two SPLC arguments, $k = 2$, with unit weights. (**a**) Prolong phase; sparse vectors are added to a zero-initialized dense buffer. (**b**) Restrict phase; the dense intermediate is mapped to a sparse resultant

3.3 Summary

Expressions involving multiple arguments ($k > 2$) can be evaluated more efficiently using k-way generalizations. In order to better characterize and compare the performance of the proposed algorithms, we introduce some diagnostic SPLC parameters which may all be computed efficiently during the SPLC list construction process. The first parameter is the *Apparent Dimension*, N_a, that is defined as the difference between the smallest nonzero entry column index and the largest column index in the resultant of the SPLC expression. The second parameter is the *Nonzero Density*, $0 < N_d \leq 1$, which is the ratio of the number of nonzero entries in the resultant of the SPLC to the Apparent Dimension. Finally, the third parameter is the number of arguments in the expression k.

The computational cost of the caching binary tree kernel is clearly independent of N_a. A worst-case scaling of the number of necessary memory reads and writes goes as $\log(k) N_d$. This cost is asymptotically favorable to that attained for example by the SparsLinC kernel which scales as $k.N_d$. On the other hand, the cost of the prolong-and-restrict dense accumulation kernel scales primarily with N_a since the prolong phase involves no branching.

4 Computational Examples

The OO lazy evaluation techniques discussed in this work are all implemented in a comprehensive thread-safe generic C++ OO AD library [16] that computes runtime sparse Jacobians using the forward mode. The Automatically Differentiable Expression Templates Library (ADETL) provides generic data structures to represent AD scalars and systems that can be univariate or multivariate (gencrically dense, sparse, or block sparse). The library handles cross-datatype operations and implements poly-algorithmic evaluation strategies. The choice of OO technique used depends on

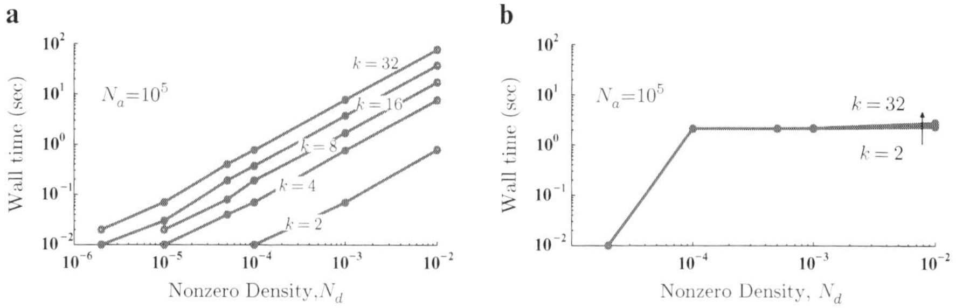

Fig. 4 Performance comparisons of several SPLC evaluations using the two proposed kernels. The test sparse vectors are generated randomly and have an apparent dimension, $Na = 10^5$. (**a**) Binary tree kernel. (**b**) Prolong-restrict kernel

the type of derivatives involved in an AD expression. The ADETL treats univariates with a direct pairwise evaluation. Dense multivariate expressions involving more than two arguments are treated using the classic ET technique. Finally, sparse and block-sparse multivariate expressions are treated with the dynamic SPLC lists and are evaluated using either of the two proposed kernels.

To illustrate the computational performance of the proposed algorithms for sparse problems, we consider a number of hypothetical SPLC expressions as well as the computation of a block structured Jacobian matrix arising from the numerical discretization of a system of PDAEs.

4.1 Model SPLC Numerical Experiments

To empirically validate the computational cost relations discussed in Sect. 3.3, we generate a number of synthetic SPLC expressions that span a portion of the three dimensional parameter space defines by k, N_a, and N_d. In particular, we execute a series of SPLC expressions $\sum_k c_k \mathbf{V_k}$ with $k = 2, 4, 8, 16$, and 32 arguments. The argument sparse vectors $\mathbf{V_k}$ and coefficients c_k are generated randomly. By freezing the Apparent Dimension, $N_a = 10^5$, we can vary N_d simply by varying the number of nonzero entries used to generate the sparse vector arguments. We consider the range $10^{-6} < N_d < 10^{-1}$ that spans a wide range of levels of sparsity. Figure 4a, b show the empirical results obtained using the binary tree and the prolong-restrict kernels respectively. The figures show plots of the wall execution time taken to construct and evaluate SPLC expressions with varying N_d. Each curve consists of results for a fixed k.

Clearly, the asymptotic behavior of the two algorithms is distinct. The prolong-restrict results show that for fairly large N_a, neither the number of arguments nor the level of sparsity N_d matter. These differences in computational cost lead to a

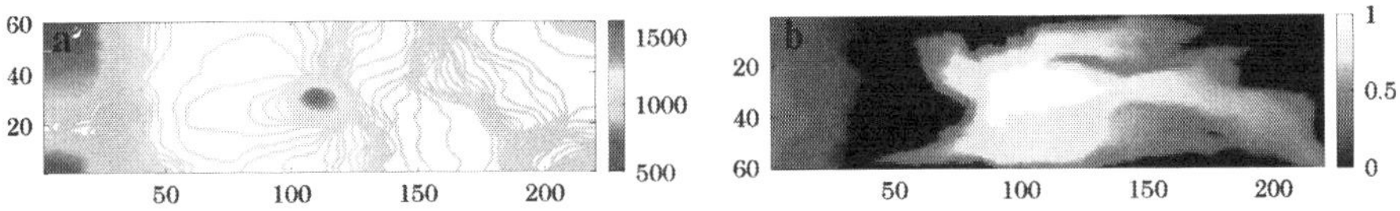

Fig. 5 Two sample state component snapshots for a simulation performed using the ADETL. (**a**) Pressure contour time snapshot (psi). (**b**) Gas saturation time snapshot

performance crossover point. The ADETL exploits this by performing install-time measurements such as those presented in Fig. 4 in order to apply a poly-algorithmic evaluation strategy that automatically selects the better algorithm for a given situation.

4.2 Model Problem Simulation Jacobian

The nonlinear residual and Jacobian evaluation routines of a General Purpose Reservoir Simulator (GPRS) are re-written using the ADETL [16]. The original GPRS code was written using hand-coded Jacobian matrices including manual branch fragments that encode a dynamic sparsity pattern. The GPRS implements fully coupled implicit finite volume approximations of compressible compositional, thermal, multi-phase flow in porous media [2]. The system of equations is a collection of PDAEs of variable size and structure depending on the thermodynamic state.

Figure 5 shows sample results obtained using the ADETL GPRS. During the course of the simulation, 735 N iterations are performed, each requiring the evaluation of the residual and Jacobian. The total wall clock times taken by the hand-differentiated and manually assembled GPRS is 238-s. The time taken by the ADETL implementation is 287-s, implying a total performance penalty of 21%. This penalty is considered minor compared to the improved maintainability and level of extendability of the new code.

5 Summary

The core kernel of runtime sparse Jacobian AD is a SPLC operation. We develop an OO implementation that combines a dynamic form of ET along with two alternate k-way evaluation algorithms. Extensive use of the ADETL in developing general purpose physical simulation software shows comparable performance compared to hand-crafted alternatives.

References

1. Aubert, P., Di Césaré, N., Pironneau, O.: Automatic differentiation in C++ using expression templates and application to a flow control problem. Computing and Visualization in Science **3**, 197–208 (2001)
2. Aziz, K., Settari, A.: Petroleum Reservoir Simulation. Elsevier Applied Science (1979)
3. Bischof, C.H., Carle, A., Corliss, G.F., Griewank, A., Hovland, P.D.: ADIFOR: Generating derivative codes from Fortran programs. Scientific Programming **1**(1), 11–29 (1992)
4. Bischof, C.H., Khademi, P.M., Bouaricha, A., Carle, A.: Efficient computation of gradients and Jacobians by dynamic exploitation of sparsity in automatic differentiation. Optimization Methods and Software **7**, 1–39 (1997)
5. Bischof, C.H., Roh, L., Mauer, A.: ADIC — An extensible automatic differentiation tool for ANSI-C. Software–Practice and Experience **27**(12), 1427–1456 (1997). DOI 10.1002/(SICI)1097-024X(199712)27:12⟨1427::AID-SPE138⟩3.0.CO;2-Q. URL http://www-fp.mcs.anl.gov/division/software
6. Bulka, D., Mayhew, D.: Efficient C++: performance programming techniques. Addison-Wesley Longman Publishing Co., Inc., Boston, MA, USA (2000)
7. Fischer, H.: Special problems in automatic differentiation. In: A. Griewank, G.F. Corliss (eds.) Automatic Differentiation of Algorithms: Theory, Implementation, and Application, pp. 43–50. SIAM, Philadelphia, PA (1991)
8. Gebremedhin, A.H., Manne, F., Pothen, A.: What color is your Jacobian? graph coloring for computing derivatives. SIAM Review **47**(4), 629–705 (2005). DOI 10.1137/S0036144504444711. URL http://link.aip.org/link/?SIR/47/629/1
9. Griewank, A.: On automatic differentiation. In: M. Iri, K. Tanabe (eds.) Mathematical Programming, pp. 83–108. Kluwer Academic Publishers, Dordrecht (1989)
10. Karmesin, S., Crotinger, J., Cummings, J., Haney, S., Humphrey, W.J., Reynders, J., Smith, S., Williams, T.: Array design and expression evaluation in POOMA II. In: Proceedings of the Second International Symposium on Computing in Object-Oriented Parallel Environments, ISCOPE '98, pp. 231–238. Springer-Verlag, London, UK (1998)
11. Kirby, R.C.: A new look at expression templates for matrix computation. Computing in Science Engineering **5**(3), 66–70 (2003)
12. Lee, B., Vuduc, R., Demmel, J., Yelick, K.: Performance models for evaluation and automatic tuning of symmetric sparse matrix-vector multiply. In: Parallel Processing, 2004. ICPP 2004. International Conference on, pp. 169–176 vol.1 (2004)
13. Narayanan, S.H.K., Norris, B., Hovland, P., Nguyen, D.C., Gebremedhin, A.H.: Sparse jacobian computation using ADIC2 and ColPack. Procedia Computer Science **4**, 2115–2123 (2011). DOI 10.1016/j.procs.2011.04.231. URL http://www.sciencedirect.com/science/article/pii/S1877050911002894. Proceedings of the International Conference on Computational Science, ICCS 2011
14. Phipps, E.T., Bartlett, R.A., Gay, D.M., Hoekstra, R.J.: Large-scale transient sensitivity analysis of a radiation-damaged bipolar junction transistor via automatic differentiation. In: C.H. Bischof, H.M. Bücker, P.D. Hovland, U. Naumann, J. Utke (eds.) Advances in Automatic Differentiation, *Lecture Notes in Computational Science and Engineering*, vol. 64, pp. 351–362. Springer, Berlin (2008). DOI 10.1007/978-3-540-68942-3_31
15. Rall, L.B.: Perspectives on automatic differentiation: Past, present, and future? In: H.M. Bücker, G. Corliss, P. Hovland, U. Naumann, B. Norris (eds.) Automatic Differentiation: Applications, Theory, and Implementations, *Lecture Notes in Computational Science and Engineering*, vol. 50, pp. 1–14. Springer, New York, NY (2005). DOI 10.1007/3-540-28438-9_1
16. Younis, R.M.: Modern advances in software and solution algorithms for reservoir simulation. Ph.D. thesis, Stanford University (2002)

Implementation of Partial Separability in a Source-to-Source Transformation AD Tool

Sri Hari Krishna Narayanan, Boyana Norris, Paul Hovland, and Assefaw Gebremedhin

Abstract A significant number of large optimization problems exhibit structure known as *partial separability*, for example, least squares problems, where elemental functions are gathered into groups that are then squared. The sparsity of the Jacobian of a partially separable function can be exploited by computing the smaller Jacobians of the elemental functions and then assembling them into the full Jacobian. We implemented partial separability support in ADIC2 by using pragmas to identify partially separable function values, applying source transformations to subdivide the elemental gradient computations, and using the ColPack coloring toolkit to compress the sparse elemental Jacobians. We present experimental results for an elastic-plastic torsion optimization problem from the MINPACK-2 test suite.

Keywords Forward mode • Partial separability • ADIC2 • ColPack

1 Introduction

As introduced by Griewank and Toint [13], a function $f : \mathbb{R}^n \mapsto \mathbb{R}$ is considered partially separable if f can be represented in the form

$$f(x) = \sum_{i=1}^{m} f_i(x), \quad (1)$$

S.H.K. Narayanan (✉) · B. Norris · P. Hovland
Mathematics and Computer Science Division, Argonne National Laboratory, 9700 South Cass Avenue, Argonne, IL 60439, USA
e-mail: snarayan@mcs.anl.gov; norris@mcs.anl.gov; hovland@mcs.anl.gov

A. Gebremedhin
Department of Computer Science, Purdue University, West Lafayette, IN, USA
e-mail: agebreme@purdue.edu

S. Forth et al. (eds.), *Recent Advances in Algorithmic Differentiation*, Lecture Notes in Computational Science and Engineering 87, DOI 10.1007/978-3-642-30023-3_31,

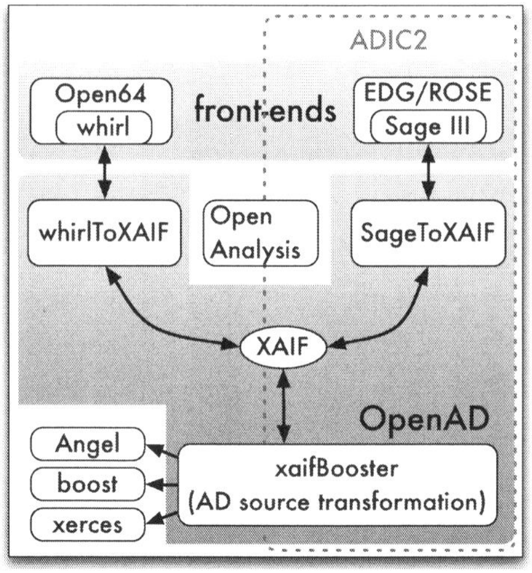

Fig. 1 *OpenAD* component structure and source transformation workflow

where f_i depends on $p_i \ll n$ variables. Bouaricha and Moré [5] and Bischof and El-Khadiri [3], among others, have explored different ways to exploit the sparsity of the Jacobians and Hessians of partially separable functions. To compute the (usually dense) gradient ∇f of f, one can first compute the much smaller (and possibly sparse) gradients of the f_i elementals, then assemble the full gradient of f. This approach can significantly reduce the memory footprint and number of floating-point operations for overall gradient computation compared with computing dense gradients.

This paper describes the following new capabilities of the ADIC2 source transformation tool.

- Pragma-guided source transformations to perform scalar expansion of the elemental components of partially separable scalar-valued functions.
- Exposing of the sparsity present in the elemental Jacobians.
- Calculation of the compressed elemental Jacobians using ColPack.
- Combining of the elementals into the scalar valued result.

1.1 ADIC2

ADIC2 is a source transformation tool for automatic differentiation of C and C++ codes, with support for both the forward and reverse modes of AD [15]. ADIC2 uses the ROSE compiler framework [17], which relies on the EDG C/C++ parsers [10]. ADIC2 is part of the OpenAD framework at Argonne, whose general structure is illustrated in Fig. 1. Briefly, the process of transforming the original source code into code computing the derivatives consists of several steps: (1) canonicalization (semantically equivalent transformations for removing features that hamper analysis or subsequent AD transformations); (2) program analysis (e.g., control flow graph construction, def-use chains); (3) generation of the language independent XAIF intermediate representation; (4) AD transformation of the XAIF representation; (5) conversion of the resulting AD XAIF code back to the ROSE intermediate representation; and (6) generation of C/C++ derivative code. The general differentiation process as implemented by ADIC2 is discussed in detail in [15]. To exploit the sparsity of the gradients of partially separable functions, we have implemented several extensions of the AD process, which are described in Sect. 3.

1.2 ColPack

When a Jacobian (or a Hessian) matrix is sparse, the runtime and memory efficiency of its computation can be improved through *compression* by avoiding storing and computing with zeros. Curtis, Powell, and Reid demonstrated that when two or more columns of a Jacobian are structurally orthogonal, they can be approximated simultaneously using finite differences by perturbing the corresponding independent variables simultaneously [9]. Two columns are structurally orthogonal if there is no row in which both columns have a nonzero. Coleman and Moré showed that the problem of partitioning the columns of a Jacobian into the fewest groups, each consisting of structurally orthogonal columns, can be modeled as a graph coloring problem [7]. The methods developed for finite-difference approximations are readily adapted to automatic differentiation with appropriate initialization of the seed matrix [2].

ColPack is a software package containing algorithms for various kinds of graph coloring and related problems arising in compression-based computation of sparse Jacobians and Hessians [12]. In ColPack, the Jacobian is represented using a *bipartite graph*. Thus, a partitioning of the columns of a Jacobian into groups of structurally orthogonal columns is obtained using a *distance-2 coloring* of the column vertices of the bipartite graph.

The coloring algorithms in ColPack are fast, yet effective, greedy heuristics. They are greedy in the sense that vertices are colored sequentially one at a time and the color assigned to a vertex is never changed. The number of colors used by the heuristic depends on the *order* in which vertices are processed. Hence, ColPack contains implementations of various effective ordering techniques for each of the coloring problems it supports.

The rest of this paper is organized as follows. Section 2 contains a brief overview of related work. Section 3 describes our implementation approach. We show experimental results for an optimization application use case in Sect. 4, and we conclude Sect. 5 with a brief summary and discussion of future work.

2 Related Work

Bischof and El-Khadiri [3] describe the approach they took in implementing partial separability support in ADIFOR. Our approach, though similar in spirit, has a number of significant differences. The ADIFOR approach assumed that the elemental functions were encoded in separate loops, while our approach does not rely on this assumption and supports partial separability when multiple elemental functions are computed in the same loop nest. To determine the sparsity pattern automatically when the Jacobian structure is unknown, both ADIFOR and ADIC2 use runtime detection through different versions of the SparsLinC library; in addition however ADIC2 also relies on ColPack to compute a coloring, which is used to initialize the seed matrix for computing a compressed Jacobian (or Hessian) using the forward mode of AD.

Järvi [14] describes an object-oriented model for parameter estimation of a partially separable function. Conforti et al. [8] describe a master-slave approach to effect a parallel implementation of AD. The master process on the basis of the partially separable structure of the function, splits the workload among the slaves, and collects the results of the distributed computation as soon as they are available. Gay [11] describes the detection of partially separable structure automatically by walking expression graphs. The structure is then used to facilitate explicit Hessian computations.

To exploit the sparsity in Jacobians of the elemental functions, we perform *scalar expansion*, which is the conversion of a scalar value into a temporary array. For example, scalar expansion can convert a scalar variable with a value of 1 to a vector or matrix where all the elements are 1 (See Sect. 3). Typically scalar expansion is used in compiler optimizations to remove scalar data dependences across loop iterations to enable vectorization or automated parallelization. This transformation is usually limited to counter-controlled for loops without control flow or function calls. The size of the temporary arrays is typically determined through polyhedral analysis of the iteration space of the loops containing the scalar operations that are candidates for expansion. Polyhedral analysis is implemented in a number of compilers and analysis tools, including Omega [16], CHiLL [6], and PLuTo [4].

Our current approach to the implementation of scalar expansion does not use polyhedral analysis. We describe our approach in more detail in Sect. 3.

3 Implementation

The changes required to support partial separability were implemented in our ADIC2 source-to-source transformation infrastructure introduced in Sect. 1.1. While this source translation can be performed in a stand alone manner, it was convenient to implement it before the canonicalization step in ADIC2. The ROSE compiler framework, on which ADIC2 is based, parses the input code and generates an abstract syntax tree (AST). ROSE provides methods to traverse and modify the AST through the addition and deletion of AST nodes representing data structures and program statements.

The translation is implemented as the following three traversals of the nodes representing the function definitions within the AST.

1. Identification of partial separability (traversal T_1)
2. Promotion of the elementals within loops and creation of a summation function (traversal T_2)
3. Generation of elemental initialization loop (traversal T_3)

In the first traversal, (T_1), the statements within each function definition are examined. If the pragma *$adic _partiallyseparable* is found, then the statement immediately following the pragma is an assignment statement whose left-hand side is the dependent variable and the scalar-valued result of a partially-separable function computation. The right-hand side of the assignment statement is an expression involving the results of the elemental function computations. The names of the variables representing the elementals are specified in the pragma.

The second traversal, (T_2), is initiated if the pragma *$adic_partiallyseparable* is found in T_1. The second traversal visits the nodes of the AST in two phases. The first phase, called *TopDown*, visits the nodes starting at the node representing the function definition and proceeds down the tree to the leaf nodes. The second phase, called *BottomUp*, visits the leaf nodes first and proceeds up the tree all the way to the node representing the function definition. Therefore T_2 visits each node twice. In both the phases information gained by visiting nodes can be passed in the direction in which nodes are visited. The following transformations occur in T_2.

1. *Scalar expansion of elementals.* In this transformation, the declaration of each of the scalar elementals is changed into a dynamically allocated array. Next, each reference to the scalar-variable elemental is modified into a reference to the array elemental.
   To allocate memory for the array, its size is determined to be the maximum number of updates to the value inside any loop nest within the function body and is calculated by the *BottomUp* phase. In the *BottomUp* phase, when an innermost loop is visited, we create a parameterized expression whose value will be the number of times that loop executes (based only on its own loop bounds). This expression is passed to its parent. Each parent loop multiplies its own local expression by the maximum of the expressions received from its children. When the *BottomUp* phase is concluded, the function definition node will contain the maximum number of updates to the elementals. This value is used to allocate memory for the type-promoted elementals.
   To modify the references of the scalar-variable elemental into a reference to the array elemental, the index of each array reference is determined by the bounds of the surrounding loops. For example,

```
double elemental;
        for(j = lb0; j < ub0; j++){
                for(i = lb1; i < ub1; i++){
                        elemental = ...     (omitted)
                }
        }
```

is transformed to

```
double *elemental;
ADIC_SPARSE_Create1DimArray(&elemental, (ub1−lb1) * (ub0−lb0));
        for(j = lb0; j < ub0; j++) {
                for(i = lb1; i < ub1; i++) {
                        temp0 = j * ( ub0 − lb0) + i;
                        elemental[temp0] = ...      (omitted)
                }
        }
```

This transformation is made possible using both phases. The expression that calculates the value of the index is created by passing the bounds of outer loops in the *TopDown* phase to the inner loops. Then, in the *BottomUp* phase, if a reference to an elemental is encountered, it is converted into an array reference and an appropriate assignment of the index-calculation expression to the array index variable is inserted at the beginning of the inner loop body.

2. *Creation of the result vector.* In this transformation, the assignment statement immediately following the pragma *$adic_partiallyseparable* is modified. This assignment statement is not affected by the previous transformation. A for loop is created to replace the assignment statement. The assignment statement itself is inserted into the body of the for loop. The for loop iterates as many times as the maximum number of updates to the elementals, which is determined in the *BottomUp* phase. For example,

```
#pragma $adic_partiallyseparable , elemental
    * scalar = temp * (elemental);
```

is transformed to

```
#pragma $adic_partiallyseparable , elemental
   for(k = 0; k< (ub1−lb1) * (ub0−lb0); k++) {
      * scalar = temp * (elemental);
   }
```

The elemental variable references within the assignment undergo scalar expansion, and the left-hand side of the assignment statement is replaced by an array reference. The dimensions of this array are the maximum number of updates to the elementals, which was determined in the previous transformation, and the loop becomes

```
#pragma $adic_partiallyseparable , elemental
   for(k = 0; k< (ub1−lb1) ∗ (ub0−lb0); k++) {
      temp_vector[k] = temp∗(elemental)[k];
   }
```

3. *Summation of the result vector.* Last, a call to a summation function is added to the code. The arguments to the summation function are the scalar dependent variable and the temporary array reference that forms the left-hand side of the modified assignment statement. For example,

```
ADIC_SPARSE_Summation(scalar, temp_vector);
```

is a call that can result from this transformation.

In the third traversal (T_3), the statements within each function definition are examined again. If the pragma *$adic_partialelemental* is found, then the statement immediately following the pragma is an assignment statement whose left-hand side is an elemental and whose right-hand side is an initialization value. Such an assignment statement is not modified by any earlier transformation. Similar to the creation of a result vector, a for loop is created that iterates as many times as the maximum number of updates to the elementals, which was determined in the previous transformation (T_2). The assignment statement itself is inserted into the body of the for loop. Finally the loop replaces the annotated assignment statements. For example,

```
#pragma $adic partialelemental
   elemental = 0.0;
```

is transformed to

```
#pragma $adic_partiallyseparable , elemental
   for(k = 0; k< (ub1−lb1) ∗ (ub0−lb0); k++) {
     elemental[k] = 0.0;
   }
```

4 Experimental Results

We evaluated the performance of the partial separability support in ADIC2 by using a two-dimensional elastic-plastic torsion model from the MINPACK-2 test problem collection [1]. This model uses a finite-element discretization to compute the stress field on an infinitely long cylindrical bar to which a fixed angle of twist per unit length has been applied. The resulting unconstrained minimization problem can be expressed as $min(f(u))$, where $f : \Re^n \rightarrow \Re$, $f(u)$ is the quadratic

$$f(u) = \int_{\mathfrak{D}} \{\frac{1}{2}\|\nabla u(x)\|^2 - cu(x)\}dx, \tag{2}$$

c is a constant, and $\mathfrak{D}$ is a bounded domain with a smooth boundary.

In our experiments, we applied ADIC2 to the C version of the function implementation after the original Fortran code was manually translated into C. As described in more detail in Sect. 3, we insert two types of pragmas to define (1) the elementals of the partially separable function and (2) the initialization of the elementals.

```
...
#pragma $adic_partialelemental
    fquad = zero;
#pragma $adic_partialelemental
    flin  = zero;
...
/* computation of the elementals fquad and fln of the
   function f and their sparse Jacobians */
...
#pragma $adic_partiallyseparable , fquad, flin
    *f = area*(p5*fquad+flin );
...
```

In the initialization portion, the value of the double constant *zero* is 0.0. After the transformation (described in Sect. 3), the last portion of the code above (the function computation) is transformed to

```
...
#pragma $adic_partiallyseparable , fquad, flin
 ADIC_SPARSE_Create1DimArray(&__adic_temp_f, ad_var_max);
 for (__adic_temp0 = 0; __adic_temp0 < ad_var_max; __adic_temp0++) {
    __adic_temp_f[__adic_temp0] = (area * ((p5 * fquad[__adic_temp0])
                                   + flin [__adic_temp0 ]));
  }
  ADIC_SPARSE_Summation(f,__adic_temp_f,ad_var_max);
...
```

where *ad_var_max* is the size of the gradient vector array for the full Jacobian.

We validated the correctness of the sparse computation by comparing it with the values produced by the analytical version. For example, for an input array size 100, the error estimated by the norm of the difference was approximately 1.5e-16, or near the limit of machine precision for floating-point computations.

We measured the execution time of the gradient computation on an Intel Xeon workstation with dual quad-core E5462 Xeon processors (8 cores total) running at 2.8 GHz (1600 MHz FSB) with 32 KB L1 cache, 12 MB of L2 cache (6 MB shared per core pair), and 16 GB of DDR2 FBDIMM RAM, running Linux kernel version 2.6.35 (x86-64). All codes were compiled with gcc version 4.4.5 with -O2 optimizations enabled.

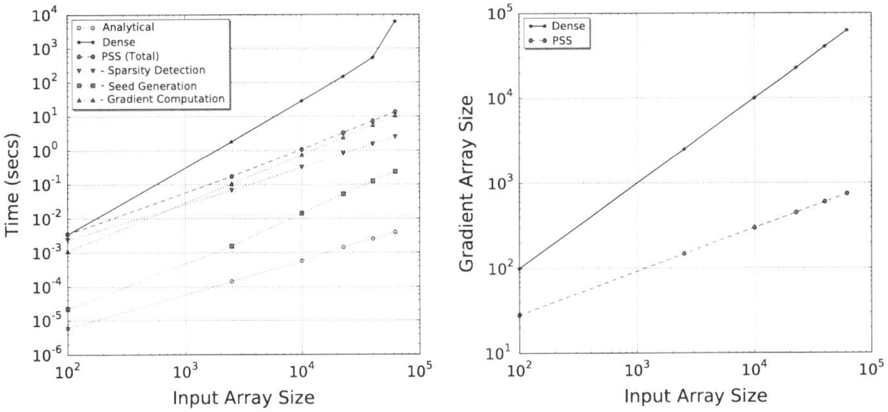

Fig. 2 *Left*: comparison between the runtimes of the three gradient versions: hand-coded analytical gradients, dense AD, and partially separable sparse AD (PSS). A breakdown of the steps in the PSS version is also shown: sparsity detection, seed generation, and gradient computation and Jacobian recovery. *Right*: gradient array sizes for the dense and PSS versions

Figure 2 (left) shows the execution times for computing the Jacobian of the function in equation (2) using three approaches: (1) manually implemented analytical derivative computation; (2) dense, forward mode AD (using ADIC2); and (3) sparse, forward mode AD (using ADIC2) with additional sparsity detection (using SparsLinC) and Jacobian compression (using ColPack). Additionally, the graph shows the time taken to perform the steps involved in approach (3): sparsity detection, seed generation, and gradient computation and Jacobian recovery. The analytic version performs best, as expected. The forward-mode dense AD version is between 500 and 3,400 times slower than the manually optimized analytical derivatives computation, while the partially separable sparse AD version achieves performance within a factor of 6 of the analytic version for small array sizes, and less than a factor of 2 slower for larger array sizes. The right side of Fig. 2 shows the reduction in memory requirements (ranging from 500 to 4,000-fold) for storing the gradients using the PSS approach compared with dense gradients. The size of the gradient array in the PSS approach is the number of colors used by ColPack to compresses the sparse Jacobian.The implication of this result is that for arrays sizes beyond a certain size, the dense approach is not feasible as it's memory requirement will exceed the available machine capacity. However, the approach exploiting partial becomes infeasible at a much larger input size.

5 Conclusion

We presented an approach to exploiting sparsity in the computation of gradients of partially separable functions, which are common in large-scale optimization. We identify partially separable computations by using pragmas, which guide our source transformation system to perform scalar expansion and generate efficient forward mode AD code for computing the gradients of the elemental functions. In addition, we exploit sparsity in these gradients by using the SparsLinC library and the ColPack coloring toolkit to enable efficient forward mode AD by using statically allocated compressed dense vectors for computing the gradients of the elemental functions. We evaluated the performance of our implementation using a case study of the elastic-plastic torsion problem in the MINPACK-2 test suite, demonstrating that (1) exploiting partial separability and sparsity significantly reduces the memory requirements of the generated code, enabling the solution of larger problems than possible with dense forward mode, and (2) the performance of the best AD version compares favorably with that of the hand-coded gradients. In future work we will extend ADIC2 to remove certain restrictions, for example, the assumption that the gradient vectors of different elemental functions are of the same size. We also plan to integrate the polyhedral analysis currently being developed in ROSE and add support for exploiting partial separability when using the reverse mode in ADIC2.

Acknowledgements This work was supported by the U.S. Dept. of Energy Office of Science Applied Mathematics Program under Contract No. DE-AC02-06CH11357.

References

1. Averick, B.M., Carter, R.G., Moré, J.J., Xue, G.L.: The MINPACK-2 test problem collection. Tech. Rep. Preprint MCS-P152-0694, Mathematics and Computer Science Division, Argonne National Laboratory (1992)
2. Averick, B.M., Moré, J.J., Bischof, C.H., Carle, A., Griewank, A.: Computing large sparse Jacobian matrices using automatic differentiation. SIAM J. Sci. Comput. **15**(2), 285–294 (1994). DOI 10.1137/0915020. URL http://link.aip.org/link/?SCE/15/285/1
3. Bischof, C., El-Khadiri, M.: On exploiting partial separability and extending the compile-time reverse mode in ADIFOR. Technical Memorandum ANL/MCS–TM–163, Mathematics and Computer Science Division, Argonne National Laboratory, Argonne, Ill. (1992). ADIFOR Working Note # 7.
4. Bondhugula, U., Hartono, A., Ramanujam, J., Sadayappan, P.: PLuTo: A practical and fully automatic polyhedral program optimization system. In: Proc. ACM SIGPLAN 2008 Conference on Programming Language Design and Implementation (PLDI) (2008)
5. Bouaricha, A., Moré, J.J.: Impact of partial separability on large-scale optimization. Comp. Optim. Appl. **7**(1), 27–40 (1997)
6. Chen, C.: Model-guided empirical optimization for memory hierarchy. Ph.D. thesis (2007)
7. Coleman, T.F., Moré, J.J.: Estimation of sparse Jacobian matrices and graph coloring problems. SIAM Journal on Numerical Analysis **20**(1), 187–209 (1983)
8. Conforti, D., Luca, L.D., Grandinetti, L., Musmanno, R.: A parallel implementation of automatic differentiation for partially separable functions using pvm. Parallel Computing **22**(5), 643–656 (1996). DOI 10.1016/0167-8191(96)00014-2. URL http://www.sciencedirect.com/science/article/pii/0167819196000142
9. Curtis, A.R., Powell, M.J.D., Reid, J.K.: On the estimation of sparse Jacobian matrices. Journal of the Institute of Mathematics and Applications **13**, 117–119 (1974)
10. Edison Design Group C++ Front End. http://www.edg.com/index.php?location=c_frontend

11. Gay, D.M.: More AD of nonlinear AMPL models: Computing Hessian information and exploiting partial separability. In: M. Berz, C. Bischof, G. Corliss, A. Griewank (eds.) Computational Differentiation: Techniques, Applications, and Tools, pp. 173–184. SIAM
12. Gebremedhin, A.H., Nguyen, D., Patwary, M., Pothen, A.: ColPack: Software for graph coloring and related problems in scientific computing. Tech. rep., Purdue University (2011)
13. Griewank, A., Toint, P.L.: On the unconstrained optimization of partially separable functions. In: M.J.D. Powell (ed.) Nonlinear Optimization 1981, pp. 301–312. Academic Press, New York, NY (1982)
14. Järvi, J.: Object-oriented model for partially separable functions in parameter estimation. Acta Cybernetica **14**(2), 285–302 (1999)
15. Narayanan, S.H.K., Norris, B., Winnicka, B.: ADIC2: Development of a component source transformation system for differentiating C and C++. Procedia Computer Science **1**(1), 1845–1853 (2010). DOI DOI:10.1016/j.procs.2010.04.206. URL http://www.sciencedirect.com/science/article/pii/S1877050910002073. ICCS 2010
16. Pugh, W.: The Omega Project web page. http://www.cs.umd.edu/projects/omega/
17. Quinlan, D.: ROSE: Compiler support for object-oriented frameworks. Tech. Rep. UCRL-ID-136515, Lawrence Livermore National Laboratory (1999). URL http://www.osti.gov/bridge/servlets/purl/793936-hdq2WX/native/793936.PDF

Editorial Policy

1. Volumes in the following three categories will be published in LNCSE:

i) Research monographs
ii) Tutorials
iii) Conference proceedings

Those considering a book which might be suitable for the series are strongly advised to contact the publisher or the series editors at an early stage.

2. Categories i) and ii). Tutorials are lecture notes typically arising via summer schools or similar events, which are used to teach graduate students. These categories will be emphasized by Lecture Notes in Computational Science and Engineering. **Submissions by interdisciplinary teams of authors are encouraged.** The goal is to report new developments – quickly, informally, and in a way that will make them accessible to non-specialists. In the evaluation of submissions timeliness of the work is an important criterion. Texts should be well-rounded, well-written and reasonably self-contained. In most cases the work will contain results of others as well as those of the author(s). In each case the author(s) should provide sufficient motivation, examples, and applications. In this respect, Ph.D. theses will usually be deemed unsuitable for the Lecture Notes series. Proposals for volumes in these categories should be submitted either to one of the series editors or to Springer-Verlag, Heidelberg, and will be refereed. A provisional judgement on the acceptability of a project can be based on partial information about the work: a detailed outline describing the contents of each chapter, the estimated length, a bibliography, and one or two sample chapters – or a first draft. A final decision whether to accept will rest on an evaluation of the completed work which should include

- at least 100 pages of text;
- a table of contents;
- an informative introduction perhaps with some historical remarks which should be accessible to readers unfamiliar with the topic treated;
- a subject index.

3. Category iii). Conference proceedings will be considered for publication provided that they are both of exceptional interest and devoted to a single topic. One (or more) expert participants will act as the scientific editor(s) of the volume. They select the papers which are suitable for inclusion and have them individually refereed as for a journal. Papers not closely related to the central topic are to be excluded. Organizers should contact the Editor for CSE at Springer at the planning stage, see *Addresses* below.

In exceptional cases some other multi-author-volumes may be considered in this category.

4. Only works in English will be considered. For evaluation purposes, manuscripts may be submitted in print or electronic form, in the latter case, preferably as pdf- or zipped ps-files. Authors are requested to use the LaTeX style files available from Springer at http://www. springer.com/authors/book+authors?SGWID=0-154102-12-417900-0.

For categories ii) and iii) we strongly recommend that all contributions in a volume be written in the same LaTeX version, preferably LaTeX2e. Electronic material can be included if appropriate. Please contact the publisher.

Careful preparation of the manuscripts will help keep production time short besides ensuring satisfactory appearance of the finished book in print and online.

5. The following terms and conditions hold. Categories i), ii) and iii):

Authors receive 50 free copies of their book. No royalty is paid.
Volume editors receive a total of 50 free copies of their volume to be shared with authors, but no royalties.

Authors and volume editors are entitled to a discount of 33.3 % on the price of Springer books purchased for their personal use, if ordering directly from Springer.

6. Commitment to publish is made by letter of intent rather than by signing a formal contract. Springer-Verlag secures the copyright for each volume.

Addresses:

Timothy J. Barth
NASA Ames Research Center
NAS Division
Moffett Field, CA 94035, USA
barth@nas.nasa.gov

Michael Griebel
Institut für Numerische Simulation
der Universität Bonn
Wegelerstr. 6
53115 Bonn, Germany
griebel@ins.uni-bonn.de

David E. Keyes
Mathematical and Computer Sciences
and Engineering
King Abdullah University of Science
and Technology
P.O. Box 55455
Jeddah 21534, Saudi Arabia
david.keyes@kaust.edu.sa

and

Department of Applied Physics
and Applied Mathematics
Columbia University
500 W. 120 th Street
New York, NY 10027, USA
kd2112@columbia.edu

Risto M. Nieminen
Department of Applied Physics
Aalto University School of Science
and Technology
00076 Aalto, Finland
risto.nieminen@tkk.fi

Dirk Roose
Department of Computer Science
Katholieke Universiteit Leuven
Celestijnenlaan 200A
3001 Leuven-Heverlee, Belgium
dirk.roose@cs.kuleuven.be

Tamar Schlick
Department of Chemistry
and Courant Institute
of Mathematical Sciences
New York University
251 Mercer Street
New York, NY 10012, USA
schlick@nyu.edu

Editor for Computational Science
and Engineering at Springer:
Martin Peters
Springer-Verlag
Mathematics Editorial IV
Tiergartenstrasse 17
69121 Heidelberg, Germany
martin.peters@springer.com

Lecture Notes in Computational Science and Engineering

1. D. Funaro, *Spectral Elements for Transport-Dominated Equations.*
2. H.P. Langtangen, *Computational Partial Differential Equations.* Numerical Methods and Diffpack Programming.
3. W. Hackbusch, G. Wittum (eds.), *Multigrid Methods V.*
4. P. Deuflhard, J. Hermans, B. Leimkuhler, A.E. Mark, S. Reich, R.D. Skeel (eds.), *Computational Molecular Dynamics: Challenges, Methods, Ideas.*
5. D. Kröner, M. Ohlberger, C. Rohde (eds.), *An Introduction to Recent Developments in Theory and Numerics for Conservation Laws.*
6. S. Turek, *Efficient Solvers for Incompressible Flow Problems.* An Algorithmic and Computational Approach.
7. R. von Schwerin, ***M**ulti **B**ody **S**ystem **SIM**ulation.* Numerical Methods, Algorithms, and Software.
8. H.-J. Bungartz, F. Durst, C. Zenger (eds.), *High Performance Scientific and Engineering Computing.*
9. T.J. Barth, H. Deconinck (eds.), *High-Order Methods for Computational Physics.*
10. H.P. Langtangen, A.M. Bruaset, E. Quak (eds.), *Advances in Software Tools for Scientific Computing.*
11. B. Cockburn, G.E. Karniadakis, C.-W. Shu (eds.), *Discontinuous Galerkin Methods.* Theory, Computation and Applications.
12. U. van Rienen, *Numerical Methods in Computational Electrodynamics.* Linear Systems in Practical Applications.
13. B. Engquist, L. Johnsson, M. Hammill, F. Short (eds.), *Simulation and Visualization on the Grid.*
14. E. Dick, K. Riemslagh, J. Vierendeels (eds.), *Multigrid Methods VI.*
15. A. Frommer, T. Lippert, B. Medeke, K. Schilling (eds.), *Numerical Challenges in Lattice Quantum Chromodynamics.*
16. J. Lang, *Adaptive Multilevel Solution of Nonlinear Parabolic PDE Systems.* Theory, Algorithm, and Applications.
17. B.I. Wohlmuth, *Discretization Methods and Iterative Solvers Based on Domain Decomposition.*
18. U. van Rienen, M. Günther, D. Hecht (eds.), *Scientific Computing in Electrical Engineering.*
19. I. Babuška, P.G. Ciarlet, T. Miyoshi (eds.), *Mathematical Modeling and Numerical Simulation in Continuum Mechanics.*
20. T.J. Barth, T. Chan, R. Haimes (eds.), *Multiscale and Multiresolution Methods.* Theory and Applications.
21. M. Breuer, F. Durst, C. Zenger (eds.), *High Performance Scientific and Engineering Computing.*
22. K. Urban, *Wavelets in Numerical Simulation.* Problem Adapted Construction and Applications.

23. L.F. Pavarino, A. Toselli (eds.), *Recent Developments in Domain Decomposition Methods.*

24. T. Schlick, H.H. Gan (eds.), *Computational Methods for Macromolecules: Challenges and Applications.*

25. T.J. Barth, H. Deconinck (eds.), *Error Estimation and Adaptive Discretization Methods in Computational Fluid Dynamics.*

26. M. Griebel, M.A. Schweitzer (eds.), *Meshfree Methods for Partial Differential Equations.*

27. S. Müller, *Adaptive Multiscale Schemes for Conservation Laws.*

28. C. Carstensen, S. Funken, W. Hackbusch, R.H.W. Hoppe, P. Monk (eds.), *Computational Electromagnetics.*

29. M.A. Schweitzer, *A Parallel Multilevel Partition of Unity Method for Elliptic Partial Differential Equations.*

30. T. Biegler, O. Ghattas, M. Heinkenschloss, B. van Bloemen Waanders (eds.), *Large-Scale PDE-Constrained Optimization.*

31. M. Ainsworth, P. Davies, D. Duncan, P. Martin, B. Rynne (eds.), *Topics in Computational Wave Propagation.* Direct and Inverse Problems.

32. H. Emmerich, B. Nestler, M. Schreckenberg (eds.), *Interface and Transport Dynamics.* Computational Modelling.

33. H.P. Langtangen, A. Tveito (eds.), *Advanced Topics in Computational Partial Differential Equations.* Numerical Methods and Diffpack Programming.

34. V. John, *Large Eddy Simulation of Turbulent Incompressible Flows.* Analytical and Numerical Results for a Class of LES Models.

35. E. Bänsch (ed.), *Challenges in Scientific Computing - CISC 2002.*

36. B.N. Khoromskij, G. Wittum, *Numerical Solution of Elliptic Differential Equations by Reduction to the Interface.*

37. A. Iske, *Multiresolution Methods in Scattered Data Modelling.*

38. S.-I. Niculescu, K. Gu (eds.), *Advances in Time-Delay Systems.*

39. S. Attinger, P. Koumoutsakos (eds.), *Multiscale Modelling and Simulation.*

40. R. Kornhuber, R. Hoppe, J. Périaux, O. Pironneau, O. Wildlund, J. Xu (eds.), *Domain Decomposition Methods in Science and Engineering.*

41. T. Plewa, T. Linde, V.G. Weirs (eds.), *Adaptive Mesh Refinement – Theory and Applications.*

42. A. Schmidt, K.G. Siebert, *Design of Adaptive Finite Element Software.* The Finite Element Toolbox ALBERTA.

43. M. Griebel, M.A. Schweitzer (eds.), *Meshfree Methods for Partial Differential Equations II.*

44. B. Engquist, P. Lötstedt, O. Runborg (eds.), *Multiscale Methods in Science and Engineering.*

45. P. Benner, V. Mehrmann, D.C. Sorensen (eds.), *Dimension Reduction of Large-Scale Systems.*

46. D. Kressner, *Numerical Methods for General and Structured Eigenvalue Problems.*

47. A. Boriçi, A. Frommer, B. Joó, A. Kennedy, B. Pendleton (eds.), *QCD and Numerical Analysis III.*

48. F. Graziani (ed.), *Computational Methods in Transport.*

49. B. Leimkuhler, C. Chipot, R. Elber, A. Laaksonen, A. Mark, T. Schlick, C. Schütte, R. Skeel (eds.), *New Algorithms for Macromolecular Simulation.*

50. M. Bücker, G. Corliss, P. Hovland, U. Naumann, B. Norris (eds.), *Automatic Differentiation: Applications, Theory, and Implementations.*

51. A.M. Bruaset, A. Tveito (eds.), *Numerical Solution of Partial Differential Equations on Parallel Computers.*

52. K.H. Hoffmann, A. Meyer (eds.), *Parallel Algorithms and Cluster Computing.*

53. H.-J. Bungartz, M. Schäfer (eds.), *Fluid-Structure Interaction.*

54. J. Behrens, *Adaptive Atmospheric Modeling.*

55. O. Widlund, D. Keyes (eds.), *Domain Decomposition Methods in Science and Engineering XVI.*

56. S. Kassinos, C. Langer, G. Iaccarino, P. Moin (eds.), *Complex Effects in Large Eddy Simulations.*

57. M. Griebel, M.A Schweitzer (eds.), *Meshfree Methods for Partial Differential Equations III.*

58. A.N. Gorban, B. Kégl, D.C. Wunsch, A. Zinovyev (eds.), *Principal Manifolds for Data Visualization and Dimension Reduction.*

59. H. Ammari (ed.), *Modeling and Computations in Electromagnetics: A Volume Dedicated to Jean-Claude Nédélec.*

60. U. Langer, M. Discacciati, D. Keyes, O. Widlund, W. Zulehner (eds.), *Domain Decomposition Methods in Science and Engineering XVII.*

61. T. Mathew, *Domain Decomposition Methods for the Numerical Solution of Partial Differential Equations.*

62. F. Graziani (ed.), *Computational Methods in Transport: Verification and Validation.*

63. M. Bebendorf, *Hierarchical Matrices.* A Means to Efficiently Solve Elliptic Boundary Value Problems.

64. C.H. Bischof, H.M. Bücker, P. Hovland, U. Naumann, J. Utke (eds.), *Advances in Automatic Differentiation.*

65. M. Griebel, M.A. Schweitzer (eds.), *Meshfree Methods for Partial Differential Equations IV.*

66. B. Engquist, P. Lötstedt, O. Runborg (eds.), *Multiscale Modeling and Simulation in Science.*

67. I.H. Tuncer, Ü. Gülcat, D.R. Emerson, K. Matsuno (eds.), *Parallel Computational Fluid Dynamics 2007.*

68. S. Yip, T. Diaz de la Rubia (eds.), *Scientific Modeling and Simulations.*

69. A. Hegarty, N. Kopteva, E. O'Riordan, M. Stynes (eds.), *BAIL* 2008 – *Boundary and Interior Layers.*

70. M. Bercovier, M.J. Gander, R. Kornhuber, O. Widlund (eds.), *Domain Decomposition Methods in Science and Engineering XVIII.*

71. B. Koren, C. Vuik (eds.), *Advanced Computational Methods in Science and Engineering.*

72. M. Peters (ed.), *Computational Fluid Dynamics for Sport Simulation.*

73. H.-J. Bungartz, M. Mehl, M. Schäfer (eds.), *Fluid Structure Interaction II - Modelling, Simulation, Optimization.*

74. D. Tromeur-Dervout, G. Brenner, D.R. Emerson, J. Erhel (eds.), *Parallel Computational Fluid Dynamics 2008.*

75. A.N. Gorban, D. Roose (eds.), *Coping with Complexity: Model Reduction and Data Analysis.*

76. J.S. Hesthaven, E.M. Rønquist (eds.), *Spectral and High Order Methods for Partial Differential Equations.*

77. M. Holtz, *Sparse Grid Quadrature in High Dimensions with Applications in Finance and Insurance.*

78. Y. Huang, R. Kornhuber, O.Widlund, J. Xu (eds.), *Domain Decomposition Methods in Science and Engineering XIX.*

79. M. Griebel, M.A. Schweitzer (eds.), *Meshfree Methods for Partial Differential Equations V.*

80. P.H. Lauritzen, C. Jablonowski, M.A. Taylor, R.D. Nair (eds.), *Numerical Techniques for Global Atmospheric Models.*

81. C. Clavero, J.L. Gracia, F.J. Lisbona (eds.), *BAIL 2010 – Boundary and Interior Layers, Computational and Asymptotic Methods.*

82. B. Engquist, O. Runborg, Y.R. Tsai (eds.), *Numerical Analysis and Multiscale Computations.*

83. I.G. Graham, T.Y. Hou, O. Lakkis, R. Scheichl (eds.), *Numerical Analysis of Multiscale Problems.*

84. A. Logg, K.-A. Mardal, G. Wells (eds.), *Automated Solution of Differential Equations by the Finite Element Method.*

85. J. Blowey, M. Jensen (eds.), *Frontiers in Numerical Analysis - Durham 2010.*

86. O. Kolditz, U.-J. Gorke, H. Shao, W. Wang (eds.), *Thermo-Hydro-Mechanical-Chemical Processes in Fractured Porous Media - Benchmarks and Examples.*

87. S. Forth, P. Hovland, E. Phipps, J. Utke, A. Walther (eds.), *Recent Advances in Algorithmic Differentiation.*

For further information on these books please have a look at our mathematics catalogue at the following URL: www.springer.com/series/3527

Monographs in Computational Science and Engineering

1. J. Sundnes, G.T. Lines, X. Cai, B.F. Nielsen, K.-A. Mardal, A. Tveito, *Computing the Electrical Activity in the Heart.*

For further information on this book, please have a look at our mathematics catalogue at the following URL: www.springer.com/series/7417

Texts in Computational Science and Engineering

1. H. P. Langtangen, *Computational Partial Differential Equations.* Numerical Methods and Diffpack Programming. 2nd Edition
2. A. Quarteroni, F. Saleri, P. Gervasio, *Scientific Computing with MATLAB and Octave.* 3rd Edition
3. H. P. Langtangen, *Python Scripting for Computational Science.* 3rd Edition
4. H. Gardner, G. Manduchi, *Design Patterns for e-Science.*
5. M. Griebel, S. Knapek, G. Zumbusch, *Numerical Simulation in Molecular Dynamics.*
6. H. P. Langtangen, *A Primer on Scientific Programming with Python.* 3rd Edition
7. A. Tveito, H. P. Langtangen, B. F. Nielsen, X. Cai, *Elements of Scientific Computing.*
8. B. Gustafsson, *Fundamentals of Scientific Computing.*

For further information on these books please have a look at our mathematics catalogue at the following URL: www.springer.com/series/5151

Printed by Publishers' Graphics LLC